The American Future

Patriots and Liberators:
Revolution in the Netherlands 1780–1813

Two Rothschilds and the Land of Israel

The Embarrassment of Riches:
An Interpretation of Dutch Culture in the Golden Age

Citizens:
A Chronicle of the French Revolution

Dead Certainties:
Unwarranted Speculations (*fiction*)

Landscape and Memory

Rembrandt's Eyes

A History of Britain I:
At the Edge of the World? 3000BC–AD1603

A History of Britain II:
The British Wars, 1603–1776

A History of Britain III:
The Fate of the Empire, 1776–2000

Hangups:
Essays on Painting

Rough Crossings:
Britain, Slaves, and the American Revolution

The Power of Art

The American Future

A History

SIMON SCHAMA

An Imprint of HarperCollins*Publishers*

HarperCollins books may be purchased for educational, business, or sales promotional use. For information, please write: Special Markets Department, HarperCollins Publishers, 10 East 53rd Street, New York, NY 10022.

First published in Great Britain in 2008 by The Bodley Head.

FIRST U.S. EDITION

Library of Congress Cataloging-in-Publication Data
The American future : a history / Simon Schama—1st ed.
    p. cm.
ISBN: 978-0-06-053923-8
    1. Nationalism—United States. 2. United States—History.
3. United States—Foreign relations. I. Title.
E156.S33   2008
973—dc22   2009358875

09   10   11   12   13   OFF/RRD   10   9   8   7   6   5   4   3   2

For Nick Kent and Charlotte Sacher, buddies on the wild ride, without whom this would all have been unthinkable, impossible . . .

History, by apprising them of the past,
will enable them to judge of the future.

Thomas Jefferson,
Notes on the State of Virginia, 1787

CONTENTS

III: WHAT IS AN AMERICAN?

IV: AMERICAN PLENTY

The American Future

PROLOGUE:
IOWA WALTZ

I can tell you exactly, give or take a minute or two, when American democracy came back from the dead because I was there: 7:15 p.m. Central Time, 3 January 2008, Precinct 53, Theodore Roosevelt High. I know this as I was regularly checking my watch, and besides you couldn't miss the schoolroom clock, its old white face the object of generations of teenage hatred and longing. I suppose a visitor from another world—London, say—might have thought there was not all that much going on in west Des Moines that evening. Minivans were pulling into the Kum & Go at the usual clip; burly men bulked up by their down jackets were stamping their feet as they fed fuel pumps to their tanks. Bags of salt were being lugged over the forecourt, the plastic gleaming in the sour orange light. After many months of maneuvering and talky self-promotion, it was time for the Iowa voters to offer judgment on who they thought should be the forty-fourth president of the United States. They would, the media hacks opined, "winnow the field," and Iowans are partial to a good winnow.

But it wasn't as if a sign was hanging from the steely sky reading "HISTORIC DAY." The sidewalks weren't carpeted with discarded election flyers, nor was every third downtown store window screaming "HUCKABEE" at you! No one that I heard was Honking for Hillary. In a couple of days of steady driving around Des Moines, the only street corner placard-waver we could find was a solitary devotee of the libertarian Ron Paul, gaunt and hairy like a midwinter John the Baptist crying his hero up in a downtown wilderness. Every so often a car would toot and the Pauline would wave his banner, and then put it down for a moment so he could clap his arms around his chest. Then he'd jump up and down a bit to keep his spirits up and the blood in his brain from going gelid.

So even though a lot of happy "we're in the limelight" waving to out-of-towners was going on from behind car windows, maybe Iowa was just too frigid for vote-hustling. So I said to Jack Judge, our crew driver, while the cameramen and director were off on the far side of the highway getting windburn as they took pretty shots of frozen corn stubble: "It's big, Jack, right?" Jack took off his beanie, pushed the mop of steel-gray hair back from his creased brow, held up the glasses he kept on a cord round his neck, gave them a big *haah* of steaming breath, wiped them with a Kleenex, and pronounced, "Way big." Jack was definitely someone to ask, seeing as he'd been a bit of a pol himself. A farm boy from Melrose, fifty miles downstate, he had had some fast growing up to do after his daddy lost his fingers in an accident with a combine. Jack had tended to the hogs, sheep, and chickens, picked the corn and beans by hand, and cut his own standing timber into fences, as best he could. "We had runnin' water," he chuckled, "the kind you had to run to fetch with a bucket." He was seventy-three now but still grittily handsome, and you only had to take a look at his open face to see a man who would do right by his family and his neighbors.

Then, in the spring of 1960, a young Boston-Irish senator came through Melrose, which was nowhere in particular but which pointed north toward Des Moines. The senator stopped long enough at a coffee shop for Jack to get a good look and take in the glamorously tousled hair, the winning freckles, the stream of wisecracks, and the merry mugging for the cameras. Surrounding Kennedy were fast-talkers in snap-brim hats, pulling anxiously on cigarettes while they shook the papers open or pushed coins into pay phones. The candidate had put in a lot of Iowa miles already, but in Melrose he poured on the happy-go-lucky like it was maple syrup on a short stack, and Jack Judge just spooned it up, signing on right there and then for the campaign. Though the senator's staccato short "a"s made it seem sometimes like he wasn't speaking English at all, leastways not the kind Jack was used to, there was no mistaking his smarts nor his appetite for action and power, which, in the normal way of things, would have raised Jack's eyebrows but for some reason this time didn't.

The Judges had all been Democrats as long as anyone could recall, raised and schooled in old-fashioned Midwest populism, the kind of country preacher-man politics that was unafraid to blame metropolitan money for small-town hardships. Nor, since it was the country's breadbasket, were they bashful about expecting the government to tide them over the rough patches with a few favors: low-rate loans, decent prices for their corn and livestock, secure markets. They were alright with the assumption that the nominee was bound to come from a quite different world just so long as he made some effort to understand theirs: the raw-knuckle mornings before dawn, the sinking misery of drought as sparse leaves on the corn stalks drooped and withered to papery rags, the heaviness of knowing they'd have to have a meeting with the bank manager before fall. For all his long cigarette holders and silky elegance, FDR had looked out for them, that much had been obvious from the AAA (Agricultural Adjustment Administration), and Harry Truman from Missouri was close to being one of them. They had wanted to feel good about Adlai Stevenson, him being from Illinois, but Adlai had reveled in his Princeton eggheadedness in such a superior way that he had proved a tough sell, especially against the war hero Ike. The thing about Kennedy, who was no less urbane than Stevenson, was his trick of making the cleverness seem down-home smart, the cocky high-school kid who could debate on Thursday and quarterback on Friday. So no one took against his can-do cheeriness, especially not when he sat down and listened to their stories of tough times and when he promised to do what he could to help them hang on to their family farms. Sure, all politicians talked that way when they were rustling up votes, but this one seemed in earnest. And he smelled, a little, of money himself, to which no one had much objection.

So Jack Judge went to work as a Kennedy campaigner and drove his battered pickup round the pothole-happy backroads to Moravia and Promise City, to Mystic and Plano, stumping for the Catholic Bostonian who was still reckoned a dark horse against Vice President Nixon. Jack listened to a whole lot of righteous bellyaching about how no one could afford a tractor now that they cost so much and how you couldn't make a go of it with less than a thousand acres, so that it was just a matter of time before they'd have to say yes to one of the big agribusinesses hungry for land. And Jack understood that people who were too proud to come right out and ask for help from the government were in bad enough shape that they wanted to hear it was on offer anyway. So he gave them a little something to hang on to: the hope that there might be someone in Washington who would pay some heed. And Jack Judge was so good at listening and knowing what to say in a neighborly way, that people started to trust him, and as the years went on, they would come up to him in seed stores and wonder out loud why in the world he wouldn't run for something himself. After years of self-effacing hemming and hawing, Jack eventually came around to their point of view and ran for office in the small way that makes democracy real, spending sixty-five dollars on "Vote Judge" signs and getting elected to chair the city council.

"How was that, Jack?"

"Oh I liked it well enough, but you know, everyone takes everything so *personal*. If some old boy had a problem with a traffic light or a neighbor's dog, he'd call me up drunk or even come around and hammer on my door like we should declare war."

Jack laughed a long happy breakfast-in-Iowa laugh.

"So why do you think it is big, this time?"

"Hell, you know that as well as I do, the country's in bad shape, don't know when I've seen it this bad."

It wasn't just the steady, muffled drumbeat of other Iowa farm boys coming home to be eulogized in memorial services on the high-school football field, or else hobbling out of ambulances with heartbreaking smiles on their faces while their moms tore themselves up inside what with trying so hard not to cry. No, it wasn't just that which had made Jack Judge furious. It was Hurricane Katrina and the pictures on his television of American *corpses* bobbing in the slick. It was the police of Gretna, Louisiana ("Small City, Big Heart" according to the municipal Web site) training their guns at people trying to cross the Mississippi bridge out of harm's way, desperate to find somewhere where they could just take a shower and get a night's sleep. It was the heavily delayed president finally choppering down into the calamity, frowning as he denied that anyone could have seen the levee break coming (many did and had said so to anyone who cared to listen). Then the president grinned broadly while he slung his arm around the shoulders of the director of the Federal Emergency Management Administration, who was supposed to be taking care of the mess, and congratulated him warmly for doing "a heckuva job." This had really turned Jack Judge's stomach. Like most Americans he got upset when things didn't work properly, including the obligations of common sense and common decency.

Right now, Jack was worried for his boy, the grandson he'd raised after the boy's mother had taken to drugs. The boy was pretty smart at high-school academics, but his real gift was on the wrestling mat, all-state caliber. He'd wrassled (as Jack sounded it) his way into all the Iowa colleges of his dreams, so that was taken care of, but Jack would sometimes wake up in the small hours worrying about what would happen afterward; whether the boy would be taken in a draft that might have to come if the country continued sending troops everywhichwhere, or what kind of job he could do in rural Iowa, where the economy in both town and country seemed on a slide.

And who from the array of Democrats did Jack think would best look after his grandson's future and America's? "Obama," he said, somewhat to my surprise. "Seems a down-to-earth guy." "Obama: *down to earth?*" I marveled, thinking of the Illinois senator's sharp threads, Harvard Law School logic-chopping mind, acquired gospel cadences, and coolcat body language. Not the kind of down-to-earth regular who perched on a lunch-counter stool in Des Moines. But then I was wrong about the kind of place Des Moines actually is. I did think that if the likes of Jack Judge had been smitten as he was in 1960, we were in for an interesting evening.

Just how different this would be from the Iowa caucuses of other election years, no one yet knew. Among both Republicans and Democrats and a whole lot of people who were officially neither, it was already a commonplace that the 2008 election was going to be fateful for the political direction of the United States, and so for much of the rest of the world. Republicans had been shaken by the inability of 150,000 troops in Iraq and Afghanistan to secure the "mission accomplished" that had decorated George Bush's premature victory speech aboard the USS *Abraham Lincoln* on 1 May 2003. As often as they rehearsed the line that after 9/11, the choice was between victory or capitulation to the terrorists, the number who truly believed that was wasting away. Some of them, like the maverick Vietnam War hero and senator from Nebraska, Chuck Hagel, for example, had broken ranks in a storm of indignation, accusing the administration of deceiving the nation by translating grief and anger over 9/11 into a war on a dictatorship that had nothing to do with the attack, and that to date had cost 4,000 American dead, five times as many gravely wounded, perhaps 100,000 Iraqi dead, a fiscal sinkhole of $10 billion a month, and that had no end remotely in sight. Even staunch loyalists recognized that the unpopularity of the war and the president had contributed to their loss of control of Congress in the midterm elections of 2006. Incumbent hard-line conservatives, like Mike DeWine in Ohio, George Allen in Virginia, and Rick Santorum in Pennsylvania—normally shoo-ins at the midterms—had all forfeited their seats, Santorum by the biggest margin in his state for twenty years.

While the Republicans could feel the ground shifting beneath their feet, they were unsure how to stand before the country. Repudiation of their own eight-year administration was not an option, though some "distancing" seemed prudent. The furthest some of the candidates, Senator John McCain, for example, would go was to concede there had been "mistakes" in planning, both before and after the invasion, but then insisting on the necessity of "staying the course," the shameful and irresponsible alternative being to "cut and run" as the Democrats proposed. Yet when, in the television debates, the candidate furthest on the right, the Texas congressman Ron Paul, laid into the president, Vice President Dick Cheney, and Secretary of Defense Donald Rumsfeld for putting one over on the country, leading it into a literally interminable conflict, and usurping unconstitutional powers to stifle dissent, you could feel the sweaty shifting of weight right through the plasma. Some of the suits did their best to change the subject to a different threat that would rile up the patriots: illegal immigration. Others, like the former Massachusetts governor Mitt Romney, threw red meat to the public by retorting to questions about closing Guantánamo that he would rather enlarge it and its powers of interrogation.

For the Democrats, the election of 2008 was going to be balm in Gilead after nearly thirty years of acute suffering. The conservative ascendancy inaugurated by Ronald Reagan had been interrupted by the Clinton victories in 1992 and 1996, but to many in the party those victories seemed pyrrhic after the midterm Republican landslide of 1994 turned control of Congress over to the party determined to correct what they thought had been an electoral freak of nature, by using all the machinery of obstruction in their power to thwart the president's initiatives. Impeachment hadn't helped, nor the Supreme

Court delivering the White House to the candidate who had lost the popular vote. They seemed doomed in the foreseeable future to be tagged as liberal losers, remote from the sentiment of the heartland. It seemed to get even worse when, contemptuous of the weakling notion that they should acknowledge the narrowness of their victory, George W. Bush and Dick Cheney proceeded to act as though they had a triumphant mandate.

Which, did the Democrats but know it (and they certainly didn't) turned out to be just what they needed. Dame Fortune turned her wheel just a cog or two. You could barely hear it creak. Enter hubris. The Republicans did precisely what they wanted and got what they sought except not exactly in the way they imagined: a deceptively swift battle that turned into an endless, unwinnable, war; a government so stripped down that it was incapable of responding with any kind of competence to natural disasters when they came along (not least because so many National Guardsmen were on tours of duty in Iraq); regressive tax cuts that replaced the Clinton surplus with deficits so colossal that the future of entitlement programs like Social Security and Medicare had been put into jeopardy, at exactly the moment when the baby-boomer generation would be claiming them. To cap it all, the party of business, long committed to deregulation, seemed to be presiding over the gravest financial meltdown since the Depression as banks wrote down or wrote off billions of dollars in failing mortgage loans.

The Democrats know that ultimately it's not enough to stand back while trying to wipe the smile off their faces and exclaiming "it wasn't us" and just cash in on the grief (though for the moment it would do nicely). Like the Republicans they felt the sea change and like them they were (and are) unsure, ideologically, which way to tack to catch the tide. But they began to speak less defensively about the integrity of public service, about recovering a sense of community in America, about a version of the nation's history and its present condition that had somehow got hidden away during the long paramountcy of raw individualism.

Sometimes, it seems to me that, for its own good, we should retire the word "narrative"—from graduate student courses, political analysts, image-doctors, from anyone who doesn't actually narrate. But perhaps, not just yet. Waiting out there to be recounted are competing histories of the United States. My hunch is that it's the most compelling story-teller, the best historian, who will be raising her or his hand beside the chief justice of the Supreme Court, come January 2009. But then I would think that, wouldn't I?

The story the Candidate was telling the well-dressed and coiffed lunchtime Republican crowd on caucus day at Kum & Go Corporate HQ (good modern art, even better coffee) was his own, since the default strategy of modern political campaigning requires that the biopic be the platform. "I am America," this line goes, "I am you" or rather "the you you want to be and the America you want to see, am I not?"

Mitt Romney needed a story in the worst way since the alternative was to have others do it for him, and then inevitably it would be about his Mormonism. But Mitt had precisely the narrative antidote to dispel suspicions, namely that of the hardheaded but warmhearted entrepreneur specializing in turnarounds (the Winter Olympics, the Commonwealth of Massachusetts) while fathering prolific, uniformly handsome boys, all of whom had gone on to do their own father-ing, resulting in the impossibly cute infants who were scooped up into the arms of Grandpa Mitt on stage and who, while looking momentarily shocked as if suddenly deprived of ice cream for no good reason they could figure out, managed nonetheless not to burst into tears. Either that was a real miracle, or else the ice cream was backstage. But you couldn't not like Grandpa, so sensationally buff for America. The lustrous crest rose darkly from his crown; the ortho-dontically immaculate smile flashed; the belted waistline advertised exemplary trimness; the sleeves were rolled up, ready for—what, precisely? Opening his own limo door? Romney spoke with self-deprecating charm and in whole sentences, a winning combination, especially in Des Moines, which is a culturally tony state capital. He even dared to quote before the Republican crowd the Harvard historian David Landes, whom he suspected (wrongly, except in the Gladstonian sense) of being a classic liberal but whose unexceptional nostrum that political cultures made economies, not the other way around, Romney thought just the ticket. His own culture was all eager enterprise, and as if in demonstration Romney moved around the dais in an economic, pigeon-toed shuffle that said "let me get AT it." Behind him the wide Tribe of Romney seemed to have mysteri-ously multiplied into yet more immaculate blondes and babies even while the rally had been proceeding, so that Mitt's turning circle seemed to be getting ever more confined until finally he seemed to be spinning in the Iowa waltz all on his own. Shuffle shuffle, turn, lean into audience, beam . . .

> Iowa, Iowa, winter spring summer and fall.
> Come and see. Come dance with me
> To the beeootiful Iowa waltz.

The crowd ate it and him up. He was their Man. He was the future. He was America. In the throng I asked the two sons (aged twelve and fourteen) of Kristin, a professor at Boston University, what they most wanted to see a President Romney secure for the American future, and they replied with frightening promptness and certainty, "Oh a balanced budget." You can see their point, even though cutting taxes to the bone as the candidate recommended didn't seem, in the short term, the best way to reach that goal. But let Ryan and Scott loose on the populace, I thought, and the Hairdo is home and dry.

"IF THE PEOPLE ARE DISINTERESTED, MOVE ON!" said the notice Scotch-taped to the wall at Hillary HQ as guidance for door-to-door canvassers. As it was by now clear that Des Moines is an intellectually strenuous place, the helpful professor in me felt compelled to point out how this "un/dis" confusion might leave the campaign open to unfortunate misinterpretation; mean-spirited souls gleefully seizing on Clinton links to sundry interest groups. So I said something, in a polite, smirk-free way. The disinterested advice was not welcomed. "Oh," said the campaign captain, eyes narrowing a little behind the steel frames, a faint but perceptible sigh of disgust moving through her sweater. "We've *already* discussed this, you know, and the dictionary says EITHER . . . ?" She gave me the rising inflection, high-school style, but this wasn't a question. Somewhere within me the pedant made a bleat of protest and retreated.

Four in the afternoon and precise dictionary definitions were not high on the agenda down at Clinton Command and Control. The hours, minutes were speeding by, and there were bodies to get out into the chill and off to the caucuses, all 1,700 of them right across the state; a transport fleet to mobilize; basic democracy enabled by

Jeep Cherokees and Volkswagens. Alexis de Tocqueville would have loved it. The one-story cinderblock building was divided into two long, narrow rooms, both of them supplied with low trestle tables, half-drunk cans of Diet Pepsi, and volunteers, most but not all of them women: student sweats and designer-label jeans; the glint of a golden scrunchy holding back glossy hair. Many of these women, elated at the prospect of a Madame President, had flown in from Beverly Hills, Connecticut, and the fancier suburbs of Chicago (where they had already networked and raised money for Hillary), and in turn got to meet her when she arrived to turn on the inspiration faucet, which, especially in a room of fifty or so, she can certainly do. The smile is authentic, the eye contact unforced, the listening attentive. I've seen it in operation, and, believe me, it makes believers.

But there was not much sign of the candidate's magnetism in the Hillary hut as light drained from the Des Moines afternoon. Not even much in the way of pictures on the walls; save a few small photos of the Candidate in schoolrooms and diners along with volunteers and senior citizens trying to look pleased at their Early Bird dinner interrupted by the campaign. What there was was last-minute toil— phone numbers crossed off, map locations checked—and a camera crew bumping into the labor with cables and mike booms, so under the circumstances it wasn't surprising that we got the cold shoulder. I would have felt much the same way. Name-dropping the BBC didn't melt the ice much either. A tall Asian American woman from Chicago, leaning against the wall elegantly earringed, legs snugly fitted into dark and shiny boots that seemed unmarked by the Des Moines frozen crud, surveyed me with amused disdain as if congratulating me on getting access to what, already, she made clear, she thought was the victorious nominee's campaign. Not everyone was quite so cocksure. "How do you think you're doing?" I asked the curly-haired, pale-faced young man who was running the press end of the campaign and who was entirely free from the expression of unearned self-congratulation that had sat on the face of Chicago Earrings. "Oh," he said, flashing a winning but rueful grin, "pretty well." But it was the prolonged pause that accompanied the continuing smile that gave the sense that there was indeed nothing more to be said, nor a whole lot more to be done. Nathan was at pains not to give the impression that it was all wrapped up. But nor did he think it was unwrapped, either.

Leaving the Hillary HQ I bumped into Lanny Davis, whom I'd seen working the phone banks in Room 2. Lanny was the epitome of Clinton loyalism through thick and thin, and the challenges he'd faced as counsel, apologist, advocate, had often come very thick. His Rodham-Clinton history went all the way back to Yale Law School where he'd first met Hillary, and looking at him powering his way through the phone calls, burrowing through the number list like a mole on speed, he seemed to be in love with her. Images flashed up of the rejected nerd while a mane of Arkansas hair and a world-conquering smile bore down on Hillary in the Law School Library. All these years had gone past, and Lanny still wanted to be Hillary's knight valiant, Sir Lanny of the Inextinguishable Lamp. He had written a book called *Scandal: How Gotcha Politics Is Destroying America,* which reflected, in a wounded kind of way, on the years when conservative pitbulls tried to bring Bill Clinton down for his ideological as well as personal libertinism, not to mention an elastic way with sworn testimony, thereby inflicting hurt on effective presidential govern-ance. After eight years of un-Clintonian incompetence, Lanny was looking forward to an era of New Politics: all high-minded debate on the Issues. Sure, I mused as I watched him drive the mobilization machinery. Let's see.

At the doorway he stuck out a friendly paw and introduced himself. As it happens, what Lanny wanted to talk to me about as he put on his lawyer's overcoat and faced the cold was ... history; not the history he was hoping to make that evening but some of the stuff I had written. Rembrandt, Robespierre, Churchill. Lanny Davis, perpetual graduate student, wanted a curbside seminar worse than another Diet Pepsi. Or was this the old Clinton flattery engine thrumming away? I had had the same experience, more than once, with Bill, the urgent inquiry into *my* work as if it were way more important than, say, tackling fundamentalists or the national debt. For a moment, confronted with the slightly blinking Lanny, him needing so badly to know about Caravaggio, I had the same warm rush of indecent egotism followed by shame and disbelief. And, just like his Master of the White House, before one had time to take a cold shower and snap out of it, Lanny was gone. There were platoons of old ladies to be introduced to Volkswagens, much work to be done before he could call it a night, and he was making his way into the parking lot, polished Oxfords stepping gingerly between the treacherous spots of black ice.

Beyond, in the Des Moines gloaming, the gusts had turned bone-cuttingly bitter. The lonely apostle of Ron Paul had given up the unequal struggle against the elements and was folding his banner into the back of a grimy truck. Dead leaves blew over the sidewalk, keeping company with jettisoned fast-food packaging, bowling down the frost-buckled flagstones like tumbleweed in spring.

Darkness brought a quickening pulse of commuting traffic. Every so often a trailer, carrying a satellite dish on its roof, would rumble by heading toward the downtown media center. But that would have happened in any year, with any lineup of candidates. That was media business as usual. So I did my best to contain a sense of theatrical anticipation. Maybe this sense I had of an American rebirth was just so much wishful thinking.

And there was not much about the caucus location that shouted "HISTORY!" The high school was named for Theodore Roosevelt. But the welcome sign, bucking bronco and waved Stetson, an allusion to Teddy's early life as Rough Rider, seemed more John Wayne, whose birthplace was in De Soto, a few miles out of town. The building was grandly old-fashioned; not the cinder block and glass constructions that sat squatly all over the American suburban landscape, but a school built of solid brick. Inside the ceilings were high; the long wood-paneled corridors were wide and smelled of floor wax. Three caucuses, two Democratic and one Republican, were meeting here, at around seven. It was six, and not a whole lot was going on in the room assigned to Precinct 53. A few tables set out with the campaign literature of each of the candidates, some of them staffed, some not; other tables with party registration forms for newcomers to fill out instantly qualifying them for the vote. Most of the action, predictably, was from television people setting up their cameras on legs pointing toward the front of the room. By 6:25, people had begun to trickle in from their offices, homes, and dinners; most of them elderly, all of them white.

Five, ten minutes later, the scene had changed. The trickle had become a stream which had become a flood, as if entire busloads of voters had suddenly pulled into the parking lot. The dozing schoolroom turned into a midwestern political souk, with champions of the candidates buttonholing voters, sitting the Uncommitted down to persuade them to their camp. In they came; a few not so white; all ages including families with small children wearing campaign T-shirts. Drawing paper and crayons had been set out on tables against the walls; some of the kids got stuck making pictures while cooler older brothers and sisters thumb-twiddled Game Boys. The place filled with talk of the war, of health care, and comradely abuse of the Bush administration and all it had wrought. (Down the corridor in the Republican caucus, there was less interest in defending the president and more in the competing styles of the contenders for his succession: the down-home intelligent candor of Mike Huckabee, the evangelical Baptist pastor whom, for some reason, the conservative powers that be suspected of closet liberalism, and the more corporate sleekness of Mitt Romney.)

By 6:45, there were no seats left in Precinct 53 and no standing room either and people were still pouring through the doors. The registration table had run out of forms, and they were being filled as fast as they could be photocopied in the school office. Party regulars came up to me shaking their heads in happy disbelief. They had never seen anything like it. A caucus that in 2004 had tallied around eighty would this time probably count three or four hundred, a pattern that would be repeated all over Iowa and all through the country as the primary season continued. Most of those I spoke to had never been to a caucus before; many of them were Independents, and all of them felt this was a year to make their voice heard. And because voting precincts were simply the political expression of residential districts, they were all neighbors, even if this was the first time they had encountered each other. They used the same stores, their children went to the same schools, they shared pews in the local churches and synagogues, and now they evidently felt joined together in an act of common citizenship.

Though the caucus nominating system was a modern invention, its roots were old and deep in American soil. Even before the revolution, the historian Sean Wilentz has noted, rowdy societies of artisans and mechanics were defying the choice of people presuming to be their betters by voting for their own nominees to city councils. Those undeferential habits would persist, eventually spawning the republican societies that made Jeffersonian democracy possible. By the 1830s, the habit of local choices and votes had moved from unruly cities into the world of the frontier. Even before it had become a state in the mid-nineteenth century, Iowa had had these kinds of local meetings in which the market and the hustings mingled together. Into town on their carts and wagons came farmers and blacksmiths with book learning, small-town lawyers who had ambitions, looking over the political wares on offer as shrewdly as the hogs and horses. Amid the traveling quacks, preachers, hot-pie stalls, people having their teeth pulled and their beards shaved, folks got to sound off about what ailed them. Bands would play, opponents get roundly vilified; one hero of the Mexican War sung to the skies, another jeered and laughed. Toughs would be on the lookout for suckers to drink or to beat into loyalty. It was frontier democracy, raucous and unruly, and it either exhilarated or horrified visiting Europeans.

Long ago there had also probably been ceremonious blowhards in beaver felt hats who liked nothing better than to Take Charge. Inevitably, then, the retired (though unhatted) schoolteacher, union organizer, and local party notable, a small, long-winded man, Jim Sutton, attempted to have the caucus Settle Down by delivering a lecture to the newcomers on Iowa's just claims to be the first state in the nation to deliver a judgment on the candidates. Iowa, he intoned, had never had a war. Canadian troopers had not, apparently, been tempted to pour across the frontier (though asking the Native American tribes might have produced a different assumption). And Iowa was, he went on, "just about in the middle of everything"—demographically, politically . . . and so on. One of the pleasures of the Democratic caucuses in Iowa was their freedom from having to listen to speeches. The assumption was that candidates' positions were already well known, and in case they were not, they were set out in the printed flyers that greeted people coming into the room. But no one had anticipated Jim and his teacherly enthusiasm. By the time that he was characterizing Iowa as America's Switzerland (minus the Alps), and proceeding with a comparative analysis, a certain restlessness was becoming audible, that could, I thought, if Jim went on, say, to comparisons with federal systems in Canada or Belgium, turn into an ugly scene, perhaps a lynch mob, even among the preternaturally good-natured citizens of Des Moines. He paid the price. A motion was made to replace Jim (pro tem chair) with an actual chair. Jim civically volunteered to replace himself with himself. This didn't go down well. Desperate, others put themselves forward, and although he demanded a recount as multitudes of hands waved to the ceiling, Jim, wearing an expression of baffled resignation, conceded, and the caucus proper, at last, was on.

At seven the doors were shut, ramping up the excited sense of expectation among the packed crowd. They were now suspended in a chamber of decision; a tiny but discrete atom in the organism of American democracy that would, ten months later, issue forth a new power. They felt the moment: Jewish grannies; teenage students; businesswomen and doctors. America. Each "preference" group identified with the candidates had been assigned a station in the room like school teams: Clinton in the right-hand corner, Obama in the left, Bill Richardson and John Edwards in the back two corners; the Rest (Biden, Kucinich, Dodd) in more amorphously defined positions. In the dead center of the room a hole was to be opened (somehow) into which the Undecideds would move, so that they could be lobbied by Persuaders from each and any of the camps toward which they might be leaning. There would be an initial count of bodies, then the Undecideds and those who had committed to candidates who were deemed to have unviable numbers would redistribute themselves to the remaining leaders. Compared to the relatively anonymous business of dropping papers into the slot of a ballot box or pulling a lever behind a curtain, and compared to the low-temperature habits of voting in Britain, the process was startlingly direct; face-to-face; voices raised, hands shaken, heads nodded in assent or dissent. It was thrilling, and the serious business hadn't even really begun.

Given the elated solemnity of the moment, the chairperson could probably have found better words to kick off the proceedings than "Let's cha-cha-cha." But it didn't matter. At 7:15, the people rose from their seats, moved from where they were standing as best as they could, in the chaotic crush, toward their "preference." For a moment the whole thing felt like summer camp: adolescent glee; lots of "OVER HERE!" shouts; self-conscious giggles. But there were also professors with urgent eyebrows, like Team Leaders, carving a route through the throng. After the immediate rush and crush, a certain ceremoniousness descended. The voters of Precinct 53 were, after all, literally taking positions; standing for something and someone. It had been like that in the local comitia of the Roman Republic too, so Cicero tells us in *Pro Flacco,* when proposals to vote sitting down had been defeated as an attempt to introduce Greek decadence into the proceedings. Citizens truly standing for something or someone stood unblushingly and visibly before their neighbors and took the social consequences. And for a moment, the Chucks and Katies seemed dissolved into the long marvelous history of civic liberty.

The "preference groups" gathered in knots around the room surveyed each other; quick head counts were taken. And then it became obvious that something startling was still going on. The left-hand corner of the room, Obama's, had not yet settled into a solid knot of supporters. It was still being fed by a moving coil of people: multi-pedal like a dragon at Chinese New Year or a slow conga line, steadily shuffling and shoving its way toward the corner, which was having to expand along the length of the left-hand wall to absorb the numbers of oncomers. Much later in the long campaign, close to the Pennsylvania primary, the Clinton team would suggest that she had done less well in the caucuses because the Obama organization had packed them with aggressive stalwarts who intimidated people into joining their camp. But that wasn't how it happened on this first night in the caucus for Precinct 53. This was the physical expression of choice. The face of the Clinton precinct captain, whom I'd seen earlier in the afternoon, as she took this in, looked bleakly exhausted. Her camp was mostly seated; there was plenty of room to spare.

It was not yet over. Supporters of minor contenders now deemed officially "unviable" were free to redistribute themselves along with the Undecideds in the middle of the room. There were no more than perhaps a dozen of them facing outward to listen to rival advocates. But a lot of cross-room appeals were being made—and looking at the ongoing carnival that was the Obama corner, there were temptations to defect to what was, in every sense, the more fun party. A blonde in her twenties, sharply dressed, curvy, a reporter with one of the news networks, had managed, by imploring, to sneak into the caucus a minute or two after the Closing of the Doors. She was barely in when a high school senior, ginger-haired and ardent, had made his move, excitedly chatting her up with merry talk of comparative health-care systems. In round one he stood with her in the Clinton corner. Now, still talking, he was moving away from the group as though he'd discovered it was the carrier of some sort of social contagion. Every so often he would peel his gaze away from the object of his infatuation and transfer it, with even greater ardor, toward the Obama corner. Finally it was too much. He broke off and started to move their way. "Stay with US," the blonde shouted. "You said you would." "Yeah, sorry but gotta do this. See you later?" "Oh SURE," she snapped back.

The final tally was done by counting off, military style; each supporter calling out the next number until the group was done. This was an economical way to count the groups, but it also made the notion of a vote—a shouted voice—powerfully literal. Thus the vox populi of Des Moines sounded: elderly aunts; high school tenors; gravelly taxi drivers; sonorous lawyers: "TWENTY-THREE," "TWENTY-FOUR" . . . By the time we got to Obama's 186 (to Edwards's 116 and Clinton's 74), the magnitude of what had just happened was inescapable. But perhaps it was just somehow the quirk of Precinct 53, where maybe Obama had campaigned more intensively than elsewhere. I went down the hall to Precinct 54, meeting in the assembly hall. They had just finished their tally and the distribution of votes was almost identical with the one I'd just seen. Perhaps this was going to happen right across the city, right across the state: Jack Judges in their hundreds of thousands willing something different in American politics; a democratic restoration.

Around the corner, the Republicans had gathered; a smaller, less hectic affair with seated caucus-goers enduring reading of statements from all the candidates—and there were a lot of them. Some, like John McCain, had mostly given Iowa a miss, gambling, shrewdly as it would turn out, on New Hampshire, where since the campaign of 2000 there had been longtime respect and affection for the senator's idiosyncratic style and beliefs. The real choice had been between Romney, the standard-bearer of the conservative notables in the party, and the unorthodox pastor Mike Huckabee. Though the numbers voting were barely a third of the Democrats, Huckabee had beaten Romney by almost the same nearly 3:1 margin that Obama had scored over Clinton.

It didn't take a genius, much less a media analyst, to figure out what was going on in Iowa: a populist rejection of political business-as-usual, of the dominant orthodoxies. The *New York Times* had endorsed Hillary Clinton. Iowa voted Obama. The conservative talk-show pundits had anointed Romney as the flag-bearer of their causes—gung-ho for the war; committed to overturning legalization of abortion; permanent deep tax cuts written in letters of blood—and had warned against the undependable affability of Huckabee. Iowa Republicans by the flockful gave him their vote. And the runner-up, also by a clear margin, was the other maverick in the pack, Ron Paul, who had also been treated by the mainstream Republican suits in the television debates like a creature from another planet who had no business, what with his outrageous attacks on President Bush (whom he wanted impeached), sharing their podium. But for the Republican faithful, who had seen their party disappear into the pockets of self-appointed oligarchs and managers, the likes of Ron Paul and Mike Huckabee were ideological catnip. I liked to think of the bearded apostle of the streets roaring happily at the television that night.

At the downtown media center: television pundits, adjusting their ties before the banks of lights as these results came in from all over the state confirming something like a little earthquake had indeed happened, stuck to their spiked guns. What was this: the joke, *Ron Paul*, taking more votes than Mitt? There was much on-camera shaking of heads and wiseacre warnings along the lines of "Senator Obama still has a very long way to go"—which was undeniable but not really the news of the night. The media corps was taking it personally, as if stung by the voters' refusal to fall in line with the truisms they had been rehearsing for months: the formidable invincibility of the Clinton campaign; the solid ranks of party notables who had declared for her; the bulging bank account; the astute campaign warriors; also—the managerial smarts of Mitt Romney; the presidential manner he exuded on the air and on the rally platform; the cross-party appeal to Independents and to the patriot corner of the 9/11 mayor, Rudy Giuliani. But all of this had, apparently, meant little or nothing. Could it be that it was precisely the parade of conventional wisdoms, the construction of inevitabilities, that was the object of Iowa voters' repudiation?

As the big screen at the media center rolled onward with its counts, it became dramatically apparent that whatever was going on was happening in numbers that had never been seen before. Rural or urban districts, it made no difference; counts, even in a politically active state like Iowa, were now up by two or three times. In other state primaries to come, voting figures would be even more staggering. In Nevada in 2004, some 10,000 had voted in the primaries; in 2008, that figure was nearly 110,000. This was the real surge, the one that mattered, of a popular democracy acting as though it could actually effect an alteration of power. And it had happened in a way that surely Tocqueville would have recognized as authentically American: a breakout from the entrapment of management; from the platitudes about the dominance of money, of television advertising; from the pet theories of the press and radio; from the cool manipulation of the campaign pros. This had happened through the recovery of directness, the transparency of neighborhood meetings, face-to-face; the shows of hands; the unapologetic sounding of voices; precisely the unapologetic demonstration of choice that was unthinkable in societies where democracy was a matter of form rather than substance, and where publicly endorsing your preference was likely to be noticed by those who might pay you back for it in currency you'd rather not have. Or it might be even worse. At the very time when the voters of Iowa were persuading each other, in Kenya voters were trying to kill each other.

On the electronically glimmering terraces of the media center, a Spanish television reporter was trying to get the words out and let her audience know the magnitude of what was happening. But they wouldn't come. Again and again, take after take, her tongue would trip over "Huckabee" or "Obama" until her verbal wheels started to spin and there was no hope of ever getting her out of the vocal ditch. She was not alone in her uncertainty. Everywhere else in the media hutch were journalists tapping frantically away at monitors rewriting the past shelf-life truisms while the atmosphere turned rank with sweaty disbelief.

That the Obama people hadn't themselves reckoned on the turn of events was clear from the complete lack of any kind of security at the entrance to the Victory Party. No gates, no frisking, nothing barring entrance to the jamboree. All candidates schedule these events, to put heart into the dispirited ("This is just the beginning"), to let the troops have a moment of exultation—or to do the "First let me thank my wife and children . . ." before bowing out. But into the brutally modernist concrete convention center flowed the full river of Obama Nation: black schoolkids in hot yellow T-shirts, ready to romp; elderly whites who looked as though they'd just come back from the Ponce de León Fountain of Youth Weekend; college students waving their arms; and a whole lot of people in between. Up the escalator came the falsetto ululations that are—peculiarly—the American cry of victory, the whoops preceding the faces and bodies. There was nothing to eat or drink at this party, not a can of Coke or a bag of mini pretzels. But the jubilant multitudes were feeding off a concentrated diet of delight.

Inside, the place was heaving and swaying, dancing and clapping. Gospel singing had turned it into the instant church of true believers, and the congregation—for Iowa is not a conspicuously black state—was just about most of America: all sizes, races, generations. When Obama showed up he seemed slighter and more sinewy than on the news, the hair coolly close-cropped as usual, dapper to show off the line of his skull as if he had the confidence that America might be ready for its contents. This bit of America certainly was.

When the riot of noise and his multiple thank-yous died away, Obama's first words immediately demonstrated the cunning of his rhetoric: "They said this day would never come," voice dipping at the end, in mock disbelief. "They" comprised everyone who indeed thought it an absurd stretch that a forty-six-year-old first-term African American senator had the remotest shot at the nomination, much less the presidency; that America had enough trouble thinking about a woman in the White House, much less a black. But "they" was also evidently reserved for all those who had assumed that whatever flowering of idealism might be at hand—the appeals to lift politics above the rancid stream of partisan demonization to propose an engagement with the actual ills that were afflicting the country—it would, sure as eggs is eggs, end up as just another naively deluded jejune footnote to the harder truths; the inexorable machine-tooled grinding of the levers of power. Obama continued to repeat "they," the people who believed "this country was too disillusioned to ever come together round a common purpose" so that, for that moment, it was the wiseacres who looked foolishly un-American. The crowd rode the moment of reaffirmation—of what? Of American democracy whose vital signs, at least on this night, were strong. But also of the living force of history.

Moving toward his peroration, Obama made sure to bring together in this big tent of hopping elation, the past with the present, memory stalking the impatiently advancing future. Into the party marched the honored ghosts: the generation of the revolution, "a band of colonists rising up against an empire"; the generation that had fought World War II, and the civil-rights generation that fueled on hope, had had the self-belief "to sit at lunch counters and brave fire hoses and march through Selma and Montgomery for freedom's cause." And at that point—for a moment—I tuned out, turned the sound right down in the arena and was somewhere else: Selma-time, 1965. I had good reason to remember its cruel havoc as if it had taken place right before my eyes, since just the previous year I'd been in Virginia stumbling into the edges of the civil-rights struggle and then I'd seen President Johnson nominated in the Democratic Convention in Atlantic City against a backdrop of agonized fury as a black Mississippi delegation tried, unsuccessfully, to unseat the white yellow-dog Democrats. Johnson's rage at the temerity and his maneuvring to make sure it would never happen was a low point. He needed the white Democrats of the South, racist or not, to cast their votes in the electoral college his way. A year later, in 1965, Johnson did something different: going on television and speaking as he said in his first sentence "for the dignity of man." But as Obama invoked the past, what I remembered most about that speech was Lyndon Johnson doing likewise, summoning those moments when "fate" and history came together— "so it was at Lexington . . . so it was at Appomattox Court House . . . so it was at Selma."

Everything contemporary seemed impregnated with history. When Obama spoke of wanting to replace the partisan division of "Red States" and "Blue States" with a recovered United States, it was impossible not to remember Thomas Jefferson's inaugural, after the bitter election of 1800 that (after thirty-six ballots of the House of Representatives) finally brought him to power, declaring that "every difference of *opinion* is not a difference of *principle*" and that "we are all republicans, we are all federalists." How surprising is it that the nation that began by wanting everything, including politics and nationhood, to be minted afresh should nonetheless need the mirror of time in which to see itself; to reach out and back to history for a sense of its own future purpose? If Gore Vidal's lament for the "United States of Amnesia" might still be right for numbers of the American public who are regularly put on the spot by interviewers asking them, mike in face, in which century, give or take a few, they think the Civil War was fought, it's equally true that for those who still think of themselves as citizens, as active participants, the habit of peering into the mirror of time to see the character of their present and future selves, dies hard.

PART ONE

I : A M E R I C A N W A R

1. Veterans Day: 11 November 2007

"America has never been a warrior culture."

Just because it was Dick Cheney saying this didn't automatically make it untrue, even on Veterans Day in Arlington National Cemetery, a year before the election. Patriotic chest-thumping from an impenitent vice president was not what anyone, least of all the veterans themselves, wanted to hear. Bodies of young American men and women were showing up regularly at Section 60, at the foot of the grassy hill. Mustard-colored backhoes stood parked in a row, steel claws raised, ready to dig. Every so often, on the hour, a soft clop of horses' hooves could be heard coming over the dips and rises of the cemetery park before a reversed gun carriage rolled into view. Most weekdays, every hour or so, those small, sad parades do the funerary honors as tourist buses are diverted to alternative routes, heading for the Unknown Soldier or JFK. But if you walk the green vales of Arlington, you can catch young soldiers of the 3rd Infantry getting ready for their next duty, operating the forklifts that hoist coffins onto the carriages. Others grab a quiet smoke beneath the plane trees before dressing the horses and getting on their ceremonials. Out in Samarra and Helmand and Mosul and Kandahar a great many more mutilated and eviscerated bodies, not American, are being tended to as best as possible without benefit of flag or drums. Only the keening sounds the same.

But at Arlington, on Veterans Day 2007, in Memorial Amphitheater there was no howling, except from small children squirming against the captivity of their mothers' laps. Cheney would utter the consolatory pieties with studied quietness, his voice falling at the end of the sentence, as if the avoidance of vocal histrionics were itself a symptom of truth-telling. Perhaps he has Theodore Roosevelt's injunction to "speak softly and carry a big stick" framed over the vice presidential desk. When, every so often, an infant would let rip with an *aaaighw*, the note bouncing off the columns, Cheney would look up from the teleprompter, sight line briefly changed and then move impassively to the next homily, like a tank rolling over a cat.

It was warm on 11 November, and the temper in the amphitheater was jocund. Sunlight falling on cherry-red caps and coats turned veteran marines into a gathering of jolly elves. The oompah from the big orchestra was classical lite, and the procession of colors into the amphitheater could have been any high-school parade but for the many years of the standard-bearers. Studded biker jackets decorated with Vietnam insignia—"Hells' Harriers," "Dragon Breath"—draped the gut-swagged bodies of old grunts, but behind the bandannas of yore they had lost their heavy-metal menace, their righteously roaring grievance. Now they were just living exhibits in the museum of stoned-age warfare, the walking wounded of the Sha-Na Na-tion. More speeches droned; more Andrew Lloyd Webber chirped; and the volunteer "service" being eulogized was rapidly turning into social granola: "veterans helping out in communities" more akin to the coast guard or the scouts; nothing to do with bombs and bullets. If Iraq and Afghanistan had turned out not to be a picnic, Veterans Day at Arlington certainly felt like one.

But America has two specified days of military remembrance; one when the leaves are fallen, the other when they spread into full spring splendor. Created after the Civil War, Memorial Day was originally known as Decoration Day from the spontaneous habit of military widows decorating graves with wreaths of white flowers. In 1868 the commander of the Grand Army of the Republic, General John Logan, decided to institutionalize a day of remembrance—for both the Union and Confederate dead—and specified the third Monday in May. For most of the country, Memorial Day is about the inauguration of warmth. Garage sales lay out their wares in driveways. America's men go through their tribal ritual firing up the grill for the first cookout. Meat meets heat, beer cans pop and hiss, and somewhere, everywhere, a microtractor is harvesting a suburban lawn. But even if the lines of spectators at the parades are thin, some remembering does get done in small-town America. In Sleepy Hollow, New York, where a statue commemorates the "honest militiamen" who caught the British spy

Major André in 1780, a dozen or so veterans, some of them octogenarian survivors of Pearl Harbor and Normandy, followed behind a
high-school marching band of big girls dressed in glossy black boots,
pleated black miniskirts, and scarlet jackets, strangely reminiscent of
the British redcoats the "honest militiamen" had thwarted. The
band murdered "Sloop John B" (a baffling selection) and "God Bless
America," and an endless procession of fire trucks from neighboring
towns followed, each bearing heraldic insignia ("Conquest Hook and
Ladder 46"), before the parade ended up at a flower-decorated "Patriots'
Park" (named for the Revolutionary War). There, amid the dogs and
babies and aunties and wives, the dignitaries did something surprising:
they connected with history. The commander of the local American
Legion, a World War II survivor, read the entirety of General Logan's
Order Number 11 from 1868, as though it had just been issued, stumbling a little over its great flights of Lincolnian rhetoric, asking for
the perpetuation of tender sentiment for those "whose breasts were
made barricades between our enemies [that is, other Americans] and
our country." The Lincolnian tone was sustained when the mayor of
Tarrytown read an abbreviated version of the Gettysburg Address,
although why he thought fit to shorten a speech that is only 400 words
in the first place was mysterious. The dead of that immense slaughter
and the president in his high hat were summoned from November
1863 to cookout day 2008, to mix and mingle with the old Vietnam
grunts in Ranger hats. But was this just an empty flourish? Was it safer,
easier, to invoke Gettysburg and Antietam than dwell on the fifty-two
American servicemen and women killed just the previous month in
Iraq and Afghanistan?

Up at the Sleepy Hollow cemetery the graves of every generation
of servicemen were receiving small American flags planted in the
earth beside them. The same had happened at Arlington National
Cemetery, where every one of the 260,000-plus graves gets decorated
and a guard posted in the fields through the Memorial Day weekend
to make sure that neither wind nor rain nor malice aforethought
might disturb them. One of those graves means more to me than
a random name and date of death. Kyu-Chay was someone whose
bright presence I can summon up a lot more easily than his death
somewhere in the dun mountains of Afghanistan. His father and
mother own a dry-cleaning store in my small town in upstate New

York, and as is the custom in Korean families, the children—two boys—stayed close. On leave from Fort Bragg, I would see Kyu helping out at the counter, handling the shirts and suits in his uniform: a big, sunny fellow, sleeves neatly rolled to his biceps, army-style, plunging energetically into the racks at the back of the store as if on patrol. That was pretty much all I knew about Kyu until, one day in early November 2006, I walked into the store and found it strewn with white flowers: on the counters, floor, propped against the walls: lilies, chrysanthemums, roses, nothing but white, a pallid shroud set down while the dry-cleaning machines throbbed on heedlessly behind the counter. Set in the middle of one of the bouquets was a photograph of Kyu in his beret, a broad smile on his big open face. Beneath the picture, a notice announced that the staff sergeant had been killed while leading a mission in Afghanistan. His father, Sam, and his mother, both hollow-eyed and stooped with grief, were still working at the shop, as much, I thought, to keep madness at bay as to carry on earning their living. They are people of great formal dignity, so I wasn't sure about the propriety of reaching over the counter, but when I did so Sam leaned forward, falling into the proffered embrace, crumpling into mute anguish, shoulders trembling.

Kyu-Chay was buried in Section 60 at Arlington, but on Veterans Day, in deference to the private sorrow of families, it was closed to visitors. A month or so later—a year after his death—I went back to find him. There wasn't enough room on the standard-issue tombstone for his story, which was, in its way, exceptional, especially for a para-trooper staff sergeant, but it was also classically American. Kyu-Chay had been born in 1971 in the ancient city of Daegu, South Korea, overlooked by Mount Palgongsan, but had grown up conscious that his city and his family had had the narrowest escapes from being overwhelmed by North Korean and Chinese forces at the Pusan peri-meter in 1950. Twenty-five years later Sam and his wife had taken their sense of grateful belonging all the way to the Lower Hudson Valley. The older brother Kyu (his younger brother shares the name) was intellectually gifted and worked hard. Upstate college and law school opened to him. But then, in 2001, after 9/11, Kyu-Chay did some-thing not so predictable for a first-generation upwardly mobile Asian American, yet something deep-rooted in the immigrant relationship with his adopted nation: he enlisted in the army. With all his smarts he was the kind of officer material the United States Army dreams of, but Kyu-Chay had something particular he wanted to do: become a cryptologist in a force where that skill was in notoriously short supply. To prepare for an Iraq tour of duty he became a fluent Arabist; and before he was deployed to Afghanistan, added Pashto and Farsi. He had pretty much everything one would want in a paratrooper staff sergeant: physical courage married to strenuous intelligence. There was a practical purpose to his learning: translated intercepts make the difference between life and death, and the glaring lack of Arabists in the CIA in the summer of 2001 had turned out to be lethal. Kyu-Chay was committed to understanding the enemy by taking the trouble to learn his language, culture, faith. But he also wanted to learn Arabic so that he might pay those who could be friends and allies the respectful compliment of learned empathy. Perhaps his greatest act of translation was to take his own complex cultural history and use it against two kinds of insularity: the American habit of assuming that if English was hollered loudly enough at a roadblock or a police station, people would eventually Get the Message, especially if that trusty old tutorial aid, a cocked rifle, was added to the instruction. Moreover, if sermons on democracy were issued at regular intervals, so the official view went, the rest of the world would one day come around to the American way of life. Equally, though, Kyu-Chay's hard-won knowledge was directed against the insularity of theocratic absolutism, a culture in which the obligation to annihilate dissent is extolled as high duty. Confronting that absolutism, he lost his life on a mountain track.

As I walked back from Section 60 through the field of stones, something struck me about them that I ought to have noticed before. Almost every soldier's headstone was inscribed on its reverse face with the name of a spouse: "Daisy His Wife, 1888–1941"; "Margaret Mayfield, 1911–1983"—although never, that I could find, "John Doe, Her Husband." Occasionally, the names of children were inscribed on the same face, although the modest format and size prescribed in the modern era precluded much in the way of an inclusive family tomb on one stone. But children, sometimes painfully young, lie in proximity to the servicemen. For historians of military death and remembrance like Drew Faust, the need to reunite military families in death, starting in the Civil War, has been a peculiarly American habit. In other more wholeheartedly warrior empires and nations, in Prussia, or Japan, severance from family was often assumed, even taken as measure of martial devotion to the Fatherland. The camp and the barracks became family, military caste overrode the sentimental attachments of hearth and home, and the dynastic commander was supreme patriarch for whom the soldier would gladly offer up his life. In Victorian Britain, regiment was family, and the apprenticeship in separation for the officer classes began as early as possible with the boarding school. In more brutal conscript societies like imperial Russia, soldiering was an extension of servitude; the delivery of the unfree into a sacrificial bondage of unlimited term.

But not in the United States, where, during much of the first half century of the nation's life, a volunteer army was a negligible presence, hardly ever more than 10,000 for the rapidly expanding continental territory of the republic. At times of emergency like the anti-excise Whiskey Rebellion of 1791 or the War of 1812, the regular army was supplemented by the mobilization of state militia and a temporary increase in enlistments. But it was only during the Civil War that millions of men were torn from their homes, stores, and farms and pitched into the muddy marches and slaughter fields, remote from everything familiar. The scale of letter-writing home by soldiers with even a bare rudiment of literacy testifies to what was felt as the unnaturalness of martial exile, the craved assumption that the separation from loved ones would be temporary. "I want to see you and the children mity bad if the war don't end vary soon I will come home on a furlow . . . ," wrote the farmer Hillory Shifflet to his wife from his camp in Tennessee in 1862. Each week, Shifflet received from Jemima back in rural Ohio not just letters, but cooked food and photographs, gloves and boots. In January of the same year George Tillotson, an enlisted man from New York, wrote to his wife: "You can't imagine how much I would give to here from home and how much more I would give to see home . . . but then I suppose the satisfaction will be all the sweeter for waiting." His homesickness was so great that though he didn't want to "insinuate that I am sorry I enlisted . . . maybe like enough I would not enlist again to be candid I don't think I would." Tillotson was lucky enough to make it home again at the end of his muster. But hundreds of thousands were less fortunate. Which is why "Johnny Comes Marching Home Again," written by the Boston bandleader Patrick Gilmore in 1863 to cheer his disconsolate sister Annie at the very moment when it was obvious that there would be no speedy reunion, remained for enlisted ranks the most poignantly felt of all Civil War songs, neither glory nor hallelujah. It was for the countless families for whom Johnny never did come marching home that the Union established war cemeteries as an act of domestic reparation. Husbands and wives, fathers and children, who had been torn apart by war, could at least be reunited in the long sleep of death.

The man responsible for inaugurating that benevolent practice— Montgomery C. Meigs, the quartermaster general of the Grand Army of the Republic—was himself buried in 1892 (aged seventy-five) on the summit of Arlington Hill in Section 2, where the cemetery began. Appropriately enough, his own family is interred in the grassy patch around the monumental, if plainly cut, tombstone. Meigs's in-laws, the Rodgerses, who came from the most distinguished naval family of the early republic, are buried nearby, making the plot a grand domestic reunion as if called to Sunday dinner. Montgomery Meigs's wife, Louisa Rodgers Meigs, faces out toward the path that tracks the brow of the hill, and in her maternal shadow lies the most startling tomb in all Arlington: that of her twenty-three-year-old son, John Rodgers Meigs, bushwhacked in the Shenandoah Valley by Confederate irregulars in the first week of October 1864. Personalized tomb sculptures are almost wholly absent from Arlington. The truism, reiterated by Dick Cheney on Veterans Day, that "we are a democracy, defended by volunteers" materializes in the featureless egalitarianism of the plainly arched low stones, each no more than a foot and a half high, that dominate the cemetery. But even before there was a prescribed regulation design, Montgomery Meigs commissioned a death-likeness of his son that would be much smaller than life-size. A mere three feet or so from head to toe, raised on a low plinth, the sculpture lies hidden from general view, tucked into the shallow space between his mother's tomb and the path.

There is something touchingly unresolved about the bronze tomb sculpture: a grieving conflict between the parents' wish to remember their son as both man and child. John Meigs, the precociously promoted brevet major of Engineers, lies just as he was found on the Swift Run Gap Road, by the edge of the woods: flat on his back, boots in the air, wrapped in his cape, Colt revolver at his side. This is the West Point cadet who, just a year before, had graduated first in his class; the officer and patriot who, his proud, stricken father wrote in his journal, "had already made himself a name in the land,"and who at "the age of nineteen, had fought with distinction at the first battle of Bull Run 21 July 1861." Now the young hero had fallen, "a martyr to liberty." But then there is the other Johnnie Meigs, the plump-cheeked youth, photographed again and again by his father in his cadet's smoke-gray uniform, a lick of hair falling over his brow, or frowning in boyish concentration as he looks at a science specimen, his mother's "darling precious John." This is the boy cut down in the flower of his years, an emblem of America's self-inflicted massacre of the innocents. Just before hostilities started, Meigs had predicted that if war was to come, it would be fought "temperately and humanely." By the time his son was killed, murdered in cold blood, he believed, by cowardly Confederate guerrillas disguised in Union uniforms, Meigs knew better.

The obverse of the tomb, the face on which the quartermaster general's own name is inscribed, looks in the opposite direction, north, toward the Doric portico of Arlington House. That house is pure Virginia history, a direct link between the two American conflagrations, the war that made America and the war that almost unmade it. The man for whom Arlington House was built was George Washington's adoptive grandson, George Washington Parke Custis. The man best known for living in the mansion, the master of the slave-tilled plantation that stretched down the hill into the valley, was Custis's son-in-law, Robert E. Lee. And it was Lee's fellow West Point graduate, Montgomery Meigs, who in the summer of 1864 made a point of turning the manorial idyll (which the Lees and the Meigses had together enjoyed as a social setting) into a boneyard. Lee, who had been the superintendent of West Point from 1852 to 1855, had, in Meigs's eyes, violated his beloved academy's code of "Duty, Honor, Country" by accepting command of the Confederate army, a treachery compounded by the fact that Lee had also been offered the same post for the Union. Other West Point graduates whom Meigs knew well—Joseph Johnston, James Longstreet, Braxton Bragg—had all followed Lee. One of the academy's roistering young bloods, Jefferson Davis, another of Meigs's former friends and mentors, became the president of the Confederate States of America. Even more iniquitous in Meigs's eyes,

Pierre Beauregard had actually left his superintendent's post at the academy in 1861 in order to join the rebels, or as Meigs always called them: "the Traitors." Had all these men not remembered that the citadel on the cliff was called the "School of the Union" and its graduates the "Children of the Union"? But it was the treason of the slave-owning Lee that most envenomed Meigs's passions. Lee had broken the house of the American union. Now Meigs would do his utmost to make sure that Lee's own house, where six of his children had been born, would be made permanently uninhabitable. Should the traitor return, he and his kin would be forced to sleep "in the company of ghosts." A student of classical literature, Meigs knew the Roman custom of sowing their enemies' land with salt to make it forever sterile (and quoted it to his superiors when, for example, they were considering how to treat the conquered port city of Charleston, South Carolina). Now he would turn implacably Roman. In August 1864 he had twenty-six Union soldiers, who had been interred near the old Lee slave quarters of the estate, brought to the portico of Arlington House like visitors about to pay their respects, and had them buried again, right beside Mrs. Lee's rose garden.

2. The fight for the citadel: soldiering and the Founding Fathers

Montgomery Meigs took Lee's treason personally because twenty-four years earlier, in the summer of 1837, the two men had roomed together by the coffee-colored Mississippi. Their task as young West Point graduates and officers of the Army Corps of Engineers had been to survey the river from the Des Moines rapids down to the new river port of St. Louis and make recommendations for improving navigation. The need was urgent because steamboats had revolutionized the possibilities of river traffic, and ports like St. Louis, then no more than 5,000 strong, were perfectly situated to capitalize on the opportunity. At the junction of the Missouri and the Mississippi, the cash staples of the lower South—cotton ginned in Eli Whitney's machines—would be warehoused and sold to buyers from the industrial North and East. Northern hardwares and manufactures would in turn be loaded on boats sailing south and west to

Memphis, Natchez, and New Orleans. St. Louis was also one of the
jumping-off points for the Conestoga covered wagon trains heading
west, carrying with them everything needed to make a new homestead America in the prairies: timber, draft oxen, saws, plows and
hoes, bedsteads, pots and pans. St. Louis had just been optimisti-
cally declared a "Port of Entry for the United States," but American
geography was notorious for failing to cooperate with the dreams of
enterprise. Upstream on the Mississippi, the rocks of the Des Moines
rapids made navigation hazardous, while downstream, debris-choked
islets close to St. Louis threatened boats with grounding at low tide
and the city with flooding should a storm surge force the water into
the narrowing channel.

　　"It is *astonishingly* hot here," Lee wrote to his wife—ninety-seven
in the shade. Hot or not, there were the two lieutenants, Lee and
Meigs, paddling their dugout canoe on the deceptively sluggish stream,
sketching its capricious course, and taking soundings while being
devoured by mosquitoes. Lee's report recommended blasting a way
through the upper reaches of the river and building dikes made from
pilings enveloped in stone and brush, which would sieve much of the
debris without forcing the current too far from its regular course. A
small dam diagonal to St. Louis would push away some of the silt,
scouring a deeper channel for the boats to navigate when the river
was low. Fastidious, beautiful maps were drawn; data collected; recommendations made for a little fleet of "snagboats" that periodically
would cleanse the passages made. The two men—of markedly different
tempers, the handsome, swart-bearded Lee even then rather grand
in his manner; Meigs, nine years his junior, six foot one in his boots,
pale and high-browed, energetic to the point of bumptiousness—were
forced into close and constant proximity. They shared log cabins, talked
with the Chippewa, made do in reeking rooms in St. Louis, where
moldy whitewash hung in limp strips from the wall, mysterious odors
defeating the cologne that the elegant Lee had brought in his traveling
bag. Though the intense, inexhaustible Meigs made Lee uneasy, a
fellowship was born. Though Lee thought the whole area, "winning
women" apart, "bloody humbug," he sportingly adjusted to it, and
the two comrades shot wild turkey from horseback, Missouri-fashion,
and caught whiskery catfish "almost three feet long," monstrously
ugly but fine eating.

Even someone as self-assured as Lee could not help taking careful stock of Montgomery Meigs, whom he declared, wryly, to be "a host [of men] in himself." After graduating fifth in his class from West Point in 1836, Meigs had been briefly assigned to the artillery but had then been transferred into the Army Corps of Engineers, regarded in every way as the military elite. He was twenty-one, still smooth-faced, but with the imperial brow and dark eyes that were to be the bane of lesser mortals rash enough to get in the way of the public virtues that necessarily came with the old name of Meigs.

The chronicle of the Meigs dynasty tracked the history of America. The patriarch, Vincent Meigs, had sailed from Dorset, England, with his wife, Elizabeth, in 1636 to the territory that would become Connecticut. It must have been the radical politics of English religion that had sent them across the Atlantic, for thirty years later, Vincent and Elizabeth took in Puritan regicides who had voted for the execution of King Charles I and who were subsequently being called to account by the Restoration courts. Montgomery's great-grandfather had been the first Return Jonathan Meigs, a name which colored the Christian sobriety of the family with a little harmless romance. In the early eighteenth century, in Middletown, Connecticut, a young Meigs had been repeatedly rebuffed by the object of his ardor, a demure Quaker. Mournfully resigned to his fate, he was mounting his horse when, as was her prerogative, the lady abruptly changed her mind, recalling him with a cry of "Return Jonathan Meigs." Embedded in his heart as the phrase that had altered his life, he felt compelled to call the first fruit of his happy union Return Jonathan, who, bearing a moniker requiring daily explanation to strangers, had no choice, really, but to become a hero.

In 1777, two years into the revolution, Return Jonathan Meigs marched to Quebec with Benedict Arnold (still the bright star of the Continental Army rather than its detested turncoat as he became in 1780). But his American regiment failed to dislodge the British and take Lower Canada. Arnold's star suddenly dimmed, and Lieutenant Return Jonathan was taken prisoner. Liberated in an exchange, Meigs lost no time vindicating his fortunes by leading 170 men in an amphibious raid on a British redoubt at Sag Harbor on Long Island in May 1777. It was the kind of guerrilla action that was the stuff of Revolutionary War

legends but which, in cool reality, seldom came off as planned. The
rare success of the raid on Sag Harbor enhanced the young officer's
reputation while that of his erstwhile general Benedict Arnold collapsed
into infamy. But the Meigs fame was richly deserved. In retaliation for
the redcoat burning of Patriot farms in Danbury, Connecticut, Return
Meigs had gathered together an enthusiastically angry company of local
militiamen, among whom were scouts knowledgeable about the swift
waters of Long Island Sound as well as the woods and fields that lined
its shores. To this troop Return Jonathan added his own company of
trained volunteers, and the little force rowed across the Sound in small
boats, taking the British napping (in some cases literally). Twelve of
His Majesty's vessels were burned and eighty prisoners taken with no
loss to the Americans. A grateful Continental Congress and General
Washington presented Return Jonathan with a sword of honor for
his welcome demonstration of both tactical competence and personal
courage. Meigs was affected enough by this official vote of confidence to
take his name seriously, returning to the fray, commanding a regiment
under "Mad" Anthony Wayne and storming the British breastworks at
the battle of Stony Point in July 1779.

Return Jonathan Meigs was, then, a paragon of all-action American
patriotism, which, once peace came, was bound to make him restless.
Unsuited to the steady round of the seasons as a Connecticut farmer,
he rode northwest to pioneer in Ohio, where he planted a new branch
of Meigses and became important enough to lay down the first homesteading regulations for incoming settlers, posted, it was said (for the
Meigses were fond of these kinds of stories), on an ancient oak by the
Ohio River. But RJ was not yet done. In 1801 he was appointed by
President Jefferson as government agent to the Cherokee Nation in
their ancestral homeland in what is now eastern Tennessee and west
Georgia. He never moved again, though the Cherokee, as we shall
see, were not so fortunate.

Inevitably there was a Return Jonathan Meigs Jr. who did what he
could both to live up to his father's dashing reputation and to make the
family name as dependable as possible, first by fighting, and then by
treating with, the Indians. The rewards for his more orthodox manner
of making his way in federal America were handsome, and Return
Jonathan Jr. became, in succession, prospering Ohio attorney, state
legislator, justice of the Ohio Supreme Court, governor (responsible

for defending the border against the British in the war of 1812), United States senator, and finally—not a job to sniff at in the early days of stagecoaching—postmaster general of the United States.

His younger brother, Josiah, took a more cerebral turn, teaching math, astronomy, and "natural philosophy" at the local college, Yale, before publishing the *Connecticut Magazine*, principally, one suspects, to oblige old classmates like Joel Barlow and Noah Webster who had literary pretensions, an enterprise that swiftly and predictably brought Josiah to the edge of ruin. Turning to the law, where his brother had done so well, Josiah took another wrong turn by defending privateers taken by the British in Bermuda, a move that got him indicted for treason, and that understandably made the quieter life of a mathematics professor seem suddenly attractive. Yale took him back, gave him years of assiduous respectability, out of which he walked yet again, migrating south to become the second president of the new University of Georgia in the town of Athens by the Oconee River, not far from the Meigs-friendly Cherokee.

So there were now Meigses north and Meigses south, and in the way of following the destiny of the nation this generous geographic distri-bution would lay up trouble for the future peace of the clan. Josiah's son, Charles, our Montgomery's father, was born in Bermuda during his father's misplaced advocacy of the maritime desperadoes, but he was educated like a good Jeffersonian democrat in Athens. Schooled in medicine at Princeton and the University of Pennsylvania, Charles Meigs then moved back south to Augusta, Georgia, to establish a practice in obstetrics, not a conventional course for a young physician but one in which he evidently found his vocation, for between anato-mizing the uterus he wrote several volumes of practical midwifery. Whatever use to the public Charles's work might have been, it must have done no harm to his own family for ten Meigs children were born, and it may be that Charles stood midwife to the birth of his own son Montgomery in 1816. Unfortunately the doctor's own consti-tution was prone to suffer from "bilious fever," which aggravated a predisposition to romantic melancholy.

But there was a southern malady that Charles's wife, Mary Montgomery, could not herself abide: slavery. In deference to his wife's strong opinions, the obstetrician and his wife moved their family back north to Philadelphia, where Monty was raised in a learned but rambunctious family home on Chestnut Street. Of the eight boys and two girls in the family, it was Monty who in every way seemed to displace more than his own weight. Large, ungainly, obstinate, he was said (by his own parents) to be "tyrannical to his brothers, very persevering in pursuit of anything he wishes, very soon tires of his playthings, destroying them appears to afford him as much pleasure as his first possession; is not vexed with himself for having broken them . . . very inquisitive about the use of everything, delighted to see different machines at work, appears to understand their different operations when explained to him and does not forget them." In short, as Lee would later indicate, Monty Meigs was, from the start, a regular handful and for all his curiosity into things mechanical, "not very fond of learning." Neither the Franklin School nor a brief spell at the University of Pennsylvania managed to rein in the persnickety temper. As an old man Meigs claimed not to see in this description anything he recognized of himself. He was wrong. What Montgomery needed, so his desperate mother and father thought, was an institution that would convert all that uncoordinated energy into patriotic usefulness. Which sounded very much like the United States Military Academy at West Point. For there was nowhere quite like it for harnessing the fidgeting of the young to the solid work of building continental America.

When Montgomery Meigs arrived in 1832, West Point, sitting 200 feet up on the west bank of the Hudson Highlands, was a scattered collection of brick two-story barracks, their conventional roof pediments the only concession to classical ornament; plus a few separate houses for the instructors. There was a small parade ground and at the edge of the cliff a gun emplacement where light cannon pointed toward the river. It was that position that had determined West Point's location and its significance. Fifty miles upstream from Manhattan, it was sited at the point where the river narrows and makes a sharp bend. The place was, and despite the nuclear reactor visible downstream at Indian Point, still is, pretty enough to get painters out of bed early on spring mornings as the pearly valley light comes up. America's first recognizable "school" of radiance-drunk artists adopted it as their very own Yankee Rhine Romance, complete with bosky islets and pairs of red-tailed hawk riding the thermals. Before the full impact of the Erie Canal had been felt, hauling livestock and foodstuffs from Ohio, the lower Hudson Valley was a region where old forests of white ash and chestnut leaf oak had been cleared for sheep and cattle pasture. In the shade of the second-growth forest that sprouted over their ruins, you can walk the lines of the drystone walls that are all that remain of that long-gone grazing and droving world. When Meigs came to West Point, modest market towns like Cold Spring were beginning to multiply churches, schools, inns, stores. Their docks were full of sailing barges and the odd belching steamboat, and their little world was hectic with America's business.

Up on its hawk roost, West Point was more or less impregnable, a fact that had not gone unremarked during the Revolutionary War. Commanding the narrow neck of river meant that its guns controlled the passage between New York and the mid-Atlantic and upstate New York, Lake George, and the route to Canada. Domination of the Hudson Gorge, though, also delivered the potential of cutting off New England from points south. Whichever side held the fort on the hill could control the destiny of America. Acutely aware of its strategic importance, George Washington posted an "Invalid" Regiment: men whose wounds or infirmity made them unfit for battle combat but who could man guns; so that the first military occupation made the place as much a convalescent hospital as fortress. Thwarted at Saratoga from cutting the American resistance in two, the British needed somehow to take West Point, and from 1779, the turncoat general Benedict Arnold, in return for a cool £20,000, offered to hand it on a plate to His Majesty's forces. In 1780 Arnold, the fabled veteran of campaigns, whose loyalty no American generals doubted, secured command of the post and would have realized his plan and perhaps succeeded in ending the revolution, had he not been exposed by the capture of the British spy Major André along with documents revealing his intentions.

The first United States Military Academy was, then, built on a site heavy with patriotic memory; one which looked to the past to create a national future. Young Meigs could not help but be aware of that during his first year as a "plebe." That was also the last year of the superintendence of Sylvanus Thayer (class of 1808), who had done more than anyone to give West Point its character as a forcing house of scientific and technical distinction. Between reveille and dusk, between the first

drill parade and lights out, the days of the cadets were remorselessly filled
with instruction on mathematics, chemistry, engineering, mechanical
drawing, and even a little geology and history. French was mandatory
but not so the cadets could steep themselves in the Pléiade or Racine
but to memorize the textbooks that Thayer had imported from the
École Polytechnique in Paris.

"Duty, Honor, Country" was the college credo, and for the most
part the cadets embraced all three, except when they escaped from the
mess-hall fare of bread, potatoes, and fat-pork beans, to Benny Havens's
establishment at Buttermilk Falls a mile south. There they could enjoy
a tankard of hot flip and flirt with the country girls. Sometimes both
were smuggled into the school, which led Thayer, in a rash moment
after a wild 4 July celebration, to ban alcohol with the predictable
result. On Christmas Eve 1826, an eggnog party in which a young
Mississippian, Jefferson Davis, was the rowdiest ringleader, was broken
up by the Officer of the Day. If he was serious, Davis and his bucks
warned, they would have to shoot him.

Davis and his fraternity had violated the honor code, which exhorted
the cadets to selfless virtue. By "Country" was meant the Union, even
for the likes of Davis, who may already have felt his true country
was the South. The college was often known as the "School of the
Union"and its cadets the "Band of the Union." But it was the first
article of the oath that was most loaded with West Point's particular ethos. For "Duty" meant the duty to respect the Constitution of
the United States, to which its graduating officers swore an oath of
loyalty, which unlike that taken elsewhere was not to the person
of a sovereign prince. That constitutional obligation to subordinate
the military to the civilian guardians of the democracy was inculcated
in each and every cadet, and it still is. It's why there may have been
eggnog rebellions at West Point but never the hatching of military
plots. Throughout much of the world—in Europe, Asia, and Latin
America—military-school solidarity has led officers to believe in their
collective superiority over civilian politicians. Not in the United States.
Though there were plenty of American soldier-presidents in the nineteenth century, many of them West Pointers, they left their swords
and their uniforms (though not their war stories) behind them when
they went on the hustings. John McCain, an Annapolis naval cadet,
would do the same. For two centuries West Point has been a sentinel

against, not on behalf of, martial power. But then that was exactly the intention of the man who founded it in 1802, and who in 1811 declared that "peace has been our principle, peace is our interest and peace has saved to the world this only plant of free and rational government now existing in it."

That same man, tall, angular, and equipped with an elegant mind, stood in the spacious hall of his Virginia villa and challenged his guest, not quite so tall but equipped with an equally elegant mind, to a guessing game. Pointing to the three busts of Worthies that lined one wall, Thomas Jefferson asked Alexander Hamilton if he recognized the identity of "the trinity of the three greatest men the world had ever produced." There was a long pause, during which one imagines Jefferson smiling as he often did when he felt superior. Preempting Hamilton's failure, the host then revealed that they were Bacon, Newton, and Locke, the patriarchs of the Enlightenment to which he had nailed his own intellectual colors. Asked who *he* thought was the greatest of the great, Hamilton took his time before replying, with pointed insouciance, "evidently . . . Julius Caesar."

Jefferson's West Point was founded to deny the United States its Caesars (of whom, Jefferson suspected, Colonel Hamilton might aspire to be the first) and to ensure the permanent victory of liberalism over militarism. Only one of its greatest, Douglas MacArthur, super-intendent after World War I, has ever flirted with martial power to the point of disregarding civilian orders, or so his president, Harry Truman, suspected. It was MacArthur who introduced systematic political discussions into the early morning curriculum at the academy, so he had only himself to blame if his students read well enough to know that a victorious general had to defer to the civilian commander in chief. Much more typical has been the other kind of West Point graduate, Dwight Eisenhower, commander of a liberation invasion, president of Columbia University before president of the United States, and who warned his country against the threat posed by the "military industrial complex" to the liberties enshrined in the Constitution. When I went to West Point to deliver a lecture not long after the beginning of the war in Iraq, the cadets and I talked about Thucydides' *History of the Peloponnesian War.* The intense debates that preceded the fatal expedition to Syracuse that mark the great cautionary climax of the work had made a deep impression. No one in that classroom wanted to be Alcibiades, the vainglorious warrior who led the Athenian Empire into self-destruction. It struck me then that West Point was perhaps the only military academy in the world programmed to have such conflicted feelings about war. But its identity as a Jeffersonian enterprise of national education—the school that imprinted itself on the young and impressionable Meigs—only came about from a fierce battle between Hamilton's and Jefferson's contending notions of the role that military power ought to play in the life of the American nation.

Caught in the middle of that debate, the first commander in chief was himself much conflicted about how a nation born in war should handle the matter in its future. Washington's baptism by fire as a young man had been in the British Army's campaigns against the French and their Indian allies, but there he had witnessed at close hand the habitual contempt that officers like General Braddock had shown for colonial militiamen and volunteers, many of whom had laid out their own money for equipment to defend the British Empire. Washington's disdain, on the contrary, had been for the mercenary regiments through whom British parliaments and governments meant to enforce unpopular laws and taxes in America. It was a common-place among American Patriot politicians, inherited from seventeenth-century English Commonwealth writers, that "ministerial armies," in the phrase of the time, were the tools of despots, servile to their masters and brutal to everyone else. To defeat them was to do good work, not only for America but for the liberties of the world. The opposite of the "ministerials" were grievously provoked citizen-volunteers who would only take to their muskets in defense of hearth and home. Such men, in their own minds, were always citizens first and soldiers second. Their fight was ultimately *against* soldiering, and only entered into for the express purpose of getting troops quartered on the citizenry out of their peaceable towns and villages. Once that had been accomplished, no further purpose was to be served by remaining in arms. But the trusty flintlock had always to be kept in working order so that the contemptible ministerials would never be tempted to try their luck again.

Hence the symbolic, rather than military, importance of the initial "shot heard round the world" on 19 April 1775, when hostilities began in earnest in the small towns of Lexington and Concord west of Boston. There, the heroic tableau of locals—farmers, smiths, and innkeepers—mustering on the village green and at Concord Bridge to thwart ministerial attempts to seize munitions was literally enacted. It was that news that brought Return Jonathan Meigs among thousands of others riding hard to Boston in 1775. So it ought to have been natural for Washington to have celebrated the Massachusetts Minutemen or their Virginia counterparts, the Shirtmen, as the patriots who won the war. But much of his experience as commander during the Revolutionary War belied that myth. Militias had been notoriously hard to discipline; quick to mobilize but even quicker to disappear. Alexander Hamilton, who was a favorite member of Washington's personal staff, the group that he called his "family," conceded in his "Federalist Paper 25" that "the American militia, in the course of the late war have by valor on numerous occasions erected eternal monuments to their fame." But, he added, "the bravest of them feel and know that the liberty of their country could not have been established by their efforts alone." Hamilton knew that Washington's strategic genius and the French alliance had ultimately counted for more than raw patriotic ardor. He had had close ties with many of the foreigners who had come to America. His admiration extended not just to the most famous of them, the marquis de Lafayette, but to figures schooled in European arms like the Prussian baron General Friedrich Wilhelm von Steuben who had published the first manual of drill and exercises for American troops. At Yorktown—the battle that ended the war—Hamilton knew from personal experience how much was owed to the excavation of mines and tunnels drawn straight from Old World texts and which had allowed American troops to move in close to the besieged British. "War," he wrote in the same Federalist Paper, "is a science." It would not be un-American to go to school to learn it.

Even some of those personally cool to Hamilton agreed with him about this. As early as 1776, a more unlikely warrior, John Adams—who, however, could see in his native Massachusetts how tough a fight the war for independence would be—made a proposal to establish a military academy, meant to train officers who might be called on in times of emergency. No one—at least no one in Congress in a position to fund the idea—paid much attention, and some attacked it as incompatible with American liberties. After the war was over, in 1783, Hamilton chaired a committee to study the new republic's military establishment in peacetime but knew that he would always run up against solid congressional opposition for anything ambitious. The volunteer army was stripped back to barely a thousand. But President Washington and more his brigadier general of artillery and first Secretary of War, the sometime Boston bookstore owner Henry Knox, continued to brood darkly on what the country might need for future preparedness. And much as they hated to admit it, American security began to face domestic as well as foreign challenges. The next five years put the reality of American federal government to the test by attacking its tax collectors and arsenals—in Daniel Shays's 1786 rebellion in western Massachusetts, and in 1791 in the Whiskey Rebellion west of the Alleghenies where excisemen attempting to collect taxes on spirits were the target. Washington called out the militia to put them down but was not very confident about the loyalty of troops from the disaffected regions. It took militia from other states to deal decisively with the rebels. The president gloomily recognized the ironic parallel with what had happened before the revolution, with his bluecoats now playing the role of oppressor. The difference, he assured himself and the country, was that this time the taxes were being levied in the name of an elected government. (But of course the same thing was being said by British parliaments in the 1760s and 1770s.)

Washington had no intention of using American soldiers against their own fellow citizens unless they had cast off their allegiance to the elected government of the United States. And he was sufficiently exercised about the threat to liberty posed by "standing armies" to hope that the foreign policy of the United States would stay aloof from Europe's wars, so that the temptation to create a large army would forever be avoided. This instinct was Jeffersonian: the belief that if somehow Americans could turn west and mind their own farms, they would forever enjoy uninterrupted blessings of peace and liberty. But the pragmatic, Hamiltonian side of Washington knew this was just a pious hope, for the Machiavellianism of the European powers was unlikely to abate just because the United States had grandly declared a Novus Ordo Seclorum (a New Order of the Centuries) on the Great Seal. Nor were the Europeans likely to confine their machinations to the old continent since there was too much at stake in the way of money in the new, where they were firmly lodged in Canada, Mexico, Florida, and Louisiana, not to mention the sugar-rich Caribbean. Even supposing the United States remained for a while in a purely defensive posture (and there were many, including Henry Knox, who felt that, ultimately, British and American co-occupancy of the land mass was unrealistic), the natural demographic increase of the country was bound to provoke friction with the other powers on the continent. Sizable French and British armies were already invested in their West Indies, attempting to put down slave rebellions and Creole discontent; their navies were still formidable presences on the oceans, capable of locking down American trade and blockading harbors should they choose.

Though financially shackled by Congress, Henry Knox was all in favor of military readiness, starting with a school that would make future generations of skilled artillerymen, engineers, and officers of horse and foot. The two colonels—Hamilton and Knox—indulged in petty rows about ranking order, but they both thought that their personal military experience made them better judges of what was needed for America's survival than the penny-pinchers of Congress. Hamilton and Washington worked together on the last speech that the president delivered to Congress in December 1796, which included his wish that both a national university and a national military academy be established. Privately Washington doubted Congress would ever fund it.

He was right. But the idea didn't vanish altogether. Already, in the early 1790s, West Point was being mentioned as a possible site for such a school. Knox had commanded the artillery bastion there, and Alexander Hamilton had actually been at the fort when Benedict Arnold's conspiracy had been thwarted. For Hamilton, then, it was entirely right that this should be the place where generations of American cadets would be instilled with the imperative of vigilance. Where better to professionalize the need for military readiness, the virtue without which Hamilton feared the republic's independence would be little more than a paper declaration? Should Congress ban the establishment and training of a peacetime army, Hamilton wrote, the United States "would then exhibit the most extraordinary spectacle

the world has yet seen, that of a nation incapacitated by its Constitution
to prepare for defense before it was actually invaded." He had a point,
but it's also true that temperamentally Hamilton was trigger-happy. As
a boy in the West Indies, he had dreamed of soldiering for the empire
and throughout his life had the impulsive streak that made his eventual
death in a duel not altogether surprising. The avuncular Washington
was not blind to Hamilton's tripwire devilry, but it was hard for him
to take against the lieutenant colonel who, bayonet in hand, had led
the storming of the British redoubt at Yorktown, a maneuver that
arguably ended the war.

Alexander Hamilton was not just dash and danger. His postwar
career, whether in office as Washington's secretary of the treasury,
or out of it, was spent thinking what kind of figure the United States
(the two words of which for Hamilton were problematically weighted)
would cut in the world. For Jefferson, the republic was supposed to be
a marvelous new organism utterly unpolluted by the atavistic habits
and warped customs of the old. It would resist the tendency that
states, even those boasting parliamentary pedigree like Britain, had
of sliding inexorably into oligarchic corruption and tyranny. And that
meant it would have no need for professional armies of any size and
should indeed always suspect their potential for political mischief. Pay
no heed to jaded lessons from the past that insisted that states, like
men, could never wean themselves from their habitual savagery. The
United States had been born to refute the cynicism that a fresh start
was not utopian, and to prove that it was entirely possible to live as a
republic of free men and yet be a moral force in the world. War was
at once the functional need and customary habit of aristocracies and
despots. Do away with the latter, and you did away with the former.
The coming of the French Revolution and Jefferson's own witness of
it in Paris only confirmed his belief that the mighty shift from despots
to democracies would obviate the habitual need for war—except as a
last resort to defend liberty.

Hamilton heard what he considered all this naive Jeffersonian optimism and rolled his eyes. Let Jefferson indulge himself in philosophical
entertainment if it amused him, but let him not do so, Hamilton
thought, at the expense of American security. It was childish folly
to pretend that the political virginity of the United States would be
sufficient protection against the predators who prowled the oceans

and swarmed across continents with armies numbering tens, hundreds, of thousands. Had not Jefferson and the gentlemen who thought
in his fashion observed what had become of the professedly peaceful
pretensions of the revolutionary *"République une et indivisible,"* the
"grande nation" that, while disclaiming conquest as the obsolete sport
of tyrants, somehow had managed to occupy—and plunder—most of
western Europe. Their war to "defend liberty" had become a transparent pretext for empire.

Hamilton urgently wanted his nation to grow up. Unlike his reluctant co-Federalist John Adams and his bitter political foe Jefferson, he
had no qualms about looking to the biggest success story of all, the
British Empire, as an exemplary model of power. What—other than
its unfortunate moment of American coercion and the tendency of
its ruler to lose his wits now and then—was actually *wrong* with the
British state? The answer was nothing! Britain had had the wisdom
to accept its defeat and concentrate on consolidating its power where
it mattered—against the French in Canada and India, on the oceans.
Good for Mr Pitt! For Hamilton, it went without saying that there
were certain instruments of economic and military heft without which
a stance as a great power was unthinkable or at any rate unfeasible: a
national debt and a bank of issue, which, as Washington's secretary of
the treasury, he resolved to establish in the United States. Hamilton
noticed as well that although many commentators characterized old-
regime France as top-heavy, the real machine of pure state power was
in Britain. Its officers of revenue and excise—resources without which
even the most virtuous of republics could not survive—swarmed over
the country, virtually an army unto themselves.

And then there was the Royal Military Academy at Woolwich—a
mean and skimpy thing as military schools went and nothing at all
to give the impression of a Britannic Prussia. But Woolwich still
offered the rudiments of instruction in the military sciences, and it
may be that Hamilton had heard of steps afoot to expand the education of arms more systematically. Or perhaps Hamilton looked at the
breathtaking aggression of the French Republic and knew that it had
nothing to do with the republican élan (as Thomas Jefferson, who
had never fired a gun in anger, fondly imagined) and a lot more to
do with the incentive of loot and power that Bonaparte nakedly held
out to his soldiers. At least, he might have said, there is one citizen-

general who doesn't speak humbug. But from his French friends in
the Revolutionary War, he would also have known that none of the
spectacular successes of the French Republic in the field could have
happened without a prerevolutionary officer class intensively educated
in military technology. Why should the United States, blessed as it was
with all manner of such practical learning, deny itself a college where
young men of aptitude could create a comparable elite of scientifically
minded officers? Meritocracy was power. On that, at least, Jefferson
and Hamilton could agree.

But Congress continued to rule that such places were inconsistent
with the "principles of republican government." (And they would cost
the nation money it could ill afford.) Instead of being an academy of
virtuous, democratically minded citizen-soldiers, such a school was far
more likely to breed a military caste, aristocratic in demeanor, separate from, and contemptuous of, civilian society. Worse, such a place
might put itself into the hands of some self-appointed hero who had
evil designs on the republic. Such horrified imaginings, which seem
so far-fetched now, were part of the hot war of principles dividing
the politicians of the young United States. Federalists like Hamilton
were unafraid of power and believed that the nation could never
survive without its vigorous and unapologetic exercise. Anti-Federalist
Republicans, champions of states rights like Jefferson, believed that
if the power of the government were not strictly confined by the
Constitution, it was all up with democracy. For both opposing groups,
the fight over the citadel on the Hudson River was a fight over the
future of America.

And then, quite suddenly, the debate became less like a seminar
where abstract theories contended for America's future, and more
like a crash course in the thorny realities of foreign policy. This loss
of innocence began with a development that should hardly have
come as much of a surprise. The Jay Treaty regularizing relations
with Great Britain, signed on 19 November 1794, had been taken by
republican France, then fighting a ferocious war against the British,
as an ungrateful repudiation of the alliance that had created the
United States in the first place. The Americans did what they could
to represent the Jay Treaty as a disentanglement. But in the French
government's view it was a shocking violation of republican solidarity. Since the directors in Paris believed that in this life-and-death

struggle for the survival of popular revolutions, all who were not with them were against them, the Jay Treaty was not just the betrayal of American promises never to negotiate a separate peace but, in effect, an alliance with their deadliest enemy.

This was a moment when what had been assumed by Washington to be America's blessing of distance did not serve diplomatic understanding well. Had the United States a better grasp of the impossibility of neutrality in what had become a world war of ideologies, it might have had some pause in its rush to disarm. On the other hand, in justice to Washington and to John Adams, who succeeded him as president in 1797, given the relentlessness of that total war, was the diplomatic freedom of the United States to be held forever hostage to those earlier engagements? If so, the Federalists pointed out, they would have merely exchanged colonial masters. Threats and bluster coming from the French in the wake of the Jay Treaty made the scales fall from American eyes. Illusions about the altruism with which France had undertaken to liberate America were now judged sentimental. Instead, that entire enterprise seemed less a disinterested expedition for liberty and more an exercise in French imperial gamesmanship.

The ways in which France then proceeded to make the United States pay for its temerity only confirmed to Hamilton, Adams, and the Federalists that they had done the right thing by signing Jay. While they were frantically attempting to build the first ships for a United States Navy, they banked on the Royal Navy getting the better of the French *marine* in American waters. This turned out to be a poor wager, for the Royal Navy was not about to put itself to the trouble of protecting American merchantmen from the attacks of the French. If it was the armed benevolence of the Crown that Americans sought, they ought not to have sundered themselves from it in the first place. What then followed was a savage yet undeclared war at sea between France and the United States, and on a scale that the United States government could hardly have anticipated. From May 1796 to March 1797 over 300 merchantmen were taken by French privateers and naval vessels.

For a while it was what came to be called the "quasi war" but the real thing seemed only a matter of time. In the patriotic furor that gripped the Eastern Seaboard, President Adams, flossy-pated and rotund, swaggered around wearing a sword through his sash like a cross between Mr. Pickwick and Napoleon. For the only time since the opening of the Revolutionary War, Adams heard the thunderous applause of American crowds. For a while the president's head was turned by the sound of bugles, and he acted accordingly. On the grounds that an immigrant nation preparing for war had better beware of spies and a fifth column, Adams and the Federalists began to take liberties. In 1789 an Alien Act gave the government rights of summary imprisonment and deportation. Naturalization time for citizenship was raised from five to fourteen years (there was more than a streak of xenophobe in Mr. Adams, who never ceased to think of Hamilton as a "foreigner"). A Sedition Act of the same year made it a criminal offense for persons to libel or even attack the United States government and its president. Not least, Adams sent a bill to Congress to finance the raising of a volunteer army. Hard-line Federalists wanted $20,000, and Hamilton $30,000. Congress gave them $10,000. In the same bill provisions were made for the funding of a modest degree of military instruction. Just four teachers were to be sent to educate the cadets already stationed at West Point in the engineering of mines and tunnels and the like. But it was a start.

Even though all of this came about through the agency of John Adams, whom Hamilton despised as an irascible egotist, he agreed it was necessary for the well-being of the country. Hamilton began to think of the new force as, in some sense, "his" army, and for the good reason that Washington had been persuaded to come out of retirement to command it, for he was the only person who could silence doubts about the army's political neutrality. But Hamilton had also managed to become Washington's second in command, which, given the great man's advanced years and uncertain health, meant that Hamilton was the general-in-waiting. Hamilton's fertile brain now began to quick march. Legions, divisions, uniforms, drills—all were set down on paper and sent to the secretary of war, McHenry. In short order, Hamilton also devised a complete curriculum for the cadets of West Point: four years, half of which would be spent in common at a "Fundamental School" (heavy on the mechanical sciences, but also with a healthy dose of history and geography) and the remainder in whichever military subdivision the cadet would be destined for: cavalry, infantry, artillery, or engineers. He had in mind around 200 cadets taught by six directors and eighteen faculty. Most important of all, officers already on active service would be required to rotate through the academy.

None of this materialized in the way Hamilton had imagined. The belated warrior Adams suddenly turned nervous about taking on a war with France. It may be that the spectacle of Hamiltonian men-in-arms springing from the field like the harvest of Jason's dragon's teeth gave him pause. At the other end of the world, Horatio Nelson's annihilation of the French Navy at Aboukir Bay at the mouth of the Nile on 1 August 1798, and Bonaparte's subsequent abandonment of his army in Egypt, followed by the coup d'état that made him first consul, evidently made an Atlantic war less of a priority. He was already facing a renewed attack from the coalition monarchies. With the threat of a French war dissipating, so did the need for Hamilton's new army. Washington's death in December 1799 put an end to it altogether. What remained, though, in early 1800, was the plan for the academy approved of by Adams, as he had always been an enthusiast of the idea. Congress, however, was less happy. There were noises about its undemocratic potential from the anti-Federalists. And for Thomas Jefferson, the whole idea smelled of Hamiltonian Caesarism.

Thomas Jefferson had been taken aback by the war crisis and what Adams and Hamilton in their respective ways had made of it. Jefferson had been the agent of the United States in Paris in the early years of the French Revolution and though he had witnessed some of the worst abuses of the Jacobin "Dictatorship of Virtue," emotionally he had never been able to uncouple the French Revolution from the unfolding history of the dawning age of liberty. It had been the French who had been forced to defend themselves against the monarchs of the coalition powers and who embodied what, in 1793, he told the French envoy to the United States, Edmond Genet: "By nature's law, man is at peace with man till some aggression is committed which, by the same law, authorises one to destroy another as his enemy." For Jefferson, that alone explained why France had reluctantly turned into a belligerent state and one in which individual liberties had been regrettably curtailed for the needs of security. Forced to choose between the British and the French, he had no doubts who were liberty's true enemies. The Anglophilia of the Federalists, especially Hamilton, he thought, was all of a piece with their design to introduce into the United States the strong-armed executive government power against which the Revolutionary War had been fought, making independence a pyrrhic victory. Henry Knox, who had founded an "Order of Cincinnati," as a hereditary association of ex-officers, Jefferson thought, was introducing a military aristocracy into the country, possibly even a monarchy with a second King George, reigning from Mount Vernon. Constitutionally entitled to be vice president to Adams (though making himself the leader of opposition to his policies), Jefferson thought the usurpation of power represented by the Alien and Sedition Acts augured the death of the free America.

Jefferson was no naive pacifist. But of all the Founding Fathers he was the most heavily invested in an eighteenth-century philosophical idealism that looked on war as the sport of tyrants. In 1775 he had been entrusted by Congress with articulating a defense of the insurrection, and much of that turned on the British king's loosing of mercenaries on defenseless American farms and shores. In his draft of the Declaration of Independence George III appeared as a Hanoverian Genghis: "He has plundered our seas, ravaged our coasts, burned our towns and destroyed the lives of our people." The British monarch and his governing toadies had planned even more heinous stratagems: the arming of Indians and, worst of all for the plantation owner of Monticello, a cynical flirtation with slave rebellion. And unlike Hamilton's war record, Jefferson had been on the sharp end of its bayonets, forced abruptly to flee Monticello in 1781 when a company of cavalry, attached (of all humiliations!) to a regiment commanded by Benedict Arnold, threatened his home.

For Jefferson, then, the revolution had been a war to end war, at least in the Americas. Later, in 1812—in the midst of ferocious belligerence in both America and Europe, the year in which Washington and President Madison's White House were burned by the British— Jefferson looked forward to a day when the Atlantic meridian would be "the line of demarcation between war and peace. On this side . . . no hostility, the lion and the lamb will lie down together." He conceded that situations might arise when it might become unavoidable, but American statesmen ought to resort to the arbitration of arms only after every conceivable avenue of diplomacy had been

exhausted. If it did become unavoidable, every effort had to be made
to moderate its barbarism. Prisoners had to be treated humanely; all
thought of torture ruled out; civilian sufferings avoided. Wars ought
to be limited in scope and time, for if they were not, they would
consume the liberties established by the Constitution. The war fever
of 1798–99 had been just such a lesson, for while the Constitution had
clearly granted to Congress the authority to declare war and make
peace, the Federalists had used their bare majority to usurp to the
executive much of that power. For Jefferson this was a dire omen.
He could imagine all kinds of trumped-up pretexts for war that would
turn America into just another squalid military adventurer with a
top-heavy, fiscally indebted government draining away its resources.
Was it for such a morally deformed America that patriots had shed
their blood?

The French war crisis had been a close thing. Hamilton's "new
army" and the war college associated with it had been a dagger pointed
straight at the heart of republican democracy. The only way to ensure
that such sinister follies did not recur was to turn the Federalists and
their president, John Adams, out of power and replace them with
Democratic Republicans, who could be relied on to avoid elective
wars and the whole vampiric apparatus of taxes and debts that went
with them. As president, Jefferson would, in fact, fight a war against
the Barbary corsair states of the North African Maghreb and send
warships to Tripoli and Algiers. But he always believed that action
was a defensive response to mortal threats to commercial shipping,
America's lifeline. The aims of the North African war were limited:
the liberation of American captives, an action that would make it clear
to the beys and sultans that the United States would not be held to
ransom by glorified pirates.

Why, then, with all this aversion to an embryonic military establishment, did Jefferson become the founder of West Point? In his
mind it all made perfect sense. By making himself responsible for
such a college, he could immunize it from the war lust he associated with the Federalists, making sure that its faculty and officers
were trustworthy democrats, sworn to defer to the civilian powers.
Instead of becoming the nucleus of an antidemocratic officer caste
within the republic, a viper in the bosom of liberty, West Point's
graduates would be something like the opposite: the tutors of

citizens-in-arms. Jefferson meant its graduate officers to go forth into the country and prepare and instruct the state militias in matters like artillery and the building of forts, thus obviating the need for a large and menacing professional army. If that made the officers sound more like teachers than soldiers, well, that was also the idea. For Jefferson, West Point was an element in his ambition to educate America for the modern world. That meant less theology and classics and more science and technology. While he was struggling to get his University of Virginia launched and funded, West Point could be created—on a modest scale—as a mini university, equipped to teach mathematics, chemistry, geology, architecture, and, of course, engineering. If a military academy could function as a modern university, it could give its students the kind of education that would equip them for other more civil vocations than the endless, self-generating pursuit of arms. And it would create a cadre of guardians who would stand against any threats to civil freedom.

That was the West Point Jefferson chartered in 1802. Its congressional funding, and therefore its initial scale, was extremely modest. There would be no Hamiltonian riding around in uniformed swagger. The cadets—all twelve of them—would be dressed in sober gray. They would be strictly tutored and held to the highest standards of academic and technical excellence and drilled in their obligation to honor the Constitution and a civilian commander in chief. In their ranks there would be nurtured no American Napoleons. Jefferson's appointment as first superintendent was Jonathan Williams, a mathematician whom he had met in Paris when Williams had acted as secretary to Benjamin Franklin. Williams had no military experience whatsoever, and this, for Jefferson, made him the perfect candidate, as West Point was meant to be more of a school and less of a war college. When Superintendent Williams said that "our guiding star is not a little mathematical school but a great national establishment . . . we must always have it in mind that our officers are to be men of science," Jefferson could only applaud. What would be the vocation of the "long gray line" of cadets? They would be nation-builders, the engineers of democracy. In the Jeffersonian mind that has always been what the American military has been *for*!

3. The Drop Zone Cafe, San Antonio, Texas, 3 March 2008

"But that's not what the military's *for*," said the retired general when I asked him if the army could have done more to repair the Iraq it had broken, by delivering a modicum of infrastructure. The odd bridge or two would have been nice. But for years after the statue of Saddam had come down, civil society in Iraq remained smashed up. Only a few hours of electricity had been generated for the cities; less oil was flowing to the refineries and out again toward the city pumps than in the time of the dictator. Roads and city streets were murder alleys; mosques were dangerous on solemn days; hospitals were without basic drugs and often without doctors. Desperately needed professionals had departed en masse for Syria and Jordan, where entire new universities had been created around the exile population. Billions of dollars meant for the reconstruction of Iraq, packed into suitcases, had gone unaccountably AWOL, and no one seemed to think this was a big deal. Newspapers of record registered righteous dismay and then shrugged their shoulders and moved on. Construction companies awarded no-bid contracts had bungled the job after pocketing front-loaded operational budgets. An American government that had begun its administration by declaring it wasn't in the business of nation-building had embarked on the biggest exercise of all. But after it had taken down the tyranny standing in its way, it turned out to have no clue about how to establish a successor state, being philosophically hostile to public administration. But acting as midwife to a democracy in a place innocent of it was "really not the army's brief," repeated the general, flashing me a high-voltage smile and taking a gulp of Sunday morning Mexican coffee.

General Alfredo "Freddie" Valenzuela and I were sitting in the favorite brunch haunt of local veterans in San Antonio, Texas, America's "military city." The brotherhood was Hispanic: men who'd been born to farmworkers on the borderland, had joined the service as teenagers and, in some cases, risen far and fast. They were tucking into chiquiles—eggs with jalapeño chilies—and in case that wasn't hot enough, upending the Tabasco to give breakfast a little more excitement. The Bloody Marys, no celery, were on the house. The good-looking, prosperous kids of the vets sat with their families while trying to keep an eye on their own small children roughhousing at the back of the room. Most of the patrons were "Jumpers" from the 82nd Airborne, and the walls of the long shacklike place were papered in photographic memories of bad places they'd dropped into: Anzio, Normandy, Arnhem, the Battle of the Bulge. In most of the old pictures they mug for the camera, arms slung around buddies, wives, or girlfriends. But a few of the jumpers stare dead ahead as though the lens had just challenged their face to a fight and they were waiting to see who would back off first. The place was loaded with tough-guy charm, handsome sunburned men who'd grown old setting their stories out on the tables like decks of cards and staying cheerful when they got trumped. Their tightness was all about where they had come from, the barrios and the pueblos; clawing their way to respect and rank; the unembarrassed pride in having, in the end, got both. Nicknames of greeting and acclaim were shouted over the heads of their wives, as more of the boys stepped through the door—"China Boy!" who sported a swinging gold Christ around his neck, arms above his head, a paratrooper Jesus, had trained Nung people as anti-Vietcong guerrillas; another of the brethren, "Jumpin' Joe" Rodriguez, had clocked over 6,000 drops as a training instructor and combat trooper. Amid the easy-over cheer, the good-natured General Valenzuela, who had taken on the FARC in Colombia and was therefore unlikely to be a pussycat, was reluctant to get back to the painful matters in hand. But he made it clear that John McCain, who in Vietnam had taken history on the body, was his kind of president. I pressed him on McCain's incautious aside that if necessary the United States needed to be in Iraq for a century. "Oh," said Freddie, black eyes merry with inside knowledge of the military man's cavalier way of putting things, "what he meant is that we can't just be up and running. A sixty-day exit strategy isn't going to work." "So you stay until when, exactly, General?" "Until they've got things better under control." "And how do we know when that day comes along? The army can't stick around seeing to the generators and the oil wells forever. You said that yourself." A big cloud briefly darkened the general's sunny countenance. "We need to withdraw with honor," he said after a long pause. "We need to rebuild our alliances. It's tough. It's tough." And then, back came the disarming smile.

The evening before, I'd seen Valenzuela surrounded by comrades at the dress-up veterans' reunion ball. Mariachi bands serenaded tables of men with prosthetic legs and hooks for hands who nonetheless took their wives off to the dance floor between the London broil and the ice cream dessert. By the cash bar, kids fresh from Afghanistan firefights with the Taliban were being inducted into the fraternity of memory by avuncular survivors of Khe Sanh and the Tet offensive. The guest of honor was General Ricardo Sanchez, one of San Antonio's own, a poor kid who had made it all the way to command of the forces in Iraq. But what must have seemed like a dream promotion for a three-star general had turned into a nightmare. Iraq couldn't be fixed. Too many people—Sunni insurgents, Shi'a militias, the Iranians—all had an interest in it staying broken. Worse, it was on Sanchez's watch that the images of unspeakable sadism of Abu Ghraib reached the world: Lynndie England pulling a naked prisoner along by a dog collar. Since he had signed a memorandum of "acceptable" interrogation techniques (designed to prevent, rather than authorize, torture), it was Sanchez who took the rap for the atrocities. Two years later, after desultory efforts had been made to send him to this post and that, Sanchez figured out he had become an embarrassment and retired, terrible odium hanging over his name. But as far as the jumpers of the 82nd were concerned, he was a brother and a son (even if not of their division), and when he proposed the toast "to the ladies" they clinked glasses and roared back.

Seeing General Sanchez joshing with the crowd lost me whatever appetite I might have had for the beef and potatoes. I expected his speech to be heavy with regimental camaraderie from which he could segue to disingenuous self-exoneration, leaning on the buddies for understanding. I was right about the appeal to martial honor and wrong about the rest. Notes of surprising disquiet crept into his remarks. No self-serving calls to circle the wagons against the Civilians Who Didn't Understand the Facts on the Ground were made. Instead there was a call to vote; to pull the lever for whoever they thought could offer the country foresight, wisdom, strength in a hard time. "Send a message," the general said; but the message was supposed to show, by the sheer numbers who delivered it, that if men in office (especially men in office who had never worn uniform) were sending youngsters to kill and to die, something more was owed the country by way of explaining the precise point of the sacrifice. Intoning "9/11" and "fighting the terrorists there before we have to fight them here" was no longer enough to satisfy the troops, either in the field or at the polling station.

And there he was again at the Drop Zone, dress uniform and medals replaced by checked shirt and jeans, shoveling his breakfast around the plate in one of the side booths. The instinct in me that wanted to talk to him fought with the instinct that screamed "are you *nuts?*" and then won. I slid into the booth and did the introduction. Sanchez is a compact man with widely spaced large, dark eyes and a broad nose, like a sharp prairie marmot in tinted glasses. He could not have been friendlier, but then this is a man who needs friends in the worst way, not to mention readers, for, inevitably, there was a book coming out, *Wiser in Battle*, the self-exoneration from which he had refrained at the ball. Even so, I was surprised at the bristling hostility to his former superiors in the Bush administration that he was prepared to lay out on the table. I hardly had begun to probe him on the failings of the planning for the aftermath of the war when he finished my sentence for me: "No real strategy, none at all, beyond getting to Baghdad." It was well known that Sanchez had barely been on speaking terms with the civilian governor, Paul Bremer (who doubtless would have his day in print), and that the two men had agreed on virtually nothing that had to do with how the military might help build infrastructure as well as tangle with the insurgents in Falujah.

"Was that your business, then, putting down the foundations of a working state? Doing the engineering?" I asked him. "How could it not be," he said, looking up from his breakfast, "seeing as everyone else whose business it was was doing such a poor job?" So Sanchez was not one of those who thought the army was just for fighting. I didn't have to tease the history out of him. Out without prompting came the honor roll of 1945, many of them West Pointers: Bradley, Eisenhower, Marshall, Clay, the generals who *did* have the strategy for peace as well as war, generals unafraid of governance. "They were visionaries," Sanchez said wistfully, "but, heck, they were professionals, real soldiers who knew what they were getting into; who knew how to make things work, a democracy, for instance." "It didn't happen this time, did it?" I added gratuitously. "It did not," he said.

So, in hindsight, was this a war that at all costs *had* to be fought?

I asked him. Aren't those the only wars for which the United States should think of sacrificing its children? He lowered his gaze, took a stab at an egg, then looked back at me and said, "Sure." I didn't know which of the two questions he was answering.

4. The trials of the Roman

The Meigses were staunch Jeffersonians. How could they not be? They were all over America—Ohio homesteaders, Georgia merchants, Philadelphia doctors. They subscribed to the vision of their country as a new polity in the world, the first true "empire of liberty" as Jefferson had put it. The difficulty of reconciling power, freedom, and justice did not dampen their patriotic energy, although Mary Meigs, Montgomery's mother, feared for the Union should her southern relatives support the expansion of slavery into that empire of liberty.

Idealism at West Point lived on in one fundamental aspect of the institution, its commitment to civil as well as military engineering. Sylvanus Thayer had resisted teaching the subject alongside the sciences of fortification and ballistics; but West Point's Visiting Board, appointed in Jefferson's spirit, had insisted on it. And the academy became America's only school of technology and engineering; the elite members of each class taking instruction from Dennis Hart Mahan, who, like Thayer, had had a European as well as American education, and then joined the United States Army Corps of Engineers. It was West Pointers who constructed America physically and materially in this period, Jefferson's dream of a westerly-stretching empire of liberty a real possibility—through the pioneering survey maps; the building of roads, bridges, and canals; the dredging of harbors; the protection of ports from natural as well as foreign threats. That sense of patriotic vocation was what made Montgomery Meigs put up with the foul smells and the ferocious heat of St. Louis in 1837: the conviction that he was America's centurion-engineer, out on the far provincial frontier, the *limnes*, creating and guarding the new Rome with as much integrity and tireless zeal as the ancients. Ten years later word would come to him of the exploits of Lee and other brother-officers from West Point in the Mexican War where

General Winfield Scott was bringing slaughter to Mexico. But Meigs reassured himself that the peaceful work he was engaged in would ultimately redound more to the happiness of his country than the annihilation of the Mexican Army, the despoliation of their people, and the annexation of Texas.

This was Meigs's West Point talking: the ethic planted by Jefferson that, for American soldiers, sustaining life, repairing damaged social fabric, and building anew was as much part of the military mission as lessons in killing. Only in America was a corps of civil engineers instituted as the highest elite of the army. On the Web page of the U.S. Army Corps of Engineers the spirit of Lee and Meigs on the Mississippi lives on, complete with images of levee restoration in Louisiana and bridge building (in every sense) in Afghanistan. In recent years, the corps has had to try to discharge its historic responsibilities with fewer resources, since it's suffered the same kind of "streamlining" (a term used without any trace of irony for the great river-minders of the nation) as other branches of the government. Ironically, then, precisely the branch of the armed services that might have made the American presence more welcome in Iraq has been the one most starved of funds, which have gone to more purely military exercises. The war-winners have been seen, until very recently, as optional auxiliaries. Similar damage to the corps has been sustained at home, where over a hundred levees, dams, and dikes for which it has maintenance responsibility have been classified as in serious danger of breach. When the corps fails to deliver on the high expectations made of it, whether in New Orleans or in Baghdad, the sense of falling short is registered with painful acuteness at the place where it began, up on the Hudson Highlands. Go into a West Point classroom, and the odds are that you will find nineteen-year-old women and men grappling with hydraulics rather than ballistics.

This, at any rate, was the kind of lieutenant that Montgomery Meigs became in the years after his expedition to the Mississippi with Lee: a master of the theodolite as well as ordnance. On his way back from St. Louis, Meigs had crossed the Alleghenies in a sleigh, had ridden the new Baltimore and Ohio railroad, and done the rest by boat and horseback. He knew exactly what it took to throw the American idea across the continent and still keep faith with it. It would never have occurred to him that the vocation of the army was not democratic nation building, beginning with his own.

This did not preclude a strong military element in his work for, so American officers believed, there was still an obstinate and vengeful enemy that wished to do the United States harm, namely the unreconciled British Empire. Hard as it might be to credit now, the Canadian frontier in the late 1830s and 1840s was an unstable and unpredictable boundary. The most militant American nationalists claimed the entirety of the northwest frontier up to the fifty-fourth latitude boundary of Russian Alaska, a presumption the British had no intention of conceding. Since the attempt in the 1812 war to take lower Canada, the British had every reason to be vigilant about American designs on the colony, especially when there were Canadian rebels actively seeking the support of American irregulars. Despite official American neutrality, skirmishes occasionally turned into real battles, seizures of ships on the rivers and lakes, raids and retaliations across the shifting frontier. As long as the border was unsettled, Congress neither wanted to stamp on the action, nor wanted to give the British provocation for a full-scale third American war. What it needed, either way, were forts, and in 1841 Congress finally appropriated funds for a chain of them across the northern frontier.

That was Montgomery Meigs' first major posting after the Mississippi. Following the work with Lee, the Corps of Engineers had returned him to Philadelphia which meant a reunion with his family. Amid the domestic comforts—gardens, songs at the piano, promenades—Meigs fell in love. Louisa Rodgers was graceful and lively rather than conventionally pretty. Photographs of her taken by her keen photographer-husband show an attractively strong face—a powerful nose and jaw, dark complexion, and thick black ringlets. Louisa was vivacious and forceful like his mother, Mary, and her grandfather Commodore John Rodgers was the most famous naval hero in American history after John Paul Jones. They married, the children came quickly and often, and in 1841, Meigs took his family northwest to the Detroit River, on the edge of the British war zone. There, Meigs spent eight years building Fort Wayne, named for "Mad" Anthony Wayne, the general under whom his great-grandfather Return Jonathan Sr. had served at Stony Point—and who had taken Detroit from the British. That enemy still seemed to be at the gates of the United States. Should its troops cross the lakes and scatter the modest frontier force, they would never, Meigs thought, be able to take Fort Wayne. Everything he had learned from Mahan at West Point—Roman and French fortification science, especially the work of Louis XIV's pet genius, Sebastien Vauban—went into the formidable structure. Built from primitive, economical materials—packed earth, fronted with thick cedar rather than masonry—its star form, taken from Vauban's classicism, allowed for projecting artillery bastions on each of the protruding points. Bearing in mind the British habit of burning and razing everything in their way, Meigs turned the barracks into an inner stronghold: its walls, made from local limestone rubble, twenty-two inches thick. The barracks, gabled and pedimented in American-Palladian grand style, still stand by the river at the end of Livernois Avenue in a tough area of the city, the property of the city of Detroit, which opens them on summer Sundays for Civil War reenactments.

So while brother West Point officers were pushing the American Empire south, carving a path in fire and blood all the way to "the halls of Montezuma" in Mexico City, Meigs became Captain Meigs, the American Vauban, unrenowned in the world, but rapidly acquiring a reputation for engineering competence and integrity as solid as Fort Wayne. In Washington, the Army Corps of Engineers knew all about the formidable Meigs: his unhelpful aversion to the bribes and kickbacks that were a routine part of frontier construction; his omniscience; his eagle-eyed passion for minutiae. Nothing doubtful got past his scrupulous inspection. It was at this time that Meigs began filling small green octavo and duodecimo notebooks with encyclopedic observations, drawings, and pasted cuttings on *everything* that came his way—topography, architectural details, load-bearing problems, the customs and appearance of this or that Indian tribe, the state of local roads and canals—all dashed down in his high-speed hand which, for someone constantly taking on yet more work, was never quite fast enough, necessitating after 1853 his using Pitman shorthand. (For the historian, Meigs's shorthand is even harder to read than his longhand.) Trust went a long way to overcome illegibility, though. Toward the end of the Civil War General Sherman signed a procurement order saying, "the handwriting of this report is of General Meigs and I therefore approve it but I cannot read it." Meigs's notebooks are full of the routine toils of supervising excavations and foundations, masonry and timberwork, roof trusses, joists, and pulleys, but they also breathe a scrupulousness rare for the time, for recording every load of material, every day's hire of work. In the golden age of the huckster, Meigs was flint: wise to the wiles of land agents, gunsmiths, timber haulers, boat captains, anyone looking to make a killing, not just from the overstretched government, but from America's homesteading immigrants, the multitudes looking to find a home that was settled and safe. It was the good faith and credit of the republic, he thought, that was at stake in such matters. And if the United States Army could not be trusted, who could?

This reputation for integrity Meigs took back with him to Washington in 1852, along with his multiplying family. He was not well off. Army pay was poor, prospects of promotion dim and slow, so he was obliged to live in the house of his widowed Rodgers mother-in-law and her daughter Jerusha on H Street. But the extended family may have helped when in the autumn of 1853, two of Montgomery and Louisa's sons died of an "inflammation of the brain" (perhaps viral meningitis); first the eight-year-old Charles and then the two-year-old Vincent, named for the family patriarch. Both parents were prostrated by grief. Louisa howled hers, and Monty, as he would again, clenched his jaw and threw himself into the work of creating Washington.

He was only thirty-six and well out of active command but in a few years would become one of the powers in the city, in part at least because he never swaggered with that self-knowledge. The army had been given power by Congress over much of the fabric of the rapidly growing city to keep it from of the clutches of the corner-cutting profiteers who battened like leeches on some of the most spectacular contract opportunities in the country. President Zachary Taylor, the insubordinate ripsnorting hero of the Mexican War, had barely taken office when stories, most of them true, circulated that members of his own Cabinet were egregiously on the take. Washington was a prime opportunity for getting rich fast since it was agreed the city needed drastic improvement. Very much a work in progress, the city was a ramshackle, chaotic, dirty, and dangerously insalubrious town of about 40,000, a quarter of whom were black, mostly free. Foreign visitors who arrived to see American democracy at work (or be confirmed in their lofty ironies about it) almost always commented on the disparity between reality and the grandeur of its original design: wide processional boulevards opening views along the mile between the President's House and the Capitol spacious enough to provoke thoughts on the arrival of a new Rome in the world. Even by 1850 the only true avenue was Pennsylvania, the rest being carriageways of dirt separated by coarse grass in which geese, cows, and hogs happily fed. After a spell of rain, everything turned into a miry bog through which ladies attempted to make a way through the ducks toward the twelve-seater omnibuses, where they seated themselves and attempted to avoid the flying sprays of tobacco juice that were a regular hazard of the American scene.

The greatest enemy of Washington's pretenses to metropolitan dignity was disease. Whatever the other virtues of Pierre l'Enfant's choice of site on the Potomac, he had failed to notice just how fetid the torpid river became in the spring and summer, and l'Enfant's ambitious plans for canals had managed to create the country's richest opportunity for mosquito breeding. The waterways of a city that l'Enfant had imagined as that American Rome, embellished with healthful fountains, a cascade falling from the height of Capitol Hill down to Pennsylvania Avenue, were choked with the remains of rats, dogs, horses, and, not so occasionally, people. Cholera, which had been an occupational hazard of anyone living in the city in the 1830s, would still make periodic visits. And in July 1850, so the coroners concluded, cholera morbus took its most distinguished victim, the president of the United States.

That Zachary Taylor had been struck down in front of the Washington Monument at the 4 July Independence festivities only made the disaster more sensational. It was a broiling day; for some reason the president was wearing a heavy coat and downed a large quantity of iced water (some said complemented by iced milk). Back in the White House he collapsed, sank into a trembling fever, then unconsciousness, waking only to declare, rather impressively, "I should not be surprised if this were to cause my death." On 9 July he was proved right. Historians have speculated that Taylor might in fact have died from heatstroke, but cholera was the official coroner's verdict. And the death of the president from drinking polluted water was the strongest incentive for the Corps of Engineers, who had been assigned the job of providing a new water supply for the capital, to rise to the challenge. When the army's first choice suddenly died—perhaps also from cholera—the work was given to Meigs, for whom it took on historical, as well as personal, significance. His hometown, Philadelphia, had become famous for the purity of its drinking fountains, doubtless welcomed by Meigs' physician father. But for Meigs, the challenge was less to equal Philadelphia's achievement than to demonstrate to his countrymen and to the sneering Europeans that a people's democracy was capable of doing as much as Rome had for its citizens. Commissioned by Congress to write a report, he boned up on Sextus Julius Frontinus, the aristocratic master of Roman hydraulics, and his great system of aqueducts. In the report "written at a gallop" and delivered in fifty-five days, Meigs declared it a scandal that "the nation's most honored citizens" had to suffer through the heat and dust of a Washington summer, slaking their thirst only with dangerously corrupted water. The remedy would doubtless be ambitious and therefore expensive, but Congress should think loftily when it came to the good of the commonweal for "water should be as free as air and always supplied by government."

What was more, a dependable supply of water would relieve the citizenry of another regular terror: fire, and their dependence on the private brigades who might or might not come to the aid of a burning household and who might or might not have enough water to douse the flames and rescue those trapped inside! If the system worked as Meigs intended, there would even be some left for the spectacular fountains that would make the city the true new Rome of the West. Capital hydraulics would show America and the world "that the rulers chosen by the people are not less careful of the safety, health and beauty of their capital than the emperors who, after enslaving their nation by their great works conferred benefits upon the city which, their treason [to republican ideals] almost forgotten, cause their names to be remembered with respect and affection by those who still drink the water supplied by their magnificent aqueducts."

The vaulting rhetoric worked. Congress appropriated the enormous sum of $100,000. In November 1852, the hitherto unknown thirty-six-year-old Captain Meigs was appointed by President Pierce's secretary of war. The erstwhile West Point hell-raiser and ringleader of the hot-flip rebellion, Jefferson Davis of Mississippi, had championed Meigs against his many critics. Meigs might have used steam pumps, but he chose the Roman way: gravity and a conduit, from the Great Falls of the Potomac. A nine-foot-diameter conduit would carry water over the Cabin John Valley and Rock Creek to a holding reservoir in Georgetown. So the city of 60,000 would receive 67 million gallons a day—one and a half times the quantity available for Victorian London!

Through the whole course of the work Montgomery Meigs was in his element, egotistically dueling in his mind with the legacy of Frontinus and the Caesars. The work *had* to be fine and enduring for "it contains my brains." Apart from the conduit itself, two stunning bridges had to be built: the first a single-span 220-feet masonry arch (then the longest in the world) with a rise of fifty-seven feet thrown over Cabin John Valley; the second an iron bridge that was both aqueduct and viaduct 200 feet long over Rock Creek (both still wonderfully extant). For Meigs, the provision of pure water was an authentic American conquest, the right kind of war to be fought. On the day of the groundbreaking of the works in October 1853, complete with ceremonial shovel, he wrote in his diary in lapidary tones: "thus quietly and unostentatiously was commenced the great work. Which is destined, I trust for the next thousand years to pour healthful water into the Capital of our union. May I live to complete it and connect my name imperishably with a work greater in its beneficial results than all the military glory of the Mexican War." Just in case it did not, Meigs had his name, Brunel-like, shamelessly stamped into the immense iron valves used on the aqueduct. On 4 January 1859 the first water was delivered to the city from Meigs's aqueduct, and he wrote in his diary: "God be thanked for making me the instrument of this much good for the city, for having given me the health, temperance, patience and skill to accomplish in the midst of attack so far so great a good . . . no more shall the houses of the poor burn in flames from want of the means to extinguish them . . . and the poor and the servant will now be relieved of the unhealthy labor of carrying water from the pumps through snowed up streets of winter." A fountain now played in Capitol Park, right before the Congress, and though he was disappointed it shot only thirty feet in the air Meigs often went to stand in front of his "jet d'eau" for "it seems to spring rejoicing in the air . . . proclaiming its arrival for the free use of the sick and well, rich and poor, gentle and simple, old and young, for generation after generation which will . . . rise up and call me blessed." It was as though he knew already he might be on the receiving end of curses.

Was it a vice or a virtue that American government began to be embarrassed by the dinginess of its accommodation, the face it offered to the world? Was the sudden appetite for splendor a sign of democratic hubris or a coming-of-age? At any rate, senators, congressmen, presidents wanted magnificence in a hurry, and it was thus that Montgomery Meigs became the Indispensable Man, and the U.S. Army Corps of Engineers the battalions of construction. It had to start with the Capitol and with the first risk: fire. In 1851, the Library of Congress, containing the great gift of Thomas Jefferson's books, burned down. The miracle was that it had not taken Congress—a pavilion block with flanking wings surmounted by Charles Bulfinch's modest wood and copper dome—with it. But the fire was taken as a sign, if any were needed, that the Capitol needed both enlargement and to become more indestructibly fireproof. Legislators were becoming increasingly aware of the scorn poured on the dome, looking, as one uncharitable critic put it, "like a sugar bowl between two tea chests."

A Philadelphia architect, Thomas U. Walter, was appointed to design and build the extended wings and a new dome, but in the spring of 1853 Meigs was assigned the work of overall supervision, which, Meigs being Meigs, meant more than just occasional superintendence; rather his own designs and concerns stamped on the work. The first— for he was still reading the ancients—was for acoustics. Meigs was prepared, at the beginning, to leave much of the exterior to Walter, but the issues posed by improved acoustics were for him, as was all his engineering, at root, like the good Jeffersonian that he was, the political working of republican democracy. Inaudibility, he thought, privileged the blowhard and discriminated against the Mr. Smiths of the nineteenth century, the little men who had been sent to Washington to give their voice with as much authority on the issues of the day as famed orators like Daniel Webster and John Calhoun. Poor or ill-considered legislation, bills compromised by being unexamined for the work of vested interests, were the result of that inaudibility! A vote *was a voice*! But transforming the acoustics of the chambers required something that Meigs anticipated would not be popular: closing the chambers off from natural light and ventilation. To deal with the objections he designed a system of steam-pumped hot air—an ancestor of common central heating—and was careful about evenly diffused gas lighting.

None of this was enough to appease those upset by the decision, especially since Meigs's enormous dome, a full hundred feet taller than the Bulfinch original, was originally pierced by windows in Walter's plans. But Meigs, who had carefully studied Brunelleschi's sections and plans for the dome of the Duomo in Florence, especially the construction of an inner shell dome, as well as Wren's St. Paul's, rapidly came to think of himself and not Walter as the true architect. A state of sullen conflict poisoned the relationship between the architect and the engineer. As long as Pierce was president, Meigs was upheld in his superior authority by Jefferson Davis. The two men, then the two families, became close, socially connected. But after James Buchanan took office in March 1857, the new secretary of war, John B. Floyd— sometime failed cotton planter and ex-governor of Virginia—turned decisively to Walter's side. Floyd had his own reasons to dislike Meigs, and they had nothing to do with architecture and almost everything to do with money. Meigs had long been a thorn in the side of the lobbyists and contractors. When the House of Representatives had wanted to remove responsibility for public buildings and utilities from the army and deliver it instead to businesses, Meigs had fought the policy and prevailed. Thwarted, a faction in Congress, mostly southern Democrats, attempted to transfer the business from the Department of War to the Department of the Interior, and this, too, Meigs contested. As soon as a new gambit was devised that favored patronage and profit over the public trust, Meigs was onto it. Inflating the scale of jobs, so that only preselected big guns could bid, was ended. Low bidders, who won contracts on that basis but who then hiked their prices after contracts were signed, became a special target of his displeasure and criminal inquiry. Meigs knew that his stubbornness in these matters, his superior refusal to play by the usual rules, earned him much hatred around Washington. Lucrative kickbacks were being lost to misguided rectitude. But grandly casting himself in the mold of the Ciceronian honest man, he thought that he had no option but to stick to his guns.

The impasse between Floyd and Walter, on the one hand, and Montgomery Meigs, on the other, became so serious that work on the Capitol stopped altogether for almost two years between January 1858 and November 1859, and when it resumed it was with much rancor. In revenge for all the obstruction, Floyd maliciously interfered with Montgomery's son John's admission to West Point. An operatic scene ensued. Meigs handed Floyd a letter implying (though pretending to hope this could not be the case) that it had been the differences between them that had led Floyd to oppose John's appointment to the academy, ostensibly in presidential gift. Floyd read the letter on the spot and went white with rage. Meigs stopped pulling punches and told him "he had many times grievously wounded me and done me great injury." Floyd said he would rather have resigned than see John Meigs go to the academy. The two could have murdered each other there and then. Instead, Meigs went straight to Buchanan, who as usual affected being much put upon and hemmed and hawed about Floyd being well intentioned. In the end, though, John was admitted, and Meigs took him straight off to the Hudson Valley.

But if he had won a skirmish, he sensed the satisfaction was temporary. "The Secretary will ruin me if he can," he wrote. "I have done my duty and he will, I trust, find that to prosecute an honest man is to bite a file against God." In October 1859 Floyd made it clear Meigs would have to leave his posts. A year later, with an election looming, it was official. Meigs was dismissed from all his great posts—the aqueduct, the Capitol, and the rebuilding of the Post Office. He was banished to the tropical fastness of the Dry Tortugas, seventy miles farther west in the Atlantic from the Florida Keys. There he was supposed to supervise work on the incomplete Fort Jefferson. The brick fort had been started in 1846 when it had been anticipated that the country might need an oceanic bastion against Spanish naval attack during the Mexican War. But that contingency now looked quaint, and the posting could hardly have been more remote. It seemed the end of the Roman's career. Floyd reveled in the humiliation. When Meigs asked for funds to complete the fort, he jeered at "the pestilent fellow who got trouble wherever he went" and how absurd it was to demand money to defend some "heap of rocks." Hearing the news at West Point, his eldest son, John Meigs, wrote with indignant teenage loyalty in defense of his pa, "this is a pretty place to send talent that has been entrusted by the Congress of the nation with the expenditure of millions of dollars."

But Meigs's exile turned out to be less of a penalty and more of a retreat where, between rapping his cane against the gun emplacements of Fort Jefferson, he took the opportunity to unbutton a little. Donning the white pantaloons of the islands, he walked the beaches, staring in rapture as "the waves splash away in great maps of light." He watched the pelicans dive for fish; filled his lungs with sea air and his ever-active mind with matters that suddenly had become pleasingly important—crabs, for example, in all their tropical variety: "Hermit crabs, fellows with bright red and purple nippers with painted legs but with leather-covered bodies," "stem crabs, fiddler crabs, soldier crabs, crabs which scarce move, a crab which darts with the speed of a spider, crabs which live on the vertical face of a wall and jump like birds from one perch to another." Sometimes he would stroll along by the mangroves that leaned over the beach, their feet in the salt water, with scores of tiny crabs crawling over his coat and shirt, tickling his scientific fancy. But the naturalist Meigs could no more take his mind off his two homes than could the banished engineer. His house in Washington he knew to be secure and awaiting his return; the wider house of the Union, on the other hand, was threatened with imminent destruction. He began to think and act in parables. One day, walking the beach, Meigs found a hermit crab with a broken shell and, in the spirit of the Corps of Engineers, decided to rehouse it. The crab was brought home together with a vacant shell that Meigs judged suitable accommodation, and then the crustacean was gently teased from one to the other. "He readily accepted the new home."

And then the world recalled him. Barely a month after Meigs sailed south, Abraham Lincoln was elected president. Early in the new year of 1861, his nemesis, John Floyd the Virginian (his state not yet formally seceded), was rumored to be diverting federal guns and munitions south to the Confederacy, and was indicted for fraud and malversation of public funds.

5. Taking sides

The joke was on John Floyd, who was indeed biting on a file. He was sweating it out in Washington, deserted by the feebly valetudinarian President Buchanan, fighting off criminal charges, worrying whether defecting to the Confederacy would make matters worse or better. Meigs meanwhile was sailing the coral reefs in his schooner, peering at the turtles with an invigorated sense of patriotic usefulness, always a hearty tonic. For if there was to be a war, and it was hard to find anyone who, once the Republican Lincoln had been elected, imagined it would not come, then federal forts in the South—from Sumter at Charleston, to Pickens and Pensacola in Florida, to the brick Fort Jefferson in the Tortugas—were all hostage to the Confederacy. It was hard to get news to the islands, but there was a rumor around that Louisiana planned on raising a volunteer force of 10,000, some of whom would sail to Fort Jefferson. But without Union reinforcement, as many men as could be packed into a mere fishing smack could take the fort, Meigs wrote to Washington.

By February 1861, he already knew of Floyd's indictment for "debasing . . . public virtue" and of his own vindication and recall. But when he was not taking zoological notes and watching over the building of Fort Jefferson, Meigs was meditating on the tragic necessity of an American war he had never imagined he would be called on to fight. How could he not? The reason to fight it was there, in the Tortugas right before him, the backbreaking labor of the twenty-five slaves who had been imported from the Florida Keys to work on the fort. Meigs had come to this conclusion slowly, reluctantly, certainly not as a fire-breathing abolitionist and not someone who was already sharpening the blade of his saber for civil war. The militant abolitionist John Brown's violent raid on the U.S. arsenal at Harpers Ferry in October 1859, which resulted in fifteen deaths, Meigs had judged as the misguided, if not criminal, adventure of a near-lunatic. It merely showed, he wrote in his diary, how a small band of fanatics can disturb or even destroy a country's peace. And Bob Lee (as he called him) had done sterling work to stamp on it. But en route to Florida in late October 1860, Meigs paid a visit to his younger brother Henry in Columbus, Georgia, and everything changed.

The Meigses, of course, had a long Georgia connection through Return Jonathan Sr. The Cherokee's and Creek Indians' former territory in east Georgia, along the Alabama line, became Muscogee County, and its first town, on the Chattahoochee River, was given the name of Columbus in 1828 when someone had finished reading Washington Irving's *Life and Voyages of Christopher Columbus*. In short order, Columbus became a center of cotton manufacturing, with its biggest mill right on the banks of the Chattahoochee, using the forty-feet falls to power flour mills as well as cotton. In the middle of the river sat the sweet islet of Magnolia, where the scent of jasmine and the abundance of shady groves and hanging Spanish moss made it an irresistible courting spot. Sweethearts would row out there of a summer night and spoon to their hearts' content. Into one sweet-scented glade stepped Henrietta Hargreaves Stewart and Henry Vincent Meigs, fifth son of the Philadelphia obstetrician who had himself once been a Georgian. After they had wed, Mr. Stewart made Henry manager of his mill by the Chattahoochee. There was money to be made in Georgia. Dollars ripened in the sun like fat peaches. Raw cotton was being shipped from the slave plantations to the factories, combed, spun, and carded. Henry Meigs turned raw yarn into fabric and off it went into America or Liverpool. Was that all to end now, Henry worried, because of some Northern madmen who understood nothing of the South? Fidgety and anxious, Henry went to Washington and confided to Montgomery his apprehensions. Monty looked at his younger brother and judged him "a sad fellow" for being so torn in his loyalties. But now Henry had evidently pitched his tent four-square in the adamant South. When the table talk inevitably turned to the coming election, Henry ranted against the "fanatics" of the North who knew as much about slavery as they did of celestial revelation and the kingdom of heaven, and against the federal government's meddling in matters that were none of its concern. If a "constitutional" president were elected who would restrain the hotheads, all might yet be well, but if Lincoln was the people's choice, then the country's fields would be "reddened with blood." As he listened to his brother, Meigs felt his mother's horror of slavery well up in him with augmented fury. How dare his brother wax sanctimonious as the Clapp Mill turned its wheels, the blood of slaves mingling with the river rush? How dare Henry write to their father, Charles, gloating that the skies over Columbus were bright with bursting fireworks to celebrate South Carolina's secession, "can a whole people be so deceived as you appear to think the South can be?"

Quite suddenly, the identity of his friends and his enemies became distinct in Montgomery Meigs's mind. They were the same as the friends and enemies of the United States of America. His wretched adversary, Floyd, he remembered, had heard the cheers of the Virginia hotheads, when as governor he had promised to embargo goods from any free state not returning fugitive slaves to their masters! Such bravery indeed! Such exercise of the public trust! Meigs had heard rumors of so many of his old West Point comrades, men who as cadets and then officers had solemnly vowed to hold as sacred the college code of Duty, Honor, Country, violating it by planning to betray the Union. What *was* it these men, these *traitors*, including his own brother Henry, imagined they were defending? Was it the constitutional right of states to go their own way? Meigs thought that the most despicable and transparent sophistry. It was what the states were seceding for that was the true issue: the American future; whether that future would live up to Jefferson's noble promises of liberty and equality, enshrined in the Declaration of Independence, or forever tarnish them with the odious hypocrisy of economic convenience. As the crisis reached a point of no return, Meigs had written that slavery "is not a thing which men brought up to look upon liberty of action, speech, thought and conscience should be called upon to worship under penalty of dissolution of their political organization and society." Sitting at his desk in the citadel of Fort Jefferson, the white hot Atlantic glare scalding the walls, there arose before the thoughtful, burdened Montgomery Meigs the weighty apparition of American history, past, present, and future; the great cause for which Meigses had fought from the time of old Return Jonathan, and would do so again, if the cause were true and worthy of sacrifice. This one was. Not long after, Meigs would call it a "holy war." He could see the two paths of American destiny stretching in front of him at this place and moment. Both were necessarily fateful. The path of blood and fire was never to be entered into without the most profound moral examination. American wars, he held, in the spirit of Jeffersonian idealism, ought never to be elective. But the path of accommodation with slaveholders, who would then forever hold the Union to ransom so they might protect their despicable institution from being swamped by democracy, was unworthy of the cause for which Return Jonathan had fought: a republic of freedom. He could never live with such an ethically debased America. "Am I to be the officer of some contemptible little state republic, some Bolivia . . . or Georgia . . . instead of the servant of a people stretching their empire from ocean to ocean and touching the confines of the Arctic as they do of the torrid zone, a people great in enterprise, science, arts and commerce and in arms [all] this because they are free?

"Is all this to end in order that slavery not freedom may have greater sway?

"Is slavery stronger than freedom? Or does the Almighty who punished Israel for desiring a king punish us for boasting of freedom yet encouraging yet upholding or tolerating even, slavery? My heart grows sick as I think of this prospect."

On 4 March 1861, the recalled and vindicated Meigs watched Abraham Lincoln sworn into the office of the presidency by Chief Justice Roger Taney on the east portico of the Capitol. The facade was still covered in the scaffolding of Meigs's reconstruction. Like the Union itself, Lincoln was in danger from the moment he mounted the steps, protected by guards supplied by General Winfield Scott. Tall, gaunt, and gawky, Lincoln seemed an unlikely man for the hour. Until that moment Meigs had no great opinion of the congressman from Illinois. Like most of the family, he had voted for Lincoln's old rival, Stephen Douglas, who had run on the northern Democratic ticket and was, they all thought, evidently the superior man. But Lincoln's great speech confirming that while the federal government would forbear from interfering with the "property" of slavery, it would not tolerate the "destruction of our national fabric" made a deep impression on Meigs. After the wretched temporizing of the Buchanan administration, it was astonishing to hear from the mouth of a politician an acutely philosophical intelligence, summoned at the behest of all possible crises, resolute in setting before the people exactly what was at stake: the life or death of the American democratic experiment. Though the new, abolitionist Meigs might have wished Lincoln more forthright on the ultimate incompatibility of slavery with that democratic Union, he agreed wholeheartedly with Lincoln's premise that "the central idea of secession is the essence of anarchy. A majority held in restraint by constitutional checks and limitations and always changing easily with deliberate changes of popular opinions and sentiments is the only true sovereign of a free people. Whoever rejects this does of necessity fly to anarchy or to despotism." Meigs was also won over by the classical eloquence of Lincoln's modesty, two qualities not usually in tandem; together with the moral craft by which he plainly set responsibility for the outcome before his fellow citizens: "In *your* hands, my dissatisfied fellow citizens and not in mine is the momentous issue of civil war. The government will not assail *you*. You can have no conflict without being yourselves the aggressors. *You* have no oath registered in heaven to destroy the government while I shall have the most solemn one to preserve, protect and defend it."

For months Meigs had been yearning for a leader who, while avoiding belligerence, would not shrink from war to save America. In Lincoln's soaring peroration evoking "the mystic chords of memory, stretching from every battlefield and patriot grave to every living heart all over this broad land," hoping for the return of "the better angels of our nature," he heard the echo of the Meigs family instinct to see American democracy in the long arc of its history. "It was a noble speech," he wrote to his father in Philadelphia, waxing almost Shakespearean himself as he was buoyed up by the solemn integrity of Lincoln's words. "No time was wasted in generalities or platitudes but he grappled at once with his subject and no man could doubt that he meant what he said. No point was omitted . . . but the disease of the body politic was analysed, its character and remedy pointed out and each sentence fell like a sledgehammer driving in the nails which maintain states. Kind and conciliatory, it still left no loophole for treason. War, I fear will come but it will be conducted humanely . . . If they bite they will bite against a file. He will defend and protect the public property . . . enforce the laws . . . and once more will freedom of speech and liberty of person be the rule of all our land and not the exception . . . The people about me applauded each sentence . . . some looked darkly and retired. Treason shrank out of sight and loyalty sat in the sunlight."

From that moment Meigs was Lincoln's devotee, restive to serve, but unsure in which capacity he might give his best. For more than a decade he had been that most anomalous thing, a Washington soldier, and one of scant rank, too, still plain Captain Meigs, for promotion was excruciatingly slow in the Corps of Engineers. But his official rank was the only inconsequential thing about Meigs. He had spoken directly to three presidents, was about to be the confidant of a fourth, carried substance in Congress and in Cabinets, and

more important than any of this, had become the personification
of public virtues that were in short supply in the capital and that, if
worst came to worst, would be badly needed: integrity, competence,
and resolution. He knew money, he knew metallurgy, and—this had
suddenly become very important—Montgomery Meigs knew forts,
North and South. He had built them, manned them, armed them,
inspected them, defended them. As more and more states voted to
secede from the Union, forts were very much on the new administration's mind. The status of Sumter in Charleston Bay was as close as
anything to being a casus belli. South Carolina had been the first to
depart from the Union but even before secession in late December, its
congressional delegation had demanded the evacuation of the federal
garrison. While the equivocating Buchanan, whose last speech to
Congress had castigated Northern "fanatics" in much the same tones
as Henry Meigs, was still in office, some sort of accommodation over
Sumter seemed possible. A meeting produced an informal standoff
arrangement by which the South Carolinians agreed to abstain from
shelling the place into submission provided no attempt was made
to reinforce it. But in January 1861, the garrison commander Robert
Anderson had imported seventy-five men from another Charleston
stronghold, an action the Carolinians decided to take as a violation of
the standoff. It would take 20,000 men to hold it, the aged commander
of the Union army, General Scott, concluded, and prepared for an
evacuation.

The humiliation was passed to the incoming president. For Lincoln
the status of Sumter and the other southern forts that would pass
to Confederate control—Pickens on the Floridian island of Santa
Rosa near Pensacola, Jefferson on the Tortugas, and Fort Taylor on
Key West—was as much a matter of national symbolism as military
strategy. He would have liked to have reinforced all of them if he
could, since a naval blockade of the South was very much part of
General Scott's "Grand Strategy" for a war of encirclement and stran-
gulation. Scott's gloomy assessment persuaded him that ultimately
Sumter was going to have to go. Lincoln made it clear that he was
not about to cede the rest as if the United States simply accepted the
fait accompli of its division. At issue was more than national amour
propre. The Confederacy was now a fact with ten states already
seceded and Virginia likely to follow. In February Jefferson Davis

had been elected provisional president and had taken a host of West Pointers with him.

If Sumter had to go, Lincoln was determined that Pickens would stay with the Union. William Seward, Lincoln's new secretary of state, aware of Meigs's expedition to the Tortugas, asked his advice. Meigs, raring to go, freely gave it in a meeting with Seward and Lincoln—a relief expedition to Fort Pickens that would land men and blockade the harbor against boat attack from the mainland. But because Washington was so insecure, crawling with spies and both the army and navy beset by daily defections to the Confederacy, the mission, Meigs thought, would have to be kept secret if it was not going to trigger a preemptive Confederate raid on Pickens. Lincoln wanted Meigs to be there in person, which gave Montgomery an opportunity to point out the indelicate matter of his lowly rank for such a trust. Promotion was put in the works.

After years of being an office officer, Meigs was excited by this call to immediate secret action. He kept the information from his wife, Louisa, who was told merely that he was on a trip to New York. Once there Meigs commandeered a steamer from the Brooklyn Navy Yard to act as warship and sailed south on a requisitioned civilian vessel. Keeping the mission from the Department of the Navy made trouble but none that Meigs couldn't cut his way through. His own ship overtook the warship and arrived off Santa Rosa Island by the second week of April. In short order he managed to station a garrison of a thousand, with another thousand standing by, the harbor now blockaded and barred against boat raids. But there was nothing Meigs could do to prevent Confederate artillery on the mainland from lobbing shells. As he was preparing to sail back to New York, he heard the sound of their fire opening up on the fort. It gave him no joy. "The opening of a civil war is not a thing lightly to be seen and though I saw my duty plainly in reinforcing this beleaguered fortress & rescuing my countrymen shut up here from the hands and power of rebels and traitors, I could not [see] the opening of the fire without great regret. It must soon come however & God protect the right."

6. Father and son

"The country is in flame," wrote Meigs in a one-line entry in his diary. So was he; passion unloosed. Louisa, who still had misgivings about the conflict, wrote to her son John that she was not enthusiastic about siding with the "extreme North . . . and such fraternity" who seemed to be dragging the country into chaos for some sort of righteous satisfaction. Nor did Louisa know what to make of the change in her familiar, dependable "Mont." First he disappeared off to who knew where without so much as a by your leave or any explanation. And then he had become fearsome to live with. "His soul seems on fire with indignation at the treason of those wicked men who have laid the deep plot to overthrow our government . . . He looks so dreadfully stern when he talks of the rebellion that I do not like to look at him." Certainly civil engineering no longer sufficed to assuage the storm of outrage that swept through Meigs when he considered what had become of his country; and what, especially, had befallen the institution to which he was most deeply attached and to which he had entrusted his eldest son: West Point. A full quarter of living West Point graduates had thrown in their lot with treason, and in the bitter spring of 1861 he felt their enmity, the collapse of their collegiate esprit de corps, everywhere he went. The commander of the battery that had fired on Fort Pickens was Braxton Bragg, just one year behind Meigs at West Point and thus well known to him in that little world. Joseph Johnston (class of '29, the same as Lee) had traded in the honor of quartermaster general of the United States for the same office in the Confederacy. And Johnston and Pierre Beauregard, the superintendent of West Point, were in command of regiments that were mustering in Virginia to threaten Washington itself; West Point traitors poised to swarm over *his* Washington, to camp on the Capitol Park, drink from the fountains he had created to slake the thirst of good republicans! Meigs took all this personally. And then there was Lee, for whose affected gallantry, honor, and all the rest Meigs had the utmost contempt, but whose shadow pursued him every day. Lee's house was on the Heights of Arlington, which if captured by Johnston and Beauregard would be able to fire directly on Washington and indeed on the President's House! A dwelling connected inseparably with the memory of George Washington would be commandeered as the citadel of treason for the express purpose of destroying what the greatest of the Founding Fathers had so painfully made: a union of the free. Compared to Washington, what was Lee? Someone who had fouled his nest. On 22 April 1861, a week after Lincoln had called for 75,000 volunteers, Lee accepted the invitation to command the Confederate forces. Meigs was implacable. Any of the officers who had violated their solemn vows taken at West Point or at the naval academy at Annapolis (founded in 1845) should be permanently deprived of civil rights, subject to the confiscation of all their property and deported. But for the renegade leaders like Lee, Johnston, Bragg, Jefferson Davis, and Beauregard, this would not be enough. They bore personal responsibility for leading the people of the South into fire and slaughter. They would forever have blood on their hands. They "should be put formally out of the way if possible by sentence of death, executed if caught."

A good piece of Meigs burned to take this fury into the field. But he also knew that it was not any reputed mastery of tactics that had recommended him as adviser to the president. Men like Seward and Lincoln both sensed in Montgomery Meigs the makings of a type that had never yet existed in the history of the United States—a war manager. However long the conflict lasted, it would take place on a scale unimaginable to the generation of Washington and Return Jonathan Meigs. The Confederates had planned for an army of 100,000, had enlisted almost that number by May, and under Joseph Johnston, their new quartermaster general, were expecting to have to establish a command economy capable of laying hands on every asset in the eleven states. When Forts Sumter and Pickens were fired on, the Union had ambitious plans laid out by Winfield Scott for seizing control of the Mississippi, cutting the Confederacy in two in the west, blockading Charleston and Savannah, but that was about all it had. Artillery and ammunition had been siphoned off south by the disloyal; customs houses, arsenals, and docks in the South had been seized. There were almost no uniforms, tents, blankets, rations, or, most important of all, animals: the mules that must pull wagons; horses for the cavalry and artillery; cattle to serve as beef on the inevitable long marches. Everything needed creating, virtually from scratch.

What was it about Meigs that made him seem the man who could rise to the challenge? He could be depended on to take no nonsense from the profiteers who were lining up to exploit the Union's predicament: syndicates who would buy ships and river transports on the cheap and lease them to the government at exorbitant rates; railroad men who would put a premium on the shipment of men and munitions; even horsetraders who would sell the army broken-down nags at extortionate prices. Meigs, it was thought, would give these rogues as short shrift as he had the contractors in Washington, call them to a severe account. He understood the engineering of war like few others: bridge-building, road-cutting, tunneling, fortification. But there was something else that Seward and Lincoln sensed in Montgomery Meigs: righteous anger translated into cold efficiency; someone who had suddenly lost all patience with the childish affectations of military gallantry; someone who seemed to know what was coming, four years before Sherman actually said it.

Not everyone shared this opinion. The secretary of war, Simon Cameron, for example, was against Meigs's appointment. For Cameron, Meigs was a jumped-up major who liked throwing his weight around; a nobody who had played with fountains and had pointlessly made enemies in pragmatic Washington. He was not someone who understood business. Cameron was overruled by the president and the cabinet. On 13 June Meigs was formally appointed to take Joseph Johnston's vacated place as quartermaster general of the Union. He was forty-six years old, and the hard work of a life was about to begin.

Meigs was a brigadier general now, but still without direct experience of fire. That was about to change. By July there were more than a quarter of a million men in the Union army and the new quartermaster general was scrambling to procure them uniforms. Any color would do—brown, blue, green, gray. (Many of the first federal soldiers went to battle wearing the identical gray as their Confederate foes.) Boots, blankets, tents, and guns were desperately needed. In every way, the Confederates seemed better prepared. Montgomery's brother Henry was supplying Southern troops with coats and shirts from his own factory in Columbus, Georgia! So the two brothers were in a uniform race. The older would win, but that conclusion was not foregone in 1861. No wonder, then, that Meigs was against any precipitous offensive into the South that might take the Union armies far from supply lines and depots that had barely begun to be constructed. Let the rebels rather come to us, he counselled Lincoln (who, surprisingly, began the practice of periodically asking his quartermaster general for strategic advice). But with Joseph Johnston established in the Shenandoah Valley with 12,000 and Beauregard less than thirty miles from Washington with another 20,000, a largely hostile Maryland to the north, and the Confederate press noisily looking forward to chasing Lincoln and the government from the capital, irresistible popular and political pressure built on Lincoln to stop the rebel offensive in its tracks. The editor of the *New York Tribune*, Horace Greeley, the most influential newspaperman in the country, urged a quick march, convinced that Confederate soldiers were an undisciplined rabble who would hardly survive their first contact with a real army. But the commander assigned to the task, Irvin McDowell, knew his opposite number well. He and the Confederate brigadier general had been exact contemporaries at West Point, and McDowell was in no hurry to advance. The troops are green, he told the veteran Winfield Scott. Theirs are green too, came the reply.

They didn't come greener than John Rodgers Meigs. On 2 July he had come home to Washington on furlough from West Point after two years at the academy. All along it had been a struggle. Montgomery recognized in his eldest son exactly the same qualities that had led his own father the obstetrician to pack him off to West Point, hoping to channel the unkempt energy into constructive achievement. It worked. But John Rodgers was, as his pa had been before him, a handful, rowdy, raucous, so resistant to family discipline that Montgomery resorted to the usual brutalities of the nineteenth-century home: tying his son to the legs of a wardrobe and denying him supper, and then whipping him when he managed to get loose. But the fact that he saw, without question, in his awkward boy the earlier version of himself only deepened the love-bond which was undeniably there. It mattered, then, for John to be admitted to West Point, and when news of the admission came through, Meigs felt his own vindication as well as his son's.

And when everything, at last, seemed smooth sailing in the autumn of 1859, and Montgomery brought John to the citadel on the Hudson

Heights, something else conspired to give father and son unexpected grief: John's scrotum. The medical examination of incoming cadets revealed something unknown to both father and son, namely a grossly enlarged spermatic vein, a varicocele "of such an aggravated character," the understandably distressed father wrote in his diary, "that it was doubtful he would be able to discharge the duties of a cadet." He was admitted, then, on probation, and sensing the acute suffering this particular problem might engender in his seventeen-year-old, Montgomery did all he could to reassure him that it was a temporary problem, not uncommon to lads his age and that would disappear in due course. Meigs was himself struggling with the shock of it, never suspecting the rowdy boy to be "disabled" in any way. Consulting anyone and everyone who might be qualified to give an opinion, beginning of course with his father, Charles, Meigs prescribed what he could for John—a suspensory truss-bandage, daily cold baths, morning and night, knowing that all these remedies might make him a figure of cruel fun among his peers. Rather sweetly Montgomery wrote John that though he might balk at dunking his member in cold water every day, "you will find that whatever is done with modest feeling is not immodest and that nothing is immodest which is necessary to health." If all that turned out to be too difficult then there was always a Dr. Pancoast, who had performed a simple surgical procedure on countless young men to remove the difficulty, and, so Dr. Charles Meigs assured him, with not a single misfortune. But all the Meigses talked about it together. Louisa wrote her son of her surprise since "you have always seemed so well and accustomed to take so much exercise" and cautioned (thinking doubtless of the attractions of Benny Havens, the cadets' watering hole) that "if you are *careful* you may outgrow it."

Dr. Pancoast's services were, perhaps mercifully, not called for. Gradually John's Trouble disappears from the family correspondence (though it was liable to flare up again in times of crisis); and probation was replaced by regular cadet status. But John was repeatedly interrogated by his father about his failure to come first in classes, about his habit of acquiring demerits, all of which Montgomery professed not to be able to comprehend though in his own time as cadet he too had been a demerit specialist. "My dear son, I am sorry to see by your letter of the . . . that you have allowed your competitors in the class to beat you in marks even in such a study as geometry in which you say you are so well prepared . . ." But then John always had his mother's letters written in a quite different vein to fall back on for comfort. On his twentieth birthday she wrote, "I am descending down the vale slowly but surely yet it seems but a very few years since that I was as young and fresh as you are. I was scarce twenty-five years old when you were born and yet I felt myself of very mature years & I remember that I actually *blushed* at the in-appropriateness of the expression when the doctor spoke of the likelihood of your being a strong and vigorous baby from the fact of your having a *young* and healthy mother . . . We advance so rapidly from one stage to another that it all seems like a dream. In a few years you will have arrived at all the dignities and privileges of manhood and the battle of life for you will commence. You must put on your Christian armor and go forth into the strife."

But Louisa meant it metaphorically. She somehow hoped John would never actually see battle. "Do not let all your thoughts be directed to making you a good soldier for this world's warfare but remember you once promised to become a soldier of Christ." A preacher, then, not a fighter? Just what Louisa's feelings must have been when she saw Montgomery helping John with his sword and sash on the eve of the battle of Bull Run, it is only too easy to imagine. Living in an army family made it not a whit easier for so evidently loving a mother. John, with his father's permission, had joined up as a volunteer aide with McDowell's army and had been assigned to an artillery battery commanded by Major Henry Hunt. "I felt a pained shocked sensation when he told me of it," Louisa wrote to her mother. There had been a difficult scene between mother and son. John told Louisa that after two years being "educated at the expense of the government" it was his duty to volunteer and that he would be ashamed to go back to West Point without serving at a time when the country needed all the men it could get. "I felt that it was the prompting of a nature, the stirring of his blood which comes from a patriotic race . . . but I felt a very motherly and womanish sinking of the heart nevertheless." On the morning of 16 July 1861 John marched out from Arlington with the troops, leaving his father proud and his mother in prayer. The war had come home to the Meigs family.

This first campaign was a famous fiasco for the Union, a shocking humiliation. So confident was Washington society of bloodying the noses of the impertinent rebels, once they knew the day on which the battle was to be joined (21 July), carriages were taken to drive to Manassas as if on a summer excursion to the country. The photographer Mathew Brady and others whom Meigs, as a keen student of the new art, knew well, were there too to record the disaster. After a hearty and optimistic breakfast Montgomery himself went, in uniform, as an observer who would report directly back to the president. Louisa was all too willing to let him go, imagining he might keep a paternal eye on John. Meigs Sr. was himself confident that an early victory was at hand, not least because in just a few months he had managed to equip a substantial army with everything it needed.

Except tactical sense, glaringly missing from the commander. McDowell's complicated plans for multiple outflanking movements suffered from confusion and irresolution and above all a failure to press home when it counted his massive superiority of numbers. Of the many regiments at his disposal, only two were ever engaged at the same time. So the advance up the hill at Beauregard's batteries stalled, and then broke under counterattack, leaving all of Meigs's mobilization—guns, wagons, animals—to the jubilant Confederates. Washington society, which had expected to be amused, was now panic-stricken, anticipating the city would be occupied by the rebels. Routes of retreat from Bull Run were clogged by a traffic jam of fashionable phaetons. Meigs overtook them riding a horse hard.

About three in the morning on Monday, 22 July, Meigs went to Lincoln in the President's House to report directly on the disaster, doubtless stressing McDowell's inability to make the best use of a finely equipped, if inexperienced, army. Lincoln received the news with melancholy stoicism and began to plan replacing McDowell with General George McClellan, who, in his particular way, would turn out to be an even greater disappointment. An hour later, at four in the morning, Meigs got home. While Louisa was relieved to see him safe, she could not stop worrying about her son whom Montgomery hadn't seen amid the chaos of battle. It was eight in the morning when a horseman galloped fast to the door of the Meigs house, dismounted, and rapped on the door. John's face was still black with smoke and powder. Montgomery was shaken from his sleep by his son announcing the scale of the rout, which his father already knew. Together they commiserated; together they resolved. Torrential July rain fell in sympathy with their somber mood. John had assumed that he would return to his regiment and protested to his father that he had come back to the city only to fetch fresh horses for Major Hunt's battery. To Louisa's relief, the father disabused the son. Meigs was proud of John's courageous service under fire, carrying communications from Hunt to other parts of the field with balls whizzing around him. Reports would commend him on the day's work, and Montgomery would egotistically congratulate his son on conducting himself so as to leave "my name unstained." But Meigs himself had been exposed enough to harm, and had had his confidence shaken about the prodigal quality of commanders, to feel that enough was enough. John would, he ordered, return to West Point, honor satisfied, country grateful. A lesson to be taken from the disaster at Bull Run, Meigs thought, was there would be no easy victory and that John might well be needed to serve again. Louisa was grateful for her husband's decisiveness. And John's grandfather, Charles, wrote the young man a burst of prose poetry that the Meigses seemed always to be able to summon for such moments.

"My dear John," wrote the doctor, "when I think of this wicked war I rejoice that I am old. When I remember you then I lament that I am more than twenty-one . . . for if I were young again I might hope to follow you . . . But after all when I do die, why may I not hope to gaze out at you from out [of] the face of some summer Moon or peep at you behind a gorgeous cloud in heaven and sympathetically move you ever safely onwards in the march for Truth and Honor."

7. The quartermaster general, 1861–64

Now the real toil began, the work that would ultimately win the war for the Union as the Confederacy would be out-supplied rather than out-fought. Meigs knew that blundering generals could lose wars, but smart, resourceful ones could never win them without consistent and swift supply. Time and again, the availability of food, clothing, draft animals, and artillery horses—as much, if not more than the munitions themselves—made the difference between success and failure, both in particular battles and whole campaigns. Take underwear and soap, for example. Diarrhea and dysenteric infections like typhoid made short work of armies, on the march and in muddy camps. Toward the end of the war Confederate armies in Virginia had no more drawers to supply men whose underthings had been reduced to foul shreds and rags. And in both armies lack of hygiene could kill more men than shells and grapeshot. And without proper footwear, there could be no victories. By the summer of 1864 the Confederates had run out of horseshoes, so they ripped the shoes from dead animals and shot any sick or broken-down horses to get at the shoes. Much of their infantry were themselves shod in rawhide moccasins, if they were lucky, or not at all. It was said that you could tell where rebel soldiers had passed by the bloody footprints on the ground. Their best hope was to take Union prisoners, for whom the first order was to get their shoes off and transferred to their desperate foes. If that happened they were in luck, for Montgomery Meigs had assumed each Union soldier would need four pairs per year, and since he anticipated (correctly) long rugged marches, specified at the outset that footwear be hand-stitched, rather than the wood-pegged shoes that could be got from factory production. This kind of provision took longer, tried the patience, but it won campaigns. In fact the battle that has been seen (not altogether accurately) as the most decisive of the war, Gettysburg, came about almost by accident when Lee's army in Pennsylvania were searching for footwear and ran into the army of General Meade! Later that year Lee actually curtailed his plan to attack Meade because of "the want of supplies of shoes, clothing and blankets . . . I was averse to marching them [his troops] over the rough roads of that region at a season too when frosts are certain and snows probable unless they were better provided [to] encounter . . . without suffering." Lee had read enough about the Napoleonic wars to know armies never won with frost-bitten feet.

On the other hand, when Sherman got to Savannah in December 1864, waiting for his army (that had been decently supplied in the first place) were 60,000 fresh shirts, drawers, and pairs of socks, 10,000 greatcoats (the assumption being that some at least of the original distribution would have survived the march through Georgia), 10,000 waterproof ponchos, and 20,000 blankets. There were also three full days' rations for each man, ready to go. Meigs had shipped all these supplies south, partly in the ironclad armed transport vessels whose fleet he had designed, and had stored them on Hilton Head Island just off the coast. Just in case Sherman made a last-minute change of plan and continued to march south, Meigs had also sent an equivalent supply to Pensacola in Florida. Against this performance, the overstretched Confederacy, for all its own miracles of mobilization, had no chance.

But back in 1861, the Confederacy—which had assumed an army of 100,000 at the beginning of the year—was, if anything, more energetic than the Union. In the North, the cupboard was shockingly bare. A chain of empty depots with no reserve stocks of anything greeted Meigs when he took over his post. And the scale of his task was almost incomprehensible. Joseph Johnston, the defector, had been quartermaster general to an army of just 30,000. Two years later twenty times that number of soldiers had to be provided for. By Bull Run there were 230,000 Union troops in the field, and after the defeat Congress was quick to authorize funds for half a million. By the end of 1862, 670,000 soldiers had been mobilized for the Union, the biggest army in military history. A department that had barely existed before 1861 needed almost overnight to turn itself into an empire of supply. Meigs had to cast his eye over the entire map of needs, from railway track and rolling stock, river and road transports, to the manufacture of munitions, clothing, and tenting (Meigs designed a two-man bivouac lighter to carry as basic equipment), medical supplies like bandages, crutches, and splints, as well as ambulances and field hospital space for over 100,000 wounded, and the sad materials of embalming and burial. It was impossible to do all this himself, and as if wanting to vindicate the honor of West Point, Meigs went back for assistants to near contemporaries whom he knew personally as loyal and competent: the good engineers. Both Robert Allen, who became chief quartermaster for the western theater on the Mississippi, and James Donaldson, who ran the Department of the Cumberland, were classmates of Meigs in the cohort of 1836. Langdon Easton, who provided for the fast-moving Sherman army in 1864, was just two years their junior. On the other hand, Meigs—and the new secretary of war, Edwin Stanton, with whom he had an instant rapport—knew a good Scottish engineer-businessman when they saw one. So Daniel McCallum, the superintendent of the Erie Railroad who had transformed its operations, was made director of military railroads.

In a matter of months, the offices on Pennsylvania Avenue just west of the White House became a hive of activity. An instant clerical staff, many of them women (the first ever hired by the government), settled in and staffed the command-and-control station from which procurement officers and inspectors were dispatched, orders placed, shipments tracked, and the all-important means of expediting them to armies and forts mapped. The unglamorous work of drafting contracts, making them legally watertight by having them witnessed in front of a magistrate, then sending inspectors and periodic audi-tors to see they were properly executed, was necessary if the army was to be protected from unscrupulous purveyors aiming to make a killing from the urgency of the moment. The quartermaster general, after all, would be responsible for the expenditure of $1.5 *billion*—in 1860s values! No enterprise in Western history to that point had ever been so costly.

As critical as it was to meet those logistic needs as quickly as possible, Meigs was not one to skimp on quality, convinced as he was that "slop shop" fabric provided at low cost was a false economy, especially for the long marches he anticipated after Bull Run. New York, Philadelphia, Chicago, and even Paris were scoured for clothing of heavy-duty quality. Revolvers as sidearms, another first, were ordered in hundreds of thousands from Samuel Colt's factory in Hartford, Connecticut, to the precise specifications laid down by Meigs. Meigs was in love with iron (he had used it on the Capitol Dome), and now he aimed to build a fleet of steam-driven ironclad gunboats that could revolutionize river transport, a means of getting essentials to the armies without running the risks of the raiding attacks that could disrupt supplies coming by wagon train or railroad. Locomotives and track that could be quickly laid where the armies needed them were paramount, but Meigs, McCallum, and the other railroad chief, Herman Haupt, all knew that just as essential were repair trains that could be sent posthaste to wherever an enemy raid had cut a line. Sherman, who was seldom free with compliments, showered them on the ability of the quartermaster's department to make good any damage to the one continuous line of track that rolled through enemy territory within hours of the damage. But Meigs could not afford to neglect traditional means of transport. Though he was always apt to quote Napoleon to the effect that a thousand men needed no more than twelve wagons, and complain that convoys were encumbered by wagons carrying nothing more than officers' baggage, he still knew that mule trains were the basic resource on which a moving army depended. Between July and September 1861, he sent thirty buyers into the field to acquire more than 100,000 mules—a quarter of all the mules then alive in the Union states. This would be an animal war—and Meigs could come on like an angry veterinarian if he thought generals and their staff were being reckless with their beasts of burden and combat: not feeding them the right mix of corn and oats; using them for prolonged periods that would critically shorten their working life. In the first buying outings in 1861, nearly 150,000 horses, used for cavalry mounts and artillery, were purchased, giving golden opportunities to dubious horse traders. When a vigilant inspector in Chicago took a close look at a buy that had been made in Pittsburgh, he discovered many horses that turned out to be "blind, swenies, spavined, stiff-shoulder, split hoof, curbed legs . . . ring-boned, deformed . . . big knee, wind-broken . . . deranged hip, stock-legged beside being too old and too young, too small and of the wrong sex."

Since it took so much initiative and around-the-clock labor to put together the needs of a huge army, Meigs wanted generals in the field to engage, rather than use up, forage, mounts, food, and the volatile enthusiasm of the soldiers in irresolute tactics, wandering this way and that. Since the generals had to correspond directly with him for their needs, Meigs was never shy about giving them a piece of his mind, or indeed lessons in tactics and strategy if he thought they could do with them—which, until the advent of Grant and Sherman, he invariably did. To Ambrose Burnside he wrote in December 1862 as if he, Meigs, were commander in chief: "It seems to me that the army should move boldly up the Rappahannock, cross the river, aim for a point on the railroad between the rebels and Richmond and send forward cavalry and light troops to break up the road and intercept retreat." Neither Stanton nor Lincoln minded these lectures on soldiering coming from their quartermaster general, for they also

knew Meigs had a way with impressing on the generals just what
was at stake. Just in case Burnside had failed to grasp this, Meigs
wrote that staying put where he was in Virginia would be "death to
our nation . . . defeat, border warfare, hollow truce, barbarism, ruin
for ages, chaos!"

Lincoln relied on Meigs for his panoramic grasp of the war—not just
arrangements of soldiers or towns to be besieged, but rather as a vast,
swarming social world to be got on the move, husbanded, treated as
the precious resources they were; deployed and expended only with
judicious intelligence—sailors and bargemen, horsemen, cannoneers,
and snipers; surgeons, sutlers; cameramen, band musicians; mechanics;
ditch diggers, gravediggers; cooks; balloonists, telegraph operators;
semaphore men. Meigs was the commander of macro and micro. He
knew where the next bridge had to be built, how to ford a river no
one else had heard of, but he could also somehow see the needs of
grand strategy. He was the omniscient Hamiltonian manager of the
biggest, most capital-intensive enterprise the United States had ever
set on foot, but it was all being done for the supremely Jeffersonian
cause: the salvation of democracy. Meigs grew heated when he thought
of the disdain of Europeans, especially the British upper classes who,
not actually living among slaves themselves, could afford to be lofty
about abolition, while imposing imperial autocracies in sundry parts
of the world. The British were fake democrats, pretending to deliver
virtue to the world while pursuing selfish imperial interests. For years
they had been jeering at Yankees for their low materialism and want
of martial spirit. Now "the same unfriendly spirit" was happy enough
to view "the dreadful carnage" while making its pronounced partiality
for a Confederate victory apparent. For the British, the bloodshed was
comeuppance, but they should understand that it was in America that
lives were being laid down for democracy; that this was a war for
the future of the world. Two years before the Gettysburg Address
Meigs wrote to his mother that after all the terrible sacrifices "the
world will be better. Liberty in all climes will take a leap forward and
future ages will rejoice in the advance of liberal ideas, in the proof,
the signal proof that the people, the true democracy, is capable of self-
government. It is a great and holy war. God is with us and who shall
be against us?"

But that war had to be won, or else America would be a squalid joke. For two years, the generals were Montgomery Meigs's bane. How could the likes of George McClellan and Ambrose Burnside squander the perfectly equipped armies he had sent them with blundering irresolution? For months they would destroy their strength through sheer inanition, wasting men (for hundreds died every day through sickness), money, mules, hay and then, almost because the standoff had become intolerable, would hurl them at impregnable positions guaranteeing thousands of casualties and no better strategic position. Was this the best of the West Point loyalists? Often, it occurred to Meigs that he could do a better job in the field himself. But he seemed to have made himself indispensable as quartermaster general, and that was that.

Meigs reserved his most withering contempt for George McClellan, whose principal cause in the war was, he suspected, George McClellan. The general in chief was always grumbling to Stanton and Lincoln about being massively outnumbered by the Confederates, using the disparity as an excuse not to engage them. Meigs thought McClellan had his own particular way with numbers, complaining he'd been sent just 150 horses a week when in fact he had received ten times that number. Antietam, the bloodiest day of the entire war, when the Union took more than 12,000 casualties, 2,000 of them dead in the cornfields of Maryland, ought to have been evidence of McClellan's fierceness when roused. But since McClellan failed to concentrate his forces against Lee's weaker numbers, and more damningly failed to fall upon Lee's retreat, both Meigs and Lincoln believed the terrible carnage to have been in vain. It had halted Lee's invasion of the North, but what should have happened was the destruction of the Confederate Army of Northern Virginia. So when McClellan had (to Meigs's mind) the gall to run for the presidency against Lincoln, whom by this time the quartermaster general idolized, Meigs characterized "little George" as "the general who, after collecting into one vast Golgotha 200,000 men, held them in that pestiferous valley and petulantly whined and scolded and complained that the horror-stricken people refused more victims to his shambles—beat a disgraceful retreat and shouted victory from a gunboat while his brave but deserted legions were battling on . . . informing him by signal of the fight he should have led in person."

In particular Meigs hated McClellan's grandiose naiveté, a function, he thought, of his remoteness from reality, in supposing that the war could somehow be fought as a tournament of gentlemen. For instance, why should McClellan put himself out to ensure that the arch-traitor Lee's wife had safe passage through Union lines to rejoin her husband in May 1862? Meigs gagged on McClellan's sentimental assumption that war was somehow a gallant business that could be conducted within the rules of decency, or as McClellan himself said after an interview with Lincoln, "upon the highest principles of Christian civilization." There should, the general announced, be no seizure of enemy property, especially land or livestock, no political executions, and, horror of horror, no thought of liberating the slaves, a measure so heinous it should not be contemplated "for a moment."

To Meigs it was incomprehensible and absurd to deny the Union anything it could take from rebel farms, towns, and plantations—horses, cattle, crops, clothing, and especially slaves who were to be made free, who were at last to become true Americans! All this talk in the South—and sections of the North—over the war being fought over constitutional principles he dismissed as so much disingenuous cant. This was the war that would finally make good on the promises held out by the slave-owning Jefferson in the Declaration of Independence, of liberty and equality. When Lincoln issued a draft proclamation in the summer of 1862, Meigs, influenced by Horace Greeley's "Prayer of Twenty Millions" published in his *Tribune*, wrote that it was something "thinking men must have foreseen since the first gun fired." Ever since his heated conversations with his brother in Georgia, Meigs had concluded that the terrible conflict had been sent by God as a chastisement to the republic for tolerating and condoning the unchristian abomination of slavery. Now some sort of expiation—a word he used often—had to be made in blood to atone for that dreadful sin. "God does not intend to give us peace again," he wrote to his mother in 1863, "until the last shackle is stricken from the wrist of the black man." So naturally Meigs welcomed Lincoln's emancipation proclamation on 22 September 1862 and looked forward even more to its enforcement on the first day of the next year. The capture of Vicksburg by Grant made him elated since it would bring the full force of this second revolution into the heart of hell, Mississippi, a state he described as being of "special malignity."

Meigs was all for a black army, aggressively recruiting soldiers from among both the slave and free African American population and equipping them for cavalry as well as infantry combat, a thought almost as inconceivable to many Union officers as it was to Confederates. But Meigs wanted to prime their anger, point it right at the enemy. "It is impossible to cast aside the millions of recruits who will offer themselves, accustomed to the climate, inured to labor, acquainted with the country and animated by the strong desire not merely for political but personal liberty." By September 1863 there were already 82,000 freed African Americans serving as soldiers and laborers; and by the end of the war, fully 10 percent of all Union troops were black. That was in large part Meigs's doing; he established black garrison forts on the Mississippi and was happy to learn that in prison camps like Point Lookout, ex-slaves were guarding their former masters. Even more radical was his prescience that emancipation would be merely a paper revolution unless the liberated slaves received land. Without it the freedman "would be at the mercy of his former master who may drive him from the acres on which his cabin is built, his family sheltered and which, by mixing with them his own and his father's labor he has acquired a natural right to possess." Five acres per family was the very least that could make a black cultivator class a reality. And before long he would raise that estimate.

As the conflict stretched on into a third and fourth year, the war had come to assume for Meigs the character of an ideological crusade, between "gallant free men" and "a barbarous people, driven by wide sweeping conscription and enforced by a merciless despotism;" a war between a perverted and an authentic version of what America was. This was the lesson he wanted to deliver to his impressionable son up at West Point. If any American wars had to be fought, then let them be fought over grave matters on which the destiny of the whole world turned. Let them be fought for decency's sake, for the cause of humanity's freedom, or let them not be fought at all. Once embarked on, though, such wars had to be prosecuted without pity toward those responsible for bringing the slaughters on. For the blood and suffering they had caused, Meigs still wanted his old comrades and mentors, Robert E. Lee and Jefferson Davis, tried, convicted, and hanged. After Lee had beaten off McClellan in a series of engagements and the fainthearted were already muttering about Lincoln's inadequacies and the need to contemplate a compromise peace with the South, Meigs wrote to his son at West Point, shortly after the 4 July celebrations, in a chilling, Tolstoyan vein: "The North must hold this wolf by the ears until it is exhausted by starvation or destroyed by the kicks and cuffs which it may yet receive. No peace in compromise with the South is possible for our industrious educated democratic people. Death or victory is the . . . necessity of our cause and I do not less doubt the ultimate victory though God for our sins leads us to it through seas of blood." For Meigs, the war was the necessary ordeal: the second American revolution.

Having outgrown its original premises, the quartermaster general's department moved around the block, into the handsome building that the banker William Wilson Corcoran had built on Pennsylvania Avenue near 17th Street, a stone's throw from the White House. Corcoran had designed the extravagant Second Empire–style building, with French mansarded gables and Corinthian columns, to be the first public art museum in America. There the citizens of Washington would contemplate the glories of the national landscape, painted by the Hudson River luminists—Thomas Cole, Asher Durand, and Frederic Edwin Church. But collecting America did not mean pledging allegiance to it, for Corcoran was a Confederate sympathizer who fled Washington and sat out the war in Paris where his son-in-law was a diplomat for the South. Into the building moved the quartermaster general, using it to store uniforms and the mountains of paperwork the department was generating. But eventually the whole staff moved in. And there, in the compromised Mr. Corcoran's paneled study, sat the beetle-browed Monty Meigs at the fulcrum of his vast, sorrowful empire of human straining. In place of pictures of dappled light falling through deep woods, the noble red man amid (for the time being) his buffalo, or flatboats gliding down the rivers, he brought maps and photographs, and those were nothing like the American pastoral. Instead, fixed in wet collodion prints, lines of the dead lay neatly assembled for disposal (and the photographer); pyramids of shells stacked by a dock; the rear ends of hundreds of mules awaiting their wagons; farmhouses and stores reduced to charred spars and brick rubble.

But those were, after all, just paper and card images. Something in the quartermaster general told him he would do his job better if he were to see matters for himself. He had absolute faith in the best of his subordinates; honest, good, serious men, he thought, were McCallum, Allen, Easton, and the rest. Together they had sent the profiteers packing, had made whole armies, prostrate with sickness, fear, and exhaustion, movements of men once more. They had believed in free black American soldiers and had put them too on horseback and by their guns. But Monty wanted to get out of town before he went crazy with the sedentary claustrophobia of it. The trouble was that Lincoln depended on him to organize the capital's defense, should Lee ever break through and threaten Washington directly; to make sure Lee never got home to Arlington Heights, from where he could launch shells at the White House—and the quartermaster's building! But Gettysburg in July 1863 put an end to the northern invasion and encirclement strategy. Meigs could move.

The late fall of 1863 found the quartermaster general at beleaguered Chattanooga. In late September William Rosecrans had lost the battle of Chickamauga, close to the Georgia–Tennessee line. Worse, as far as Meigs was concerned, the West Point engineer, of whom better things were expected, had done a McClellan and abandoned his army toward the end of a disastrous day, leaving his chief of staff to hold off the Confederates. Now he had taken the Army of Cumberland, the prize force, comprehensively equipped by Meigs, back to Chattanooga, where he was in a state of siege. Chattanooga was at the hub of rail lines, crucial to denying Lincoln his strategy of cutting the Confederacy in two, west and east. Rosecrans was supposed to have taken it. But it had taken him instead. Lincoln removed Rosecrans and sent Grant, Sherman, and General Joe Hooker to relieve the siege. But he also sent Meigs to sort out logistics on the spot. He would be the fixed point in the shakiness. When someone yelled for more artillery, more horses, more grape, more rations, Meigs would know whether it could be supplied and, if it could, get it there quickly. He could see right away that the situation of the Army of Cumberland was bad. Men had been living off weevily hardtack for too long, and now even this was in short supply and the army was in real danger of being starved into surrender. What would rescue it? Carpenters, Meigs thought, and telegraphed the secretary of war to that effect. They would build supply boats and they could bring in food supplies for both men and animals along what the soldiers called "the cracker line." Getting this done on the spot made Meigs happy, though he covered his elation with his usual mask of sobriety.

But when he slept under the cold November stars, in a Meigs blanket stitched to make a sleeping bag and filled with hay for warmth, and gazed at the thousands of lights set by campfires on the hill, Meigs felt the strength that comes from the instinct of rightness. He felt close to the thousands of men; to their fear and their courage; to the vermin that were crawling their backs; to the liquor they needed to get through; to the homespun letters they were scratching to their homes in Illinois and Vermont and New Jersey and his own ancestral Ohio. And then on the morning he stood with Hooker's soldiers in the fog that shrouded their advance up Lookout Mountain and thought that such a blessed concealment could only be the work of God, and he continued to think this as the men shot and charged their way uphill and took the ridge. And then he watched as Union soldiers swept their way up Missionary Ridge—impulsively advancing beyond Grant's orders but protected from deadly cannon fire by the misplacement of the guns at the distant summit of the ridge rather than the highest point from which they could be effective. This too, Meigs thought, was an intervention of the Almighty in blue serge. But then it was Braxton Bragg who was commanding the enemy. Bragg who had graduated a year after Meigs from the academy, who had fought against the Seminole and the Mexicans; who was undoubtedly brave and undoubtedly dim. A stickler for the rules, serving as quartermaster and commander, Bragg was reputed to have made requests from himself for certain guns and then declined them. But he had not noticed the top-of-the-hill problem. And so he lost Chattanooga. And Chattanooga began to lose the Confederacy the war.

Once the situation was stabilized, Meigs rode with Sherman's army a little ways into Georgia. Crossing the line stirred something painful in the recesses of the quartermaster general's family feeling. Georgia was Meigs country; the place of Return Jonathan Sr.'s autumn years with the Cherokee; the place where his grandfather Josiah had been university president; the state where his father, Charles, had begun his practice; the state where Montgomery Meigs himself had been born! What if his father had stayed in Augusta? How different would his life have been? Would he, heaven forbid, have been a Confederate, just as Lee and Davis had obeyed instinctively the summons of their place? Somewhere to the south, on the Alabama line, was his estranged brother Henry, the Confederate quartermaster, still, for all he knew, turning out the coats and trousers that Montgomery's and Grant's men had taken aim at on Lookout Mountain and on Missionary Ridge. There was nothing to be done about this family misfortune. As far as Montgomery was concerned, Henry had put himself beyond the pale and was kin no more. But their father, Charles, was tormented by this war of brother-quartermasters and whatever turn the war took had continued writing to his son in Columbus. Knowing Montgomery was with the Union army, Charles implored him not to cut his brother off entirely. Perhaps he might even bring himself to attempt some contact, for their mother's sake. Monty was hard-bitten, not easily moved by such appeals. Family feeling was patriotic, or it was contemptible. Henry, he wrote back to his father, had taken himself clean out of the company of his friends and family by being "false to his country's interests . . . He has taken a post with rebellion and civil war." But then, Meigs added a sentence which suggested that he believed Henry might not, after all, be irredeemable. Perhaps he had been Led Astray by the Georgia heiress. And he could yet be brought out of the land of iniquity and back to his senses. "I only hope that the advance of our army may catch him and send him north, out of the infamous company which has corrupted his good nature." In Ringgold, Georgia, Meigs made inquiries, of the citizens, of captured Confederate soldiers, of anyone who might know a Mr. Henry Meigs of Columbus. Nothing was known. But Montgomery could not quite give up, so he left a letter for his brother in the drawer of a washstand in the lodging house at Ringgold, enclosed within another asking any person who might be in a position to send it forward to see it reached his brother's hands. What the letter said, whether it held out an olive branch or a rod of imprecation, we can only surmise.

Six months later, in Virginia, Meigs wandered among scenes of extreme distress at Fredericksburg, the town that had been designated by General Meade a collection center for the wounded. With the idea

of grinding Lee down, making his deprived and exhausted army incapable of defending Richmond, the Confederate capital, Ulysses S. Grant
had done the opposite of McClellan: had sought engagements wherever he could; had attacked frontally with superior force and suffered
terrible casualties. Those from the savage battles at Spotsylvania Court
House and the Wilderness, at least 14,000 of them (Meigs put it more
like 20,000), had overwhelmed the meager capacity of hospitals and
had been thrown on the streets, where they lay with dirty bandages
and raw, sometimes suppurating, stumps. Cornelia Hancock, one of
the extraordinary "angels of mercy" who tended the wounded, and
who thought the aftermath of Gettysburg had inured her to the worst,
nonetheless gagged on the horror at what she saw in Fredericksburg
in the spring of 1864. "It seems like the Day of Judgment. I went into
a dark loathsome storehouse, the floor smeared with molasses, found
about twenty wounded who had not had their wounds dressed for
twenty-four hours and had ridden some fifteen or twenty miles. O
God! such suffering never entered the mind of man or woman to
think of." Private houses were desperately needed to get the sick
and wounded off the streets, and to contain infections, but much of
the population of the town had disappeared on the arrival of federal
troops, and those who had stayed barred their doors out of fear or
hatred. Horrified by the scale of the suffering, Clara Barton, the leader
of the nurses (and founder of the American Red Cross) and Senator
Henry Wilson (the commander of the 22nd Massachusetts Volunteers)
went to Lincoln to ask him to send Meigs directly to Fredericksburg
to make the situation more tolerable and get food to men who might
as easily die of starvation as gangrene.

So there Meigs was, doing what he did well: commandeering houses
for the wounded, if necessary turning inhabitants out of doors if that's
what it took, using churches as hospitals, finding food other than hardtack and tea for the sick. He had seen American war from the remoteness of his office, had coolly inventoried its wants and its damage; had
come close to fire at Bull Run and the fog-girt Lookout Mountain. But
now, in Fredericksburg, its appalling truth confronted him: maggots
half an inch long crawling in open wounds; men screaming as they
were jolted into town aboard one of Meigs's two-wheel ambulances;
women, brave beyond any imagining, opening the mouths of the

wretched for water or—more safely—lemonade; men crying for their wives and mothers or for death. All of that Cornelia Hancock could bear just so long as they did not ask her to write home. That she could not manage without tears.

Through all this, Meigs found himself thinking, from time to time, about his son. Since his return to West Point after the battle of Bull Run, John Meigs had continued his eventful career at the academy. While slaughter was proceeding in Maryland and Virginia in 1862, John managed to get himself arrested and court-martialed for fighting with a fellow cadet who, he told his father, had insulted him with dishonorable remarks. Not only that, but John had compounded the first offense by exploiting his father's closeness to Abraham Lincoln, turning a moment intended for the exchange of courtesies into an attempt to have the president of the United States, then visiting the academy, commute his sentence of summer detention. John even had the cheek to thrust a prepared card at Lincoln which read: "the punishments of Cadets J. R. Reid [a friend who had been involved] and J. R. Meigs are remitted." Lincoln was all considerateness, turning him down in the gentlest way: "Well Mr Meigs, I should be very glad to do so if it would not be interfering with the authorities here." Surprisingly, the father does not seem to have berated the wayward son that severely, since both he and Mary Meigs made a point of visiting the delinquent during his detention. They were rewarded for their faith by John Rodgers Meigs graduating first in his class.

By late June 1863, the twenty-one-year-old brand-new Lieutenant John Meigs was directing fortifications for a major part of Baltimore, then threatened by Lee's advance north into Union territory. He was barely out of the academy when he showed himself very much his father's son, suggesting modifications of armored railroad cars, designed to protect the line, so that fieldpieces could be mounted inside them. Meigs then had his own little fleet of five of those ironclads with which he patrolled the lines, taking the fight to enemy raiders with his howitzers, rebuilding track, telegraph, and even bridges that had been destroyed as he went. Each of the cars had pet names given by John after Union victories—*Vicksburg, Antietam,* and so on—and some of them he had disguised as unarmed store wagons. "When Johnny Reb comes up to them and sees the little shutters drop out of their ports and the whole thing suddenly transformed into an inapproachable blockhouse, I think he will be astonished," the excited engine-riding Johnny Meigs wrote to his father.

8. John Rodgers Meigs, the Shenandoah Valley, summer and fall 1864

Just a year since he had worn the smoke-gray uniform of the cadet! Just a year. He could scarcely believe it. Even in that very little span John had noticed changes in the men. "At the beginning of the war," he wrote to his mother, "soldiers wrote letters to their friends and made little arrangements about their watches etc. before going into action. Now they march upon an expected battlefield with little more emotion than an ordinary parade ground." Had he himself become hardened, impassive? He had seen, done, pretty much everything the war could throw in the way of a young lieutenant of the Engineers. It had not just been joyrides aboard his iron railroad cars, howitzers at the shutters. John Meigs had become, just like his father, indispensable: the man to go to, for scouting terrain and topography; map-making, fort-building, bridge-building (or destroying), path-clearing, road-making. His reputation grew apace. Ulysses Grant asked for him in the slogging campaign to Richmond, something that delighted Montgomery, but he was told he absolutely could not be spared.

Like his father, John let it be known when the generals missed opportunities. Or worse. He was at the rout at New Market in the Appalachians and thought it had come about through General Franz Sigel's "poor management of his troops." He had been in hand-to-hand combat; had come close to capture when his scouting company had been ambushed in the West Virginia woods. He had lost his horse, had sprinted through the undergrowth for his life. He could act the man very well, the managing engineer; the "Lieutenant General" as the soldiers jokingly called the twenty-one-year-old with the tousled hair. "Hand full of maps and head full of plans always," his proud father wrote in his diary. When John's blood was up he could be every bit as fearsome as Montgomery. During the debacle at New Market, John had tried to rally troops fleeing to the rear and a Virginian Union soldier (there were some) saw the stocky little lieutenant "cut down a straggler with his sabre" when the trooper refused to fight. But he was also the boy who needed to hear from home; who survived on letters as much as rations. "You have not written to me since last April though I have sent several letters to you," he upbraided his father in a wonderful role reversal (Montgomery having been suddenly busy with commanding the defenses of Washington). Every contact with home was emotional bonus rations. One day his maternal uncle Robert arrived with his regiment "or the remains of it" at Harpers Ferry in West Virginia looking, John drolly reported, "as the saying is 'demoralized' that is as if he had not shaken hands with the paymaster for some time." Just as the son reproved the father, the nephew offered the uncle money to spruce up a bit, "but he said no he had borrowed $10 and was alright for a while." The two of them lay flat out on buffalo hides on the ground in the gloaming and spoke of "the dear ones at home," John throwing his arm around his uncle's neck as they lay there or taking his hand as they went through all the aunts and brothers and sisters and girls and chums who had laughed their way through the parlor in Washington. "The song you have often heard commencing 'Do they think of me at home,'" he wrote to his mother, "expresses a feeling which we often experience. It sometimes seems not improbable that the constant danger to which we are exposed may make our friends as it does ourselves forget that we are running such terrible risks."

So while he was capable of saber-slashing a coward or putting the torch to farms and barns, riding rebels down, and—a great satisfaction to a West Pointer—burning and sacking the rival Virginia Military Institute (from which John extracted a bust of Washington to be sent to his old academy as a trophy), John Meigs kept his tender streak. It was the women and children who moved him most. He hated the way the Confederates used "curly-haired" children as pickets to scout for Union soldiers, taking advantage of their high spirits to put them in harm's way. Then there was the "poor mother" who had come all the way from Boston to Virginia to find her thirteen-year-old boy who had enlisted with a regiment, though too young to serve. (The Confederates in their extremity filled ranks with fourteen- and fifteen-year-olds.) Meigs helped the mother to find the officer who had recruited the "child" and get him arrested. But there was no happy ending for her. On the eve of taking him home he "slipped off to another regiment and in his last letter informed her he was having a splendid time." On the Monocacy battleground in July he found a mother and daughter, "Miss Alice, a lovely and accomplished girl," who had taken to their cellar in the thick of a terrible firefight. Alice—for whom John evidently felt something—"declares she did not feel very badly frightened though the muskets were popping out of the windows and the balls rattling against the walls," until a shell crashed through the wall of the dining room and burst just over their heads with only a thin flooring between, the first of seven to hit the house.

The more he saw, the more John felt, and what he felt most keenly was the mystery of survival or immolation; the peculiar randomness of death's choice. He deeply admired the ferocious Sheridan and the affection was evidently returned, the bond of two short heroes. Sheridan was known as "little Phil" to his men so in no time the general called his chief engineer and aide-de-camp "little Meigs." But John thought Sheridan and General Tolbert, his second in command, altogether too reckless in exposing themselves to danger; always discussing maneuvers right behind the skirmishers who usually were the first to advance. Near Charlestown, a Dr. Rulison on Tolbert's staff was having just such a talk with the general. John was watching, and in his way, half man half boy, was worrying. It was late August, Dog Star days and harvest moons. The discussion was taking place in a grassy field, when John saw a man walk right before them "so coolly I thought he must be one of our men who was wounded." And then that same man "stopped . . . raised his gun and fired." John caught the whistle of the bullet as it came past and its stop as it struck something, someone. He looked hard at Tolbert and Rulison. For a moment there was silence, so John supposed the ball must have hit one of their horses. "But then I saw the doctor put his hand to his side and someone cried out that he was wounded." "I'm shot through," Dr. Rulison said in John's hearing, with the same incredulity any target of gunshot might have. "My God my God I'm dying." In less than an hour, John told his father, "he was dead." It was all such a mystery.

Two weeks later on 18 September, John was preparing for battle the next day and as usual felt the need to write to his mother. Sometimes when he had finished everything he meant to say, he felt he must carry on anyway. "I still feel like writing on and talking a little while longer." That was it. It was like talking.

"We are going to have a terrible fight tomorrow and will have to be up at two o'clock in the morning. I feel that the chances are in our favor and if our troops behave well we must win the day and a glorious victory it will be.

"Still God only knows what the result will be and if it is not for us I am afraid it will fearfully be against us.

"Give my love to all."

9. Montgomery Meigs and Louisa Rodgers Meigs, October 1864–December 1865

"SEND MONTGOMERY HOME. HIS BROTHER IS DEAD." So read the telegram, bald and brutal, that the quartermaster general sent on 4 October 1864 to his brother Emlen in Philadelphia, with whom Monty Jr., the third Meigs son, was then staying. Struggling with his grief, attempting to see in the killing the unanswerable will of the Almighty, Meigs attempted to take solace in making an inventory of his firstborn's virtues: courage, resourcefulness, selfless patriotism; the usual list. There had been deaths in the family before, the two little boys in 1853, a stillborn girl. When Louisa's cousin George died in a naval attack on Charleston Harbor, Meigs had written philosophically: "and so goes on the work; one after another the nation's dearest jewels are laid upon the altar of sacrifice." But as many deaths as Montgomery Meigs had reckoned in his office, this one was inexpressibly terrible. He dug deep for composure; managed it by trying to discover exactly how his son had met his end.

There was no doubt at any rate about how he had been found. An orderly who had escaped the attack had made it to General George Custer's headquarters on the morning of 4 October. A major sent to the Swift Run Gap Road had discovered "the body of my son," Meigs wrote in his unnaturally calm, official voice, "where he had fallen un-rifled, the left arm raised (as if he had got off a shot), the right extended at his side. He lay upon his short cape or cloak, a bullet through the head"—just below the right eye in fact—"and another through the heart." And this much more was agreed on: that John and two orderlies were riding along in the autumnal rain, making their way back toward camp. Ahead of them they saw three other riders dressed in raincoats that covered their uniforms. Meigs and the two orderlies rode on in file, assuming the men were Union troops. But when they reached the unknown men, who were riding abreast across the road, according to the orderly the men suddenly wheeled around, grabbed the bridles, and opened up, firing the shots that killed John Rodgers Meigs. This happened, so the orderly said, in spite of young Meigs surrendering in a loud voice. The conclusion, accepted by Sheridan and by Montgomery Meigs, was that John had been "bushwhacked"—killed in cold blood by disguised Confederate partisans, probably irregulars belonging to John Singleton Mosby's band, famous for their brutality in the Shenandoah Valley. Sheridan believed this enough to order the burning of every farm and barn within a five-mile radius of Swift Run Gap Road. For a while, the fate of Dayton itself hung in the balance. Sheridan wanted vengeance for "little Meigs." But he spared Dayton.

Three days later John's body was brought to the Meigs house in Washington. A day after that he was taken with full military honors to Oak Hill Cemetery in Georgetown. Both Abraham Lincoln and Edwin Stanton stood, hats removed, heads bowed, as he was laid in the chapel. On 10 October, a week after he was shot in the Virginia rain, John was buried beside his two younger brothers and the infant who had been too little to acquire a name when she died. Attending this second ritual were just father and mother, young Monty, and Louisa's brother William. "We planted an ivy at the foot of the oak under which he lies & left him alone in his glory," wrote Montgomery in his diary.

But the quartermaster with the fine high dome of a head and stately beard could not leave matters there. Convinced his son had been murdered in cold blood, he hired a detective, Lafayette Baker, to try to verify the orderly's story and put out a reward of $1,000 for information leading to the Confederate killer's capture. But the more the sleuths dug, the shakier the first account seemed to be. When eventually it was safe for the men who rode abreast that evening to emerge from nervous obscurity, they told a different and more credible story about the last minute or so. They were, in any case, not bushwhackers but

Confederate soldiers whose uniforms were simply hidden by their rain-coats. Trying to get back to their own camp, they realized that these Union soldiers were between them and safety, briefly talked it over and decided to attempt a capture by surprise. But, so one of them said, taking his pistol from beneath his poncho meant revealing the telltale gray at which point John, who always kept his own revolver handy, aimed a shot at this George Martin. It struck him, and it was the shots that were got off in return that killed the young lieutenant.

Who knows? But reading John's letters so full of gumption, the Confederate story rings truer than the one that had him surrender right away. And why should his father not want to believe a version that had his son gunned down fighting for his life and his country? But Meigs believed the Confederacy capable of anything and preferred the "cold blood" story. From time to time, after the war was over, he would get letters from the wives of his West Point comrades before the fall—especially from Varina Davis, imploring his intercession with the government for the release of her husband. Iron had entered Meigs's soul. As far as he was concerned Robert Lee, Jefferson Davis, Braxton Bragg, Pierre Beauregard, and Joseph Johnston had all killed his boy with the murder and mayhem they had unloosed on the republic. And they had murdered his beloved friend and president to boot.

For on Good Friday, 14 April 1865, after going to church and writing in his diary that the country was "drunk with joy" at the peace, Meigs returned home. Around ten in the evening he was told that Secretary Seward, his mentor, had been the victim of a savage knife attack. Seward's house was just three blocks away. By the time Meigs got there, blood was everywhere and Meigs held his hand to the wound to staunch the flow. Astonishingly, Seward would survive; Lincoln of course did not. It was while he was attending to Seward that the horrifying news of what had happened at Ford's Theater arrived. Meigs went from one horribly wounded man to another. Lincoln, who had been taken from the theater to a house across the street, was already unconscious when Meigs arrived. Giving himself something to do in the dumb horror, Meigs made himself the gatekeeper, deciding who should and who should not gain admission. He himself kept vigil as Abraham Lincoln lay dying, expiring at 7:22 in the morning. At the funeral five days later, Meigs rode at the head of two battalions from the quartermaster general's department.

None of this did anything to lessen Meigs's resolve that Arlington House and the 1,000-acre estate around it should be a national military cemetery. In June 1864, three months before John's death, he had already ordered the first Union bodies buried close to the Lee house. They would be the advance guard, he thought, for thousands of those who had perished because of the infamous treason of the general (whom he still wanted to see tried, judged, executed) along with Davis. But when Meigs went to look over the site, he was displeased to discover that the bodies had been interred in the old slave burial ground of the Lee estate, close by their living quarters. Orders were reissued to the effect that bodies ought to encircle the mansion, coming as close as possible to the Doric portico itself. And this time, Meigs went to Arlington Heights, the site from which he had always imagined Confederate shells raining down on Abraham Lincoln's house and on his Capitol Dome, and made sure that the shovels struck; that Lee soil was purified with the bones of the blessed dead. By the end of the war there were 16,000 bodies buried on the Lee estate.

It was some while before he had John taken there and had the perfect bronze of the boy-man made for his memory. By this time Montgomery had come to believe in the story he had rejected in the grieving aftermath of the shooting. Now he wanted John to be the sweetly dauntless patriot more than he wanted a prosecution for murder. So the bronze revolver by his side exposes an empty chamber to indicate the firing of a bullet at his rebel assailant, and Montgomery could go and visit his son on the brow of Arlington Heights and mourn his lost patriot.

But that was not how the mother felt or what the mother wanted. Not for Louisa her husband's submission to the ways of Providence. She wanted him back from the grave. "My darling precious John," she wrote to her sister Ann, the "Nannie" John had loved, "I seem to hear his step in the hall and I see his bright happy face as I last saw him. I crave to be alone, to sit & think of the past, those sweet happy days at West Point how they return and what bright memories they bring of young forms and faces that will never meet there again . . . Letters from every quarter come to me and Mont to assure us how much he was loved & what a reputation he had already achieved. It seems an increase of agony to know what a brilliant future was before him. All that he was and all that we have lost. I have at the foot of my bed his trunk, his poncho and his camp stool and short cloak he wore that fatal night. The hat and coat are splattered with mud. You know how it rained that evening. The cloak shows where he lay upon the ground and the jacket is pierced through with a bullet just over the heart. These cast-off garments seem to tell me the whole story. I take them out and look at them again and again and kiss and hold in my embrace the dear hat which still retains the perfume of his hair. I seem to feel that I have him here again with me. My darling precious boy . . ."

A month later Louisa had made a shrine to John in her own bedroom, framing his drawings, cap and sash hanging over her bed, a trunk filled with his things "the altar where I most love to pray." A year later Andrew Johnson was president, and there was talk of amnesty for the Confederates, something which still drew bitter reproof from Montgomery. But Louisa could not shake off her grief "which trembles at my heart . . . and which will terminate only with my own life." She went about the business of the household, just as Monty continued to manage the quartermaster's department, demobilizing the troops, standing down the immense machine he had made to save the Union; meeting with his scientific friends in Washington. But when she was by herself Louisa was overwhelmed by loss and wandered the rooms absentmindedly. "He seems to have left his footprints everywhere in this house, traces of his hand in books or work of some kind I encounter every day. He has left such a void, such an aching void in Mont's heart and mine that we must go down to our graves sorrowing . . . Mont never dwells on this sorrow, he seldom speaks of our dear boy. I know it pains him to do so but he could not find indulgence for his grief as I do but it has entered his inmost soul, I found him alone in the parlor the other day looking at a bust which he has lately made of Monty & as I came in he said how much he wished he had one of our dear John. 'As the war is over and as time wears away I seem to miss the boys more & more. I *want* them,' he said in such a tone that I could scarcely refrain from bursting into tears."

Montgomery Meigs strode into the room, and it was as though I'd always known him: the good general. Presumptuously, I told him as much. "I've been living with your great-great—how many greats?—uncle." "Three," he replied without having to count on his fingers. The Meigses knew their genealogy, and this one had a doctorate in history. He smiled as he said this, and it crossed my mind that the quartermaster would not have been so easy on first introduction. Suddenly, the grungy little green room at NBC Television seemed populated with Meigses: Return Jonathan the guardian of the Cherokee; Josiah the restless professor; Charles the gynecologist; Johnny laid out, eyes to the sky on the Shenandoah road. Was I imagining there was a Meigs look, for the current Monty seemed to wear it? Like his ancestor he held his long-limbed height straight up, a West Point bearing that could be informally unfolded into a chair. Present Monty offered a bright and open face, generously inviting engagement, whereas Past Monty, in the beautiful Mathew Brady portrait photograph, is locked off behind the whiskers of authority, answering the calls of severe contemplation. Full-length, three-quarters profile Meigs stands as if simultaneously present and unavoidably engaged elsewhere with the look, as Brady must have imagined it, of that oxymoronic thing: living history. The upper part of the head was uncannily identical in the two Monties: big fleshy ears, deep-set dark eyes beneath a slightly overhanging brow, the nobly domed cranium whose curvature I suddenly realized I had seen many times that week in Washington, as the Capitol cupola; the American legislature configured as the thought-full skull of Montgomery Meigs, architecture as self-portraiture.

I had just been watching General Meigs (now retired) speak on cable television with a British brigadier he'd known during his command of the Stabilization Force in Bosnia. Persuading Serbs and Bosnians to communicate across their ancient tribal and religious loathings and terrors had made him canny about what similarly needed to be done in Iraq if American troops were ever to depart with honor. "Have them do their deals as they know how to do them and stay the heck out of the way," he said of a lesson learned in Bosnia, making light of his skills as an arbitrator of decency. He had learned the hard way the indispensability of social understanding and political acumen to soldiering. "Did they teach you that at West Point?" I asked. "They did not. Something Americans don't do well," he added ruefully, "understanding other cultures."

Comparative anthropology hadn't been much on his mind either, not at the start. It had been hard to escape the Meigs tradition and young Monty hadn't especially wanted to. His father had been a World War II tank commander; grandfather in the navy, great-grandfather ditto. He had been taken to see the model ships at the museum in Annapolis Naval Academy, and there he gazed at the past and saw the future. There had been a time when he'd thought he might be a doctor but at Colgate University, right in the middle of the 1960s, when American history, *especially* the history of American wars, was deeply unfashionable, somehow ancestral memory and present vocation resolved themselves into clarity. Meigs went to West Point and then on to Vietnam as an infantry officer in the most dangerous outfit of all—reconnaissance. He was a Jeffersonian idealist; there were such types in the rice paddies. "I thought it was important to protect South Vietnam from being conquered by the North . . . we did nothing wrong; no atrocities" (I hadn't asked). But then on the summit of Hamburger Hill, with his company taking appalling losses for no particular objective that he could understand, something ugly began to pick at Monty Meigs's conscience: that the whole war was "a strategic error of horrendous proportions"; a war that should never have been waged. At Georgetown University these days Meigs teaches a course on "Why presidents go to war when they don't have to."

The disenchantment bit deep. For some time he thought he'd get out of the military, but then he couldn't. "I looked in the mirror and thought, no. I'm a soldier; that's what I am." A command position in Europe followed, where, so long as the Cold War continued, so did the rationale for American troops, along with that education in comparative culture. But there were no simple outcomes. Desert Storm in 1991 was justified by Saddam Hussein's invasion of Kuwait, but in forty minutes at Medina Ridge, Meigs was in the battle that incinerated Iraqi cavalry inside their tanks. The NATO command in Bosnia—trying to separate the sides—was, evidently, altruism meets pragmatism; a dash of Jefferson, a shot of Hamilton. We spoke of those two founders and their respective philosophies of American war. Jefferson had, he thought, the luxury of picking his fights and keeping a skeleton army of professionals; making West Point an academy of engineers, as the young country was without immediate land enemies, and the conquests were of geography and the Natives. There have been moments when the Jeffersonian commitment to fighting only wars that defended liberty were realized—the Civil War; World War II—but since 1945, the military had been Hamiltonian; a vast permanent corporate institution. Every so often the West Point "stars" divided on their allegiance. Omar Bradley had been pure Jefferson; the ex-superintendent Douglas MacArthur, the incarnation of Hamilton. And every so often a general who ought to have been one kind turned out to be another. It had been Dwight Eisenhower, the deepest embodiment of the West Point ethos of command in World War II, who at the end of his presidency had sounded *exactly* like Thomas Jefferson, warning against the threat posed to American democracy by the "military-industrial complex." But, Meigs thought, in the end the scale of Cold War preparations had meant that a Hamiltonian mind-set had, for better or worse, prevailed; the self-generating momentum of military preparation dominating serious discussions of the cause for which treasure and blood would be spilled. He lightly rubbed his chin as he said that, not exonerating himself from what had happened.

It had been "preparedness" that had persuaded the army to train officers and men for a second war in Iraq even though the decision (at least officially) hadn't been taken or even properly debated. The imperative of offensive preparation had been just another form of self-fulfilling prophecy. Reflexive instinct. You can fight the wrong war with the wrong enemy and inadvertently make new ones, the general said. Another smile, this time of regret, a pause, "This had nothing to do with al-Qaida." The decision had been taken after he had retired as a four-star general and while he was occupying the most paradoxically, or possibly penitentially, endowed Lyndon Baines Johnson Chair of World Peace at the University of Texas.

Knowing your true enemy; that's what the quartermaster general had done. Did the general ever think of his great-great-great-uncle? He did. He understood perfectly the importance of being a pain in the neck. He knew exactly what it had been like for Montgomery to have faced down the kickback artists and the array of businessmen who had not expected to have to bid for their contracts. When he had run the Joint Task Force on Improvised Explosive Devices (road bombs), and, like the good engineering Meigs that he was, was concentrating on what could be done to defeat them, both politically as well as technically, he discovered that "there are still people in town interested in noncompetitive contracts. It makes you draw a line and say no, we're not going to do that." Could that be said for whole wars? It had better be, he thought. You couldn't miss the Meigs inheritance. He told the top brass what they didn't want to hear, namely that they were going after effect when they should be going after the cause, identifying and penetrating the networks that produced the IEDs rather than just catching up with the latest cell-phone detonators after the fact. But he was warned off. This was politics. This was none of Meigs's business. The commander of Central Command, John Abizaid, felt moved to remind Meigs that he was officially retired. "Look, Monty, you're not helping, the way you're going about things." Meigs persisted; Meigses always do. "HEY, look," Abizaid exploded, "this is not your fucking war to fight." Meigs declined to retreat; Meigses seldom back off. "You know the family has a characteristic of that flinty obdurate nature," he remarked, looking sunny as he said it. "I don't see it in you," I said cheekily, thinking this is one of the most decent men I have met in a long time. "Oh it's there," he replied, looking back at me with a straight face. "You can't see it, but it's there."

11. Hamilton resurrexit

On 11 March 2006, General Montgomery Meigs walked into a briefing room in the White House. It was breakfast time. On a side table were coffee, bagels, the usual. On the long table in the middle of the room was an array of Improvised Explosive Devices. The horrifying casualty rate in Iraq from these bombs and mines was proving a textbook case in asymmetric warfare. On the other side of the table were President Bush, Dick Cheney, and Donald Rumsfeld, all of whom wanted some good news. From Monty Meigs, though, they got the tough truth that so many people didn't want to hear. There was no magic shield that dollars could buy. The answer lay in attacking the insurgent networks from within. Two days later Bush spoke about the problem and put a brave face on it. It was a big problem, no doubt. But "we're putting the best minds in America on it."

The room where Meigs had briefed the president is called the Theodore Roosevelt Room. On one wall hangs a portrait of the Rough Rider president, who believed a nice little war was just the moxie to revivify an America enervated by its foul cities: filthy lucre, even filthier slums, polluted air, and corrupt plutocrats. Americans needed to restore the national manhood by getting out more and taking potshots at their enemies. In 1906 the president who had declared with his customary candor that "no triumph of peace is quite so great as the triumph of war," received the Nobel Peace Prize in Oslo. The immediate reason for this improbable act of recognition was that at a conference in New Hampshire, TR had negotiated an end to the Russo-Japanese War. Perhaps when the presenting orator, Gunnar Knudsen, said that "the United States of America was among the first to infuse the ideal of peace in political practice," he was thinking of Thomas Jefferson, who had indeed wanted it to be so (even if he found himself at war with the Barbary States on the Maghreb). But at the moment that he received the prize, Roosevelt's administration was trying to suppress a lengthy and brutal guerrilla insurgency in the Philippines. The president had claimed in 1902 that war was over, but it would not be until 1907 at least, and after 4,000 Americans and tens of thousands of Filipinos died, that the rebellion was pacified. For Mr. Knudsen to have asked the American ambassador to convey to President Roosevelt the gratitude of the Norwegian people for "all he has done in the cause of peace" must have called on all his skills at producing a Scandinavian straight face.

Every chance he got, Teddy Roosevelt sounded off about the tonic invigoration of belligerence. "All the great masterful races" (among which he meant Americans to number), he boomed, "have been fighting races." If Jefferson and Hamilton had pointed the United States in alternative directions of destiny as a world power, there was no question where Roosevelt's preference lay. Jefferson he despised as a remote intellectual and a weakling in matters of war and peace, one of the very worst of presidents. Alexander Hamilton, on the other hand, he revered for his frank passion for power, his vision of strong central government, and his unapologetic determination to make the United States a player on the world scene, admired and feared for its military prowess. So it was no accident that it was at the Hamilton Club in Chicago in April 1899, during the first year of the war against the Filipino resistance and with an election not far away, that the then vice president spoke of "the Strenuous Life."

Even by TR's standards the speech was an astonishing performance, a warning that if the United States did not wish to become another China "and be content to rot by inches in ignoble ease within our borders," it had better embrace strife and battle. "When men fear work or fear righteous war . . . they tremble on the brink of doom . . . thrice happy is the nation that has a glorious history. Far better is it to dare mighty things, to win glorious triumphs even though checkered by failure than to take rank with those poor spirits who neither enjoy much nor suffer much because they live in the gray twilight that knows not victory or defeat." The enemy within was "the timid man, the man who distrusts his country, the over-civilized man who has lost the great, fighting, masterful virtues . . . the man of dull mind whose soul is incapable of feeling the mighty lift that thrills 'stern men with empires in their brains'—all these of course shrink from seeing the nation undertake its new duties; shrink from seeing us do our share of the world's work, by bringing order out of chaos in the great fair tropic islands from which the valor of our soldiers and sailors have driven the Spanish flag . . ." What would the enemies of this war have the government do? Deliver the Philippines to people who "are utterly unfit for self-government?" "I have scant patience with those who fear to undertake the task of governing the Philippines . . . or [who] shrink from it because of the expense and trouble." But Roosevelt had even less patience for those who "cant about 'liberty' and 'the consent of the governed' in order to excuse themselves for their unwillingness to play the part of men." Let the naysayers in Congress be warned. Should any disaster befall the troops, it will have been the fault of those weak-kneed, lily-livered legislators!

Playing the part of Man was at the core of Theodore Roosevelt's self-making. Now that the trophies on the walls of his Long Island house, Sagamore Hill, and the reflection he caught in the mirror declared him to be a fine specimen of American manhood, he felt he ought to perform the same invigoration for the American republic. Predictably, Roosevelt had been a child with poor eyesight and sickly frame, who through his bullying father's admonitions and his own formidably precocious will had turned himself into an example of American masculinity. When his first wife, Alice, died, he took his grief west on to the prairie, farmed cattle, shot at Indians, captured rustlers, so that when the time came to enter politics he could imagine that he personified the authentically American spirit of conquest precisely at the moment when there was no more America to conquer.

Driven by this restless sense of physical self-realization, Teddy Roosevelt was a man of his times as well as a man who imprinted his pugilistic personality upon them. Although he invoked Lincoln and Grant in his Chicago speech as Americans who never shrank from conflict, it was in fact the passing of the Civil War generation and their memories that fed the craving for imperial muscularity. Montgomery Meigs had gone to his rest at Arlington in 1892, by which time more than 300,000 Civil War dead had been interred in seventy-three national cemeteries around the country. After he had demobilized a million and a half men in arms, he remained quartermaster general to a shrunken army of no more than 70,000—the vast majority of whom were finishing off Native Americans as the railroads, mining companies, and cattle ranchers consumed what was left of the open West. So as well as attempting to ensure adequate pensions for the veteran survivors, Meigs reverted to the career that had sustained him before the war: architecture. In particular he designed the astounding Pension Building, a brick-and-terra-cotta galleried temple of immense scale and grandeur. Decorating the facade is a great frieze by the sculptor Caspar Buberl, representing scenes from the war, which include more than tough infantrymen on their march, but also supply wagons and their teamsters, at least one of whom Meigs typically specified must be a liberated slave. Seen in profile cracking a whip over his mule team, it's one of the great images in American public sculpture.

A year after Meigs's death in his grand house on Vermont Avenue (also built by him), the young history professor Frederick Jackson Turner stood up at the World's Columbian Exposition at Chicago in July 1893 and declared that the frontier "has gone and with its going has closed the first period in American history." It was not a happy moment to be making such a proclamation. Five hundred banks failed in that same year, tens of thousands of businesses went under, millions were thrown out of work, and massive strikes were dealt with roughly, tearing apart any sense of shared national purpose. What might restore it; perhaps a blue-water destiny?

A transoceanic imperial presence as a remedy for American exhaus-

tion, saturated domestic markets, and a sudden unwonted sense of
territorial claustrophobia, had been promoted before the depression of
1893. Benjamin Harrison, who had got to the White House despite losing
the popular vote to Grover Cleveland in 1888, and who had the urge to
overcompensate for his dubious victory by exercises of national assertion, beat the drum for the expansion of both the army and the navy.
In 1890 the son of Dennis Mahan, the professor who had taught both
Meigses at West Point, Alfred Thayer Mahan published *The Influence
of Sea Power on World History*, and with its lessons in mind Harrison
persuaded Congress to fund the construction of sixteen battleships.
In 1893 the fleet sailed into New York Harbor as an iron regatta in an
attempt to take the city's mind off economic catastrophe, coming too
late to save the election for Harrison. Grover Cleveland, who got his
revenge for 1888, was cool to talk of island empires, rejecting the annexation of Hawaii, which under Harrison had seemed a sure thing.

But the sea change was literal, and in the end, unstoppable. What had
happened was Herbert Spencer, the kaiser, and Joseph Chamberlain.
Charles Darwin had actually taken Spencer's phrase about "the survival
of the fittest" for his own evolutionary theories, but Spencer returned
the favor by popularizing a theory of bio-social struggle in which the
weak were weeded out and the strong inherited the earth. Thus it
was with species, thus it was (to the satisfaction of the likes of Andrew
Carnegie) in business, and thus it was, thought the young Theodore
Roosevelt, with nations and empires. Put Mahan and Social Darwinism
together and look hard at the British Empire and the challenge it
faced from imperial Germany, whose kaiser was a devoted reader of
Mahan, and the conclusion was inescapable: either the United States
had to embark boldly on naval and military renewal, and territorial
expansion across the sea, or else it was doomed to become, in TR's
strange obsession, "China."

Had he wished, Roosevelt could have invoked Jefferson as well as
Hamilton for, as war-averse as the founder of West Point had been, he
did make it clear that if the American future was to be commercial as
well as agricultural (and thus be able to import manufactures), it had
to ensure its shipping was always free to sail the ocean. Should there
be any threat to that freedom, it had to be resisted, if necessary, with
force. A century later, there was a strongly developing sense that if the
Pacific as well as the Atlantic were to be kept free for American trade,

this required staking out a chain of island possessions that could act
as the guardians of that liberty. And if China was indeed a sinkhole of
power, into the dangerous vacuum would inevitably come competitors:
the Japanese, the Russians, the British, and the Germans, who would
take the space and leave none for the United States.

The China of the West was Spain: decadent, superstitious, anachro-
nistically monarchical, and sitting on an empire that was in the process
of disintegration. The issue for policymakers once Cleveland had been
succeeded by William McKinley in 1897 was how far to help push that
empire in Cuba and the Philippines into terminal decomposition. And
when that happened, how exactly should America profit? As McKinley's
assistant secretary of the navy, Roosevelt more or less made his own
Mahanian policy, helped by the complaisance of the official head of
that department, John Long. And the policy was to make sure, in the
event of a Spanish-American War, that there would be American arms
in both the Caribbean and the Pacific. Rebellions in both Cuba and
the Philippines helped the cause, allowing the yellow press, Joseph
Pulitzer and William Randolph Hearst competing with each other in
Hispanophobic headlines, to whip up campaigns of indignation against
concentration camps and starvation inflicted on helpless natives by
the cruel and decadent Don. The USS *Maine* was sent to Havana,
where it blew up (probably, but not certainly, by accident), killing
hundreds of American sailors and making the hue and cry for war
irresistible for McKinley, especially in a year of midterm elections. For
the same reason congressmen were not about to stand in the teeth of
a gale of patriotic hysteria as the bodies from the *Maine* were brought
back to the mainland, the journey filmed by Edison's Vitagraph and
watched in thousands of nickelodeons all over the country. War was
duly declared, and Teddy Roosevelt resigned from his post at the
Department of the Navy to raise a regiment of cavalry for Cuba. The
Rough Riders, with Roosevelt's friend Major General Leonard Wood
(personal doctor to McKinley and gung-ho imperialist) as commander,
were filmed galloping around in training at Tampa, but thereafter only
rough-marched, there being no adequate transports to convey their
horses to Cuba. No matter, the battle of San Juan, such as it was, and
TR's part in it, created a military glamour, an aura of virile zeal, that
he could convert into votes.

So when Admiral Dewey annihilated the Spanish fleet in Manila

Harbor, effectively ending the war, and the Philippines suddenly dropped into America's lap, what was to be done with them? The story told by the government and by the newspapers about the war in Cuba was of a benevolent and disinterested liberation. In the spirit of the Founding Fathers who had risen against an imperial power, America had come to Cuba to strike the chains from the island and deliver it to its own people. An act of Congress preempted any thought of annexation and required the government to respect Cuban independence as a nonnegotiable clause in any peace treaty with Spain. And that was how matters unfolded, notwithstanding the insistence of the Americans that they, rather than the victorious Cuban rebels, take the surrender; an act that did not sit well with the new authorities. And then there was the ubiquitous Major General Leonard Wood, who turned himself into a kind of proconsul in Cuba, delivering public health to Havana and sundry other blessings of American civilization.

Respecting, more or less, Cuban rights only made the Filipino rebels—who had been fighting their own war against Spain and might well have succeeded without any help from the Americans—assume that something of the kind would follow for their islands. Had not McKinley himself declared that the annexation of the Philippines would be an "act of criminal aggression?" Yes, he had, but believe it or not there was yet another election coming up before too long, and he was running to keep the White House. And aside from the organs of Pulitzer and Hearst, the *New York Tribune*, the *American Review of Reviews*, and the entertainments in the nickelodeon where audiences saw Admiral Dewey on his battleship and the gallant volunteer lads preparing to sail, there were powerful voices urging him to act: Senators Albert Beveridge of Illinois and Henry Cabot Lodge of Massachusetts. "God has not been preparing the English-Speaking and Teutonic Peoples for a thousand years for nothing but vain and idle self-contemplation," boomed Beveridge. "No! He has made us the master organisers of the world to establish system where chaos reigns. He has made us adepts in government that we may administer government among savage and senile peoples." More encouragement to do the bold thing came from Britain, where Rudyard Kipling wrote "take up the white man's burden" to influence the decision. The "best you breed" had an obligation to civilize "your new-caught sullen peoples / half devil and half child."

Just in case McKinley needed it to overcome his doubts (and earlier convictions), he called on a higher authority to help him out. "I went down on my knees and prayed to Almighty God for light and guidance," a practice that has become so routine in the White House in the past century that carpets must suffer undue wear and tear. God was in. And He spake to the president thus, saying 1) you're not going to give the Philippines back to the Spanish, 2) to walk away from responsibilities now would be cowardly and dishonorable, and 3) those Filipinos aren't ready for self-government and never will be without a prolonged dose of decent strong American administration. So "there was nothing for us to do but to take them . . . and to educate the Filipinos, to uplift and civilise and Christianise them." Apparently McKinley hadn't noticed that the Filipinos were actually already Catholic, or perhaps that didn't count. At any rate all was well for "I went to bed and slept soundly."

The motion to annex went through the Senate, although not before opponents expressed their pain and rage at what the government was making of America: a ruthless empire indistinguishable from the British or French. George Frisbie Hoar, the other senator from Massachusetts, was eloquent in a way that sounds through the generations: "You have no right at the cannon's mouth to impose on an unwilling people your Declaration of Independence, your Constitution and your notions of what is good." But his voice and that of a newly formed Anti-Imperialist League were drowned out in the jingo. Furious at their betrayal by the Americans, the Filipino Republicans, led by Emilio Aguinaldo, decided on resistance. Inevitably, hostile confrontations turned into a shooting war. Once blood had been shed and a call had been made for 70,000 volunteers, America was swept by a wave of patriotic fury. There were more movies featuring Americans against Filipinos, although much of the footage was shot in a back lot in New Jersey with the National Guard dressing up in white pajamas to pretend they were the enemy. Shoot, duck, RUN! In the presidential election campaign of 1900 Teddy Roosevelt traveled over 12,000 miles by train, accusing William Jennings Bryan and the Democrats of being the heirs to the "Copperheads" who wanted to make peace with the Confederacy; pounding tables on behalf of the gallant volunteers who were already sacrificing themselves in battles with the ungrateful Filipinos.

McKinley liked to think that God was watching over his work, but even God had the odd day off, and on one of those days in September 1901 the president was assassinated by an anarchist. TR rushed to Buffalo, where he was sworn in. Though the war in the Philippines had become something no one in Washington had anticipated—a wretched slog with American troops taking heavy casualties from determined insurgents and unable to maneuver in any kind of conventional military manner—President Roosevelt was not about to reverse course. He sent General Arthur MacArthur (West Point) to the Philippines, and the conflict settled down into a horrifying slaughter: Filipinos picking off infantrymen; Americans wreaking revenge by burning villages and crops, and treating villagers, whom they usually called "niggers"or "gu-gus" (after their coconut oil shampoo), as subhuman. Since it was said to be difficult to distinguish between native guerrillas and noncombatants, the massacre of villagers in any area thought suspicious became commonplace and even expected. Torture of prisoners to extract information, especially the "water cure," became routine. Water was poured through a funnel in quantities to distend the stomach and give the prisoner a sense of drowning. If the torturers didn't get to hear what they wanted, soldiers would jump or stamp on their prisoners' stomachs to induce vomiting, and the process would start all over again. Naturally, photographs were taken of the torture. In one of them a soldier stands watching from a few feet away while his comrades administer the cure. With one hand he leans on a pile of rifles; the other is on his hip at his belt. His left leg is crossed jauntily over his right, and on his face is the unmistakable beginning of a smile. Albert Gardner of the 1st Cavalry actually specialized in songs that made the business into a jolly rigmarole: "Get the old syringe boys and fill it to the brim / We've caught another nigger and we'll operate on him."

America was not morally dead to the atrocities. By early 1902 anti-imperialist writers had gathered enough information on torture and indiscriminate massacre to publish a report titled *Atrocities Perpetrated Against the Civilian Population.* Senate hearings were called during which the junior and senior senators from Massachusetts divided exactly down the middle on the issue: Henry Cabot Lodge making sure that much of the testimony appeared behind closed doors to the "Insular Affairs Committee," while his nemesis George Hoar conducted the hearings as best he could in public. In a post–Abu Ghraib world the defenses are familiar: we're told not much of this happened, and when it did the army conducted its own thorough investigations and discovered that most of the accusations were unfounded or exaggerated. Yes, there were bound to be a few rotten apples, but these were either the occasional American soldier driven mad by tropical fevers, or else a handful of native allies, the Macabebe. And in any case, it could not be expected in such a place that war could be played by the rules that obtained among the civilized. These were savages who would cut your throat as soon as look at you. Either America wished to win this war, or it did not. If it did it must be expected to take savagery to the enemy. Anything else was a dereliction of duty.

This was, pretty much, the view of the most important living American, President Roosevelt. But it was not the view of the second most important—or at any rate famous—American living, namely Samuel Clemens, aka Mark Twain. As much as TR, Henry Cabot Lodge, William Randolph Hearst, Edison, and Henry Adams were one kind of American voice, loud with their Hamiltonian sense of global destiny, Twain and Henry Brooks Adams and W. E. B. DuBois had the other kind of patriotic voice, which saw in military intoxication a perversion of everything the American democratic experiment was supposed to stand for.

Twain was no pacifist. He had been in Vienna when the Spanish-American War began, on yet another of the trips meant to right his perpetually shaky fortunes. Predictably he had been lionized but equally predictably found himself having to defend American intervention in Cuba. As best he could he insisted that this would be no war of imperial annexation, but a disinterested assistance to the Cuban revolutionaries, who would reap the benefits of victory. There would be a free Cuban republic with its own constitution; everything would be hunky-dory between benefactor and protégé, and they would all live happily ever after.

By the time Twain returned to New York on the SS *Minnehaha* on 16 October 1900, American foreign policy had become, in his mind, tragically indefensible. He was the country's greatest celebrity; his white whiskers and mischievous eyes known through the many photographs, right across the nation. And almost the first thing out of his mouth after he had walked down the gangplank was an attack on the Philippine annexation and war. "I have read the Treaty of Paris [between Spain and the United States] carefully," said Twain to the reporter from the *New York Herald*, "and I have seen that we do not intend to free but to subjugate the people of the Philippines. We have gone there to conquer and not to redeem. It should, it seems to me, be our pleasure and our duty, to make those people free and let them deal with their own domestic questions in their own way. And so I am an anti-imperialist. I am opposed to having the eagle put its talons on any other land." Ten days earlier he had already sounded off along the same lines for the *World* in London. "It [the Philippines regime] was not to be a government according to our ideals but a government that represented the feeling of the majority of the Filipinos, a government according to Filipino ideas. That would have been a worthy mission for the United States. But now—why we have got into a mess, a quagmire from which each fresh step renders the difficulty of extraction immensely greater. I'm sure I wish I could see what we were getting out of it and all it means to us as a nation."

Instantly Twain became the voice (and vice president) of the Anti-Imperialist League. In 1901 he published a withering attack on the sanctimonious pretensions of missions to the uncivilized: "To the Person Sitting in Darkness." The tone was acid, caricaturing the civilizing mission as a rum business: "Extending our blessings to our Brother who Sits in Darkness has been a good trade and has paid well on the whole and there is money in it yet." American policy had been enthralled by the British, by Cecil Rhodes and the appalling Joseph Chamberlain, and the Boer War. That's what drove the Philippine War: naval envy. "It was a pity, it was a great pity, that error, the one grievous error, that irrevocable error. For it was the very place and time to play the American game again. And at no cost. Rich winnings to be gathered in again, too rich and permanent, indestructible; a fortune transmissible forever to the children of the flag. Not land, not money, not dominion—no something worth many times more than that dross: our share, the spectacle of a nation of long-harassed and persecuted slaves set free through our influence; our posterity's share, the golden memory of that fair deed." And Mark Twain—the embodiment of everything American, the scourge of humbug—ended by directly attacking two of his countrymen's most cherished objects, the uniform and the flag: "our flag—another pride of ours, our chiefest.

We have worshipped it so; and when we have seen it in far lands—glimpsing it unexpectedly in that strange sky, waving its welcome and benediction to us—we have caught our breath, and uncovered our heads and couldn't speak for a moment for the thought of what it was to us and the great ideals it stood for. Indeed we *must* do something about these things . . . We can have a special one . . . just our usual flag with the white stripes painted black and the stars replaced by the skull and crossbones."

Twain paid dearly for his temerity. He never ceased being seen as the greatest of all American writers although he was now also regarded as a bitter and unpatriotic eccentric. Society invitations were fewer and far between, not that Twain cared all that much for them (though he did care some). But honors still came his way, including an honorary degree at Yale in 1901. Theodore Roosevelt happened to be there on the same day. The assassination of McKinley meant that the president was kept away from crowds but he must nonetheless have heard the roar of applause that went up for Twain as he received his degree. Later in private he said that "when I hear what Mark Twain and others [meaning the Anti-Imperialist League] have said in criticism of the missionaries I feel like skinning them alive." It had become personal. Twain wrote vicious attacks parodying Roosevelt and his worldwide popularity. The president was "the Tom Sawyer of the political world of the twentieth century always hunting for a chance to show off . . . in his frenzied imagination the Great Republic is a vast Barnum and Bailey circus with him for a clown and the whole world for an audience." And he let it be known that he thought the president "clearly insane" and "insanest on the subject of war." But as the public hostility mounted there were moments when he chose to pull his punches. When a particularly horrifying case of the water cure involving a priest, Father Augustine, came to light and the league asked its vice president to write something appropriately damning, he retreated. He was sixty-seven; he was tired; there was only so much he could do. Understandable. Sad.

TR affected to brush all this off. The last thing he wanted to be thought of was bookish. But he was a writer and a prolific one at that, and in other circumstances Mark Twain was *his* kind of writer: the voice of the people. (In fact it was Roosevelt who had the sententious voice; Twain the genuine rasp of American comedy.) But nothing stopped Teddy. On Memorial Day 1902, he went to pay his respects to the gallant dead at Arlington National Cemetery. He passed (as we all do) through the McClellan Memorial Gate that Montgomery Meigs had erected, presumably in a fit of forgetfulness of all the aggravation "little George" had caused him. And there, arrayed before him, some stooped, many gray-bearded, their medals hanging from lank frames, were the veterans of horror: from Bull Run, from Antietam, from Gettysburg, from Lookout Mountain, from the Wilderness. "Oh my comrades," the president shouted, much overcome, "the men who have for the past three years patiently and uncomplainingly championed the American cause in the Philippine islands are your younger brothers, your sons."

It was no fault of those men, off in Mindanao, but no, they weren't. The old boys had fought an American war for freedom. The young boys were fighting to extinguish it.

If only Teddy Roosevelt had been able to sit in on Monty Meigs's course on "Why presidents go to war when they don't have to."

12. American war: Rohrbach-lès-Bitche, the Maginot Line near Metz, 10 December 1944

Sonofabitch, if it was this cold then you'd think the mud would have frozen. But it got loose enough to clog up the caterpillar treads; slowed the whole damned thing down; made the battalion sitting ducks for antitank guns, 88s coming in every which way. No wonder the 4th had had it after months of mud, taking it on the nose. Now it was the turn of the 12th, his outfit, most of them still green, snotnoses, never under fire before; him too if he thought about it, nothing Fort Benning prepared you for. You got used to it pretty damned quick though. What was he doing commanding a tank battalion anyway? Pop was navy; Grandpa too; Annapolis folks, like the Rodgers. A good place Annapolis, the piebald plane trees right down to the water; a good place for the baby to come while he was off in Lorraine trying to finish off the Reich. Maybe there was snow on the ground in Maryland. Was any of this America's business, his business? Liberating the French? Hell, they had built the damned Maginot Line to stop the Germans, much good that had done them in 1940, and now here's where the

Panzers had dug themselves in, sixteen-year-olds and old guys, they'd said, but it sure didn't sound like kids and pensioners. The line wasn't what it had been in 1940, they'd said. No electricity after we cut it. Not much of anything for them to survive on. But for an army with no prospects they were putting on a hell of a show. So why was he here? Ah yes, "Duty, Honor, Country," all that West Point horseshit. Or was it? It was a West Point war now with Bradley and Eisenhower making the decisions. That had to be good. He'd been a kid, wet behind the ears, seventeen, for Christ's sake, when he went up the hill, put on his plebe's gray. The family, they'd wanted him in the navy, where Meigses went lately, but he wanted something else. Maybe it had something to do with all that hurting, the broken neck; the iron brace that had pinned him; made him grit his teeth and be damned if he wouldn't make his own way. Obstinate, the Meigses. Maybe they had all been a tad ornery, the quartermaster and his wire ropes. There they were up on Arlington Hill, the old boy and his son with his boots pointing to the stars. He wasn't ready to join them, not yet awhile, not with the baby coming. But he had a record of getting into trouble; those motorbike accidents at Fort Benning that had nearly done him in. Dumb. It was late, how late? Needed to take a look at the maps again; figure out a way through. Better get some sleep, though; this the second night in a row; not good to be drowsy; needed to be sharp to lead the attack tomorrow, everyone in the 23rd depending on him, had better find a way to punch through to Rohrbach, get rid of the artillery, give the guys on the ground a fighting chance. Clean it out, get into Deutschland, finish them off, good guys win, bad guys, very bad guys, lose. Patton, Ike, all of them happy. Europe free of the lousy Nazis. Go home. Go home to baby Meigs. Suppose it was a boy? They'd have to call him Montgomery. Baby Monty. Why not? That's just how it was.

On the morning of 11 December, Lieutenant Colonel Montgomery Meigs led the first wave of attacks on German positions at Rohrbach. The night before the battle he had argued against a frontal attack on enemy guns hidden behind twenty feet of concrete. He lost the argument. He was killed instantly while standing in the turret as the tanks advanced. The following day the badly mauled 23rd Battalion, supported by other units of the 12th Cavalry, took the objective. For his "utter disregard for his own life in leading his battalion," Meigs was posthumously awarded the Silver Star and Purple Heart. His widow worried about whether his body should be brought back to the United States but opted in the end to have him rest in the American Cemetery of Lorraine at Saint-Avold. There is, however, a grave marked with his name in Arlington.

Exactly a month after his father's death, on 11 January 1945, his son was born. His mother named him Montgomery C. Meigs.

PART TWO

I I : A M E R I C A N   F E R V O R

Listen to me, says Obama, listen to me and you will catch the American future. But I pay attention and hear the American past, not a dragweight on "change"; just the solid ground beneath the high-sailing dirigible of his rhetoric. The American future is all vision, numinous, unformed, lightheaded with anticipation. The American past is baggy with sobering truth. In between is the quicksilver Now, beads of glittering elation that slip and scatter, resisting prosaic definition. Obama wants to personify all these tenses. So he takes his listeners to the next promised land via Selma, Alabama, and the 1960s, Gettysburg and the 1860s. His effort to rekindle a sense of national community suggests another Great Awakening, but he knows all about the first spiritual revival in the eighteenth century and the second in the nineteenth; upheavals of the soul that changed the country. This attachment to the past is not just cultural exhibitionism, a guaranteed vote-loser in America. Rather, it's the grace note in Lincoln's "mystic chord of memory"; the sonority without which appeals to invoke American spirit in tough times are just so many sound bites.

Listen to me, says Obama, check out my Cicero, my measured cadence, now legato, now staccato, the latter delivered with narrowing eyes, lips slightly pursed between the calculated pauses; the head still and slightly cocked to one side, as if awaiting the promptings of ancestry. Now whom do you hear? You hear my warrant, an even bigger, deeper, preach: Martin Luther King. I am the fruit of his planting; the payoff of his sacrifice.

But when I see millions registering to vote for an African American, American African, a Kenyan Kansan from Honolulu, the *Harvard Law Review,* and the Chicago precincts, I'm reminded of someone quite different who made Obama's nomination possible. I hear a big black woman from the Delta, preacher's daughter, eyes wide with resolution, brow furrowed with passion; a gospel pair of lungs on her, sweating in her print dress on the Atlantic City boardwalk. I hear her start on up, the voice hoarse from many choruses sung in peril, a voice that took people like a handclasp. "Go Tell It on the Mountain" she sings, "Over the hills and everee whe-ere," and over the oncoming cop cars it goes, over the saltwater taffee vendors, past the brush-cut politicos in limp seersucker jackets hurrying to their appointment with history, clutching their important attaché cases; all pretending that she wasn't there, this embarrassing, overwhelming, ungainly, fired-up woman with her half-closed eyes, wet curls plastered on her forehead swaying a little as she gives it all she's got. The song gets louder, and less harmonious, now swelled by the chorus of students and civil-rights workers, black and white, not all blessed with perfect pitch, lining up behind her along the wall of the Convention Center. Are the men with the attaché cases hearing it inside? The sound seems to bounce from the building and drift over the green, slightly soiled Atlantic Ocean . . . "That JEEsus Christ is born." That's who I remember: Fannie Lou Hamer.

13. Atlantic City, August 1964

Fannie Lou seemed like a rumpled saint to me, but then what did I know? I was nineteen, editing a reticently titled undergraduate maga- zine, *Cambridge Opinion*. In 1964 we had opinions to spare on pretty much everything from Harold Wilson to Wilson Pickett, but editori- ally we confined ourselves to a single topic an issue. Once a term the magazine would opine on, say, the State of Prisons or the Look of British Modern Art (for there was some) always solemnly, defensively, capitalized. This particular issue was to cover the trifling matter of the fate of the United States in an election year. My friend, co-editor, and business manager and I sailed forth on the MS *Aurelia*, which had been, not so very long before, a tender for U-boats, and had now converted into an Italian student liner. Scheduled to make the crossing in about ten days, roughly the same time it took Dickens a century earlier, MS *Aurelia* sailed straight into a frisky gale that played havoc on obligatory efforts to promote Goodwill Among Nations: no stabilizers, no "Kumbaya." As the ship wallowed in the trough, discarded streamers bearing uplifting internationalist messages bobbed away in the wake. A counselor patrolled the Games Room looking worried, asking if anyone had seen the Belgians. No one had. The remnant of scarlet Italian meals flowered the decks, much to the irritation of the Neapolitan stewards, who would halt before these small sad deposits and confront those who they thought might have been responsible, asking accusingly, "YOURS?" What the guilty parties were supposed to do about it was unclear, but aggressive swabbing ensued.

We thought we knew America, but what we actually knew was Malamud, Bellow, Baldwin, which was something else entirely. Beyond the elegant museums on Fifth and the tonier stretches of Park and Madison, where, if we played our cards right and affected an Oxbridge accent, we'd be sure to run into Holly Golightly, New York seemed lurid and jumpy. As the thermometer climbed into the nineties, sidewalkers slowed to a gasping shuffle as they made their way to or from Grand Central, dripping into their poplin. The city was gamely attempting to put on its welcome face as a World's Fair opened on Flushing Meadows. Modernist pavilions, steel and pine, celebrated the Achievements of General Electric or the dawning of the jet age. Much of it was free. Investigate the Charms of Norway, and a braided blonde would greet you hospitably proffering brisling on rye. *Takk*, Solveig. In the Ford Pavilion the brand-new, dangerously sexy Mustang was being unveiled by peppy young men in blazers. But out in swelterland, N.Y., the Long Island Expressway was a parking lot and drivers were aggravating their ulcers, leaning on their horns and getting testy with the kids.

The unsurprising truth was that although JFK had gone to his grave at Arlington, the wound that had ripped open the body politic on that merciless day the previous November obstinately refused to heal or even scar over. Robbed of the boyo Kennedy grin that somehow promised all would be well in America, much evidence to the contrary notwithstanding, the body politic had a lesion that had gone bad. The America that had loved Kennedy (and much of it had not) was doing its best to become reconciled to Lyndon Johnson, but it was having a hard time. Some of this was just East Coast nose-holding, an incredulousness that Camelot had fallen to a Texan from the banks of the Pedernales with horse shit on his shoes. In the Century and the Knickbocker clubs, members didn't expect to see Pablo Casals showing up at 1600 Pennsylvania Avenue again anytime soon.

More serious were the apprehensions of black America. When Kennedy had gone down, their leaders initially thought that his civil-rights agenda had fallen with him; that that must have been the reason why he had been taken out. But although there was no love lost between Lyndon Johnson and Bobby Kennedy, for the time being still attorney general, the president had taken up the civil rights cause as if it had been his own crusade in the first place. He had even (between gritted teeth) helped launch Kennedy's Senate campaign in New York, introducing the man he detested as "dynamic," "compassionate" and "liberal." In the Senate, Kennedy figured to be less of a thorn in Johnson's side than in the Department of Justice. Leaning on whomever had to be leaned on in Congress, Johnson pushed through legislation that outlawed segregation in education and any public spaces: no more separate lunch counters, soda fountains, or schoolrooms. The Civil Rights Act became law on 2 July 1964. The story goes that as he signed the bill, Johnson said that the South would be lost to the Democrats for a generation.

Starting when, exactly? Not, Johnson hoped, in the upcoming election, in which he would be running against the conservative firebrand from Arizona, Barry Goldwater, who had been singing just the kind of music that yellow dog southern Democrats wanted to hear: that civil-rights legislation violated the sanctity of "states rights"—since the Civil War a euphemism for institutionalized racism. So although the president talked a good talk about "ending poverty and racial injustice in America," it was uncertain just how far Johnson would go to enforce the new legislation. And there was an immediate issue not addressed by the provisions of the Civil Rights Act: black voter registration. The Fifteenth Amendment to the Constitution, ratified in 1870, had prohibited any obstruction to the right of citizens' voting based on race or color. (The Fourteenth Amendment had already made anyone born in the United States, or naturalized, a citizen.) And for a while, in the states of the defeated South, more blacks exercised their right to vote than whites. But all this changed with the election of 1876, in which the Republican Rutherford B. Hayes won fewer popular votes than his Democratic opponent Samuel Tilden, but managed nonetheless to secure the electoral college. There was, however, a price to pay for President Hayes, namely the end of enforcing the Fifteenth Amendment. On his inauguration in 1877 Hayes shamelessly promised to protect black civil rights while knowing that he had agreed to withdraw all federal troops from the South. It would take almost a century for that betrayal to be reversed. In the meantime, blacks in the South, apart from suffering every kind of discrimination in the workplace and public places, had been kept off the voter rolls by spuriously complicated questionnaires designed to test their knowledge of the state constitutions with questions few whites could have answered. In Mississippi in 1963, just 7,000 blacks were registered out of an eligible voting population of 450,000. A supplementary Voting Rights Act had been promised, but black leaders less trusting than Martin Luther King weren't confident that the president would be willing to alienate what was left of his southern base by pressing too hard for the measure. Thus was born a campaign of practical action, the Mississippi Summer Project, known to those who became part of it as the "Freedom Summer": a thousand volunteers, from North and South, black and white, working in a climate of violent hostility, to get African Americans registered in Mississippi and to test the power of the Civil Rights Act. In the third week of June, three of those volunteers—James Chaney, black and Mississippian; Michael Schwerner and Andrew Goodman, both Jewish and white from New York—were arrested in Neshoba County for speeding. Held briefly in jail in a small town called Philadelphia, and given a supper of spoonbread, peas, and potatoes, they were released and told to leave the county. Heading to Meridian, they never arrived. Two carloads of Ku Klux Klansmen caught up with them, shot the two Jews in the heart, beat Chaney to a pulp, and then shot him in the head.

Their bodies would not be found for months. The governor of Mississippi protested at the fuss, saying for all he knew the three could be in Cuba. But the ominous disappearance of the Mississippi Summer Project workers was taken, as intended, as a declaration of war on the Civil Rights Act by the segregationist South, including Democratic senators like the plantation-owning James Eastland. Unreconciled to what the president, whom they now wrote off as a race traitor, had done, most of Mississippi's Democrats declared their support for Goldwater. That defection gave civil-rights workers an opportunity to propose that the exclusively white delegation to the Democratic party convention in Atlantic City be replaced by delegates from a newly founded Mississippi Freedom Democratic Party (MFDP), which, for the first time, would represent all the people of the state. At an impromptu convention in July, sixty-eight delegates were named, among them as vice chairman of the new party, Fannie Lou Hamer, a cotton picker from Ruleville, near Greenwood, the hub of the nascent civil-rights movement in the Delta. By the summer of 1964, she had already got used to the daily death threats made against her and her family for her temerity in getting blacks to the polls. Bullets came through her living-room windows. Forcibly sterilized without her knowledge when she was young, Fannie Lou's body had been violated again in Winona, Mississippi, after she had attended a Voter Registration Workshop. "We're going to make you wish you was dead," the sheriff had said as Fannie Lou was savagely beaten. But though the slightly hooded eye we saw at Atlantic City was the result of one of these assaults, she never did wish that. Fannie Lou reckoned this was what Christians went through for the Lord's cause. And she went right on singing and being a regular nuisance.

In the third week of August, Fannie Lou Hamer took the long bus ride from the Delta to Atlantic City. I made the much shorter bus trip from New York. My bus was air-conditioned; Fannie Lou's was not. I could smell trouble, though, and dashed toward it. My friend and I were kitted out with press credentials: small blue plastic badges bearing, without a trace of irony (so we hoped), the legend CAMBRIDGE OPINION. How did we manage this? Through an Irish American political wizard, then assistant secretary of labor in the Johnson administration who also happened to be Daniel Patrick Moynihan. A Harvard sociologist (and, years later, senator for New York), Moynihan was an old friend of my Cambridge history professor, J. H. Plumb, who airily told me that if I really wanted to see the inside of American politics, then I should write "Pat" a letter. Sure, I thought, dismissing the possibility that two undergraduates would be taken seriously by the author of *Beyond the Melting Pot*, at the time the last word on the fate of America's immigrant dream. But what did I have to lose? From the coziness of my great-uncle Joe Steinberg's Brooklyn row house I wrote, "Dear Mr. Moynihan, do forgive the intrusion on your busy agenda but Professor Plumb wondered whether there was the slightest possibility that . . ." Faster than Road Runner, back came a heavy cream envelope summoning us to an audience in Washington.

It was one of those blistering D.C. days when you expect to see camels bearing tourists down the Mall rather than buses. Heatstroke was a possibility just from crossing Pennsylvania Avenue to the Department of Labor, one of the neoclassical masonry monsters built at the turn of the last century to give American government an air of paternalist inevitability: outside, scalding limestone; inside, polished granite and gloomily stained walnut. The Department of Labor, then under one Willard Wirtz, had assumed an unexpected air of fresh importance following LBJ's declaration of a "war on poverty," although an ominous conflict with North Vietnam was the immediate war on Washington's mind after American ships had been fired on in the Gulf of Tonkin. In the vapor of patriotic fury that had predictably followed, Johnson had used the occasion to secure a resolution from Congress giving him unprecedentedly broad and undefined war powers. There was something fishy about the whole business. Exactly what were those American gunships doing there anyway? my friend and I wondered out loud in a room full of Georgetown students. This turned out to be a bad idea. The students declared themselves invested with similar war powers, and came within an inch of using them on us.

So it was good to be welcomed by the outstretched hand of Daniel Patrick Moynihan. His face was the color of a summer carnation, in the middle of which was planted a roguish grin. He had been born in Oklahoma but had grown up in Hell's Kitchen in New York, where he shined shoes for quarters while his mother kept a four-ale bar. With his flashing dark eyes and cupid's bow lips, Pat looked every bit the fallen cherub who liked a nip now and then. (When he was ambassador in New Delhi legends spread of his breakfast treble Scotches followed by a freshening dip in the embassy pool, after which the Moynihan wits were razor sharp for business.) The voice matched everything else: a merry lilt that was so rolling and rounded it seemed to come from a mouth permanently filled with humbugs. We made our little speeches to him as he beamed back at us, our new uncle in the wily world of Washington, and then he told us that he was arranging for our press credentials to the convention, but perhaps it would be a good idea to go to the platform hearings of the party, then in full swing in a Washington hotel. What were those? we wondered. "Oh," said Pat,

"that's where interested [he drawled this last word ironically] organizations make their views known to the party on whatever ails or inspires
them, and then, from the fruit of such deliberations, a committee writes
the party platform for the convention." This seemed like a good idea.
We were grateful, duly attended, listened to speeches on race relations
and civil rights, education, labor conditions and reported back to our
mentor. "How did you find it all?" he asked. Informative, we said,
omitting the qualifier "numbingly" lest it seem ungrateful. I noticed
a thick white document sitting on his desk bearing DEMOCRATIC
PARTY CONVENTION 1964 on the cover page. "Is that a document I
should read before the convention?" I cheekily asked our mentor. "Oh
I suppose so," he replied, flashing one of his most impish smiles, "it's
the party platform." Confused, I began the sentence "But I thought
you said . . ." and never got to the end. The Killarney grin from the
assistant secretary of labor assumed Cheshire cat dimensions.

In Atlantic City, I met Liz Moynihan, Pat's wife and managerial
minder, a Texan, and thus a crucial link between the Kennedy-Irish
connection and Johnson's Austin crowd. The importance of that bridge
between two hostile camps became more evident when Pat got me
into a reception that was ostensibly a fund-raiser for the Kennedy
Memorial Library that was planned in Boston, but to anyone with
eyes and ears, was Camelot at cocktails. There was the court historian
Arthur Schlesinger, dapper in the usual bow tie; there were Bobby
and Teddy not talking. The long wake for the fallen hero was still
going on. People choked up in front of the photo of JFK and John-
John on the beach. But between the hors d'oeuvres the air was also
thick with plans for Act Two, to be launched with Bobby's campaign
for the Senate. Pat, obviously signed up as spearbearer, was deep in
conclave mode, surrounded by estranged courtiers. But then, while
nodding vigorously and beginning one of his full-cheeked colloquies,
he spotted me, paused, and gave me a conspiratorial wink.

I left the party heady with precocious insider wisdom. Strolling
back along the boardwalk, past the entrance to the convention hall,
I heard, for the first time, that voice. "We Shall Overcome" it was
singing, and around it a chorus of great majesty had transpired; as if
Mahalia Jackson and Aretha and Odetta and pretty much every voice
that had ever been raised in painful hope had somehow gathered for
rehearsal in Atlantic City. The Beatles were due in after the conven-

tion had left town, and I was shamelessly riding the craze. "Do you know Lennon?" I'd be asked. "Loike a bruhther," I'd reply, my voice as glottal as I could get it. But now they could keep "A Hard Day's Night." This was the music I wanted to hear.

But who were the bad guys? A crowd was gathered around the shell of a blue Ford that had been trucked from Mississippi all the way to the boardwalk. The car was burned out, Naugahyde seats still acrid with scorch. This was what the Klan had done; what the Freedom Democrats were up against. They were also up against the power of the presidency. For Johnson—who thought he had already stuck his neck out enough for the civil-rights people—was incensed by the challenge of the Freedom Democrats. Fannie Lou, their most unstoppably vocal champion, had made it clear she would make the case public to the Credentials Committee, and if she secured just eleven of their votes, would have enough to take it to the Convention Floor. Each state delegation would then be asked to give their vote on whether the Freedom Democrats should replace the regular Mississippi delegation. The embarrassment such a scenario would present to Lyndon Johnson's claim to embody a post-traumatic American Coming Together was a nightmare. What had been planned as coronation might turn into chaotic farce.

It only got worse when Fannie Lou made the case in public before the Credentials Committee. The moving ruckus on the boardwalk; the split in the civil-rights movement, much to LBJ's chagrin, had got media traction. When she rose to speak, giving her name and address and pointedly adding the name of Mississippi senator James Eastland who had defected to Goldwater, every network camera was trained on her. The big beautiful voice was sternly resolute, tragically impassioned, as she told her life story. "If the Freedom Democratic Party is not seated today, I question America," she said. "Is this America where we have to sleep with the phone off the hook because we be threatened daily, just 'cause we want to register to vote to be first-class citizens?"

Beside himself, LBJ called an impromptu press conference on an inconsequential matter, just to get the cameras off the woman (he described her more coarsely) who had now become not so much an inconvenience as a personal nemesis. Minions were sent to Atlantic City to sway the MFDP into being Reasonable. An offer was made. Under no circumstances would the party leadership consider a full

MFDP delegation, but two black members could join loyalist white regulars. Senator Hubert Humphrey, a champion of civil rights since 1948 and one of the favorites to join LBJ on the ticket, put the offer to Fannie Lou. She wouldn't hear of such a thing. Had all the work, the danger, the suffering been for so meager a crumb? Exasperated, little Humpty Hubert, his egg-like dome glowing pale, asked the big black woman from Ruleville, "What *is* it you want, Mrs. Hamer?" "Why, Mr. Humphrey," she said, looking sweetly back at Humpty, "don't you know? The Kingdom of Jesus; that's what I want."

Fannie Lou didn't get it. LBJ did some more heavy leaning; threatened the zero-funding of a poverty program here, a school's budget there, and lo and behold, the numbers on the Credentials Committee sympathetic to the Freedom Party began to waste away. Martin Luther King was disinclined to be seen in their company. Could you blame him? There was an enemy to be fought, and its name was Goldwater. If everyone just behave there would be a Voting Rights Act on the books the year after the election. Why rock the boat? The obligations of pragmatism hung ominously over the boardwalk. The famous Atlantic City numbers game began to be a count of those singing along with Fannie Lou. One morning she showed up and the massed choir had become a chamber ensemble. Pressed once more to be reasonable, she stubbornly persisted in her rejection of so demeaning a proposal. That was it; offer withdrawn. The coronation could now proceed. On nomination night, Johnson, hitherto concealed from public view like a mysteriously veiled bride, ascended by hydraulic lift to the stage to be hailed by the roars of conventioneers while tiny plastic cowboys on parachutes rained from the ceiling. A week later in the Convention Center, the Fab Four sang "I Should Have Known Better" in front of orgasmically shrieking multitudes, while, an hour's drive away, North Philadelphia burned in the first ghetto riot of the year. Fannie Lou went back to Ruleville to see how Jesus was doing.

14. Saved

Hubert the Happy Warrior may have been thunderstruck or entertained by the naiveté of Fannie Lou's answer to his presidential offer, but she was in deadly earnest. She did want the Kingdom of Jesus in America.

Did people not talk that way in Humphrey's home state of Minnesota? (Actually, they did out on the prairie.) And they certainly talked that way in the Deep South when wrongs had to be righted. There were, to be sure, young militants of the SNCC (Student Nonviolent Coordinating Committee) like Stokely Carmichael whose disenchantment with the Democrats after Atlantic City led not to Jesus but to Malcolm X and to the Black Panthers. But the core of the civil-rights movement still thought of itself as a ministry. Take away the preaching from Martin *Luther* King, and you can have no idea of the might of his eloquence to shame America into living up to the precepts on which the country had been founded. If the unfulfilled promise of the Declaration of Independence that "all men are created equal" was routinely invoked against segregation and racism, so was St. Paul's Epistle to the Galatians (3:28) in which the apostle pronounced that "there is no Jew nor Greek, neither slave nor free man, there is neither male nor female for you are all one in Christ Jesus." It was precisely because the black churches dared to insist on color-blind Christian fellowship that their sanctuaries had become targets of arson and bombing. The reason for the selection of Chaney, Goodman, and Schwerner as murder victims was that they had been snooping around the burned-out ruins of the Mount Zion Methodist Church near Philadelphia, Mississippi, the place that had been designated a center of civil-rights education and voter registration by the Summer Project. Getting niggers all riled up about things they ought not to be bothered with was not what churches were for, the Klan thought. Hell, those preachers were all communists in dog collars anyway.

But that was exactly what the churches were for. Getting African Americans in the South, in fact all over the United States, to raise their voices, to brave things they had long been too reluctant or too intimidated to dare: sitting on the wrong side of lunch counters, riding in the wrong seats of the bus. But putting their lives on the line to do all this was inconceivable without the exhortation of the ministry. When another of the band of reverends, Fred Shuttlesworth, whose house had been dynamited by the Klan, was advised to get out of Birmingham, Alabama, as quickly as he could, he responded typically, "I wasn't saved to run." Neither were others of the ministry—King himself of course but also Ralph Abernathy and Joseph Lowery, also Alabamians. When they resolved to be staunch for freedom, they saw themselves as the inheritors of the long history of the black churches, from the clandestine converts of the slave plantations, to the itinerant preachers of the antebellum South and North; the militant abolitionists who called America to a Christian accounting with the original sin of the Republic; and the churches that had provided succor and solidarity through almost a century of Jim Crow segregation.

Which is why, when Barack Obama found himself under fire for associating with the confrontational Jeremiah Wright, his longtime pastor at Trinity United Church of Christ in Chicago, his immediate response was historical rather than polemical. In a speech delivered in Philadelphia, refusing to run away from the issue, Obama attempted an explanation of the union of race and religion in America; of the place of unseemly passion in the black church. When he first experienced the shouts and clapping of black worship, he remembered, it was a reclamation of "a moment we didn't need to feel shame about"; the recovery of "trials and triumph, at once unique and universal, black and more than black." The anger embedded in those memories, he said, was real even if often unproductive. "To wish it away ... without understanding its roots only serves to widen the chasm of misunderstanding that exists between the races." And suddenly the moment in American history seemed bigger than a political adjustment; and more like a call to reestablish moral community in the United States. Taking religion seriously, Obama seems to say, is not something that ought to divide the country more deeply but something that might actually bring it together; something white America could feel as intensely as black America. And then he went on to address sympathetically what he knew to be *white* anger. Put the two passions together, and a transformation might happen, he says.

But white rage isn't much in evidence at Woodstock, no not *that* Woodstock, the one about thirty miles northwest of Atlanta, surrounded by lovingly tended fairways. First Baptist Church sits at the end of a long driveway and is approximately the size of your average provincial airport terminal, only much better appointed. And if your idea of a house of worship involves damply smelling limestone and worn prayer hassocks, you had better go home to Barchester. For First Baptist is fragrant. Fresh-cut flowers stand at the Welcome Desk; the floors are polished tile and stone. Soaring glass walls are tinted subtly enough to let in light without heat. Escalators silently convey congregants (and there are 7,000 of them at this morning's "traditional" Sunday service) toward discreetly chiseled gold wall inscriptions attesting to the merciful love of God.

And why not? Pastor Johnny Hunt explains to me that just as people these days want a choice of mall, they are going to shop for their church. "That's just the way it is," he says through his honeyed baritone, an unapologetic smile flashing from the dazzling ortho-dontics. We're at Bible class before the main event, and it's filling up fast with tall men in chinos and Ralph Lauren golf shirts and perfectly groomed women in pastel suede and cashmere. The scent that hangs over discussions of Micah is Chanel, not incense. Pastor Johnny, gamecock zesty, with wavy silver hair, a lime-green tie, and that sock-it-to-you smile knows exactly what he is doing. He is a full-service provider. First Baptist Woodstock is, in fact, a small town that works. Its revenues are secure, its accounting transparent, its mission clear, its outreach benign, and its spirits buoyant. What corporation or for that matter medium-size sovereign state could make the same claims? First Baptist is the government the well-heeled of Greater Atlanta thought they didn't want or need. It comes complete with schools, a college, medical services, social workers; entertainment (Christian rock is a multimillion-dollar business), retirement facility, and mortician. What Pastor Johnny understands is that for all the blowhard professions of rugged individualism you hear on right-wing talk radio, middle-class Americans are lonely; heartsick at the loss of community. Even if their parents and grandparents couldn't wait to hightail it out of the immigrant districts into the verdant suburbs, they were surprised to discover that what they needed, even more than the 8,000 square feet of McMansion, the four-car garage, the life membership of the country club, and the Viking Range kitchen, was fellowship, a laying on of hands; the comfort of social connection in a headset universe. They want this whether they have been busting their buns at the gym or busting their balls at the office. And they want this so *much* more than they want an evangelical at 1600 Pennsylvania Avenue, devoted, insofar as the Supreme Court and the Constitution will allow, to banning abortion and gay marriage.

Pastor Johnny understands American solitude, and the high-wattage smile tells his flock that when he deplores transgression, he knows whereof he speaks. Not that long ago (a few decades) Johnny Hunt, half Lambee Indian, sharp in the pool room, his life measured out by the beer bottles, was tomcatting around bars and alleys. Then—isn't this always the way?—he was rescued by Janet the Cheerleader and her twirling batons. ("I noticed how pretty her baton was.") Janet is at Bible class too, her self-evident sweetness not completely masked by pancake; hair impeccably streaked. Later in his sermon, Johnny would publicly celebrate her as his personal redeemer, right up there with Jesus, without a twitch of embarrassment. But Janet must be used to this by now. Before Micah and Isaiah get started, there are the Announcements, which amount, in effect, to First Baptist's order of business: a prayer intercession for Rhonda and Mark, who together have to deal with another round of chemo; support for the mission in Argentina (evangelical Protestantism a hit with the gauchos); volunteers signing up for Church on the Street, which goes into the tougher neighborhoods of Atlanta looking to help the homeless and the addicted. I ask Johnny about the strings attached to that help. "None," he says flatly and without any defensiveness. "If they want to accept the Lord, as I did, they are most welcome, but we give what we can anyway." And you believe him because, even though every fiber of your agnostic-skeptical brain is screaming NO, the rest of you is recognizing that this is one decent (and snappily dressed) vicar.

The signs on the seats of the vast amphitheatrical "chapel" were marked "SAVED," which was nice for a Jew like me, but not, I thought, guaranteed. As the faithful trooped in there was a hubbub of rumor. Mike Huckabee, Baptist preacher, ex-governor of Arkansas, local saint to Bill Clinton (the other famous Arkansan of unsaintly fame), a victorious veteran of a hundred-pound battle with girth, a perfect emblem of the Republican commitment to shrinking government, was coming to First Baptist. If you were a committed Christian and couldn't quite get over your anxieties about the Book of Mormon, then Huckabee was your man for the White House, and since Iowa he was running strong. On the strength of the rumor the lady next to me had driven sixty miles that morning for a Huckabee sighting. From the front of the chapel, gently graduated steps, as if rising to paradise, ascended to a stage, flanked by two huge video screens on which slides smoothly scrolled, rather like local commercials at the movie theater. But instead of Country Joe's Barb-B-Cue and Tom's Tune-Up 'n' Lube, they were: "MUSLIM BIBLE DAY; PREMARITAL COUNSELING;

ADDICTION RECOVERY; OBEDIENCE THROUGH BAPTISM; PRE-BIBLE CLASS." Between the screens were, on each side, two electronic portals, glowing cerise or celadon, and at the center, a pale skin-tone shimmer in which was set a single glowing, silver cross. It was fabulous.

On came the choir, overwhelmingly white, a mere 150 of them, greeted by a wave of happy sound from the 400-piece orchestra: Andrew Lloyd Webber does the hymnal. "He CAME, he DIED, he ROSE again on HIGH . . . ," they sang, the colossally amplified sound crashing over the flock, which nonetheless did not rise from their seats but stayed put, oddly passive like children in a classroom, uncertain of their allotted quotient of eagerness. As if in supplication, one of the cassocked choir would every so often slowly lift both arms, palms upward, trembling, like a marionette worked by a celestial puppeteer. But there was no general joining in, for the choir was just too far away, across the vast wooden prairie of the stage. To the front of that space on each side of the niftily kitted choirmaster strode four women in their twenties and thirties. They were glossily shampooed and dressed just barely the respectable side of enticing: calf-high boots, cowgirl pleats, shredded buckskin blousons. Spotlit, they belted out the delight.

It was a hell of a warm-up for Governor Huckabee, whose amiably gangling form now loped on to the stage. Arriving a little while before the service, he had been surrounded by believers, eager to shake the preacher's hand, pledge allegiance, touch the raiment, which in this case was a dark gray suit that flattered the newly trim lines of the ex-chubby. "I *have* come to campaign," he winsomely confessed (the disingenuous official line at Woodstock being that he was just another worshipper who happened by in his helicopter), "I have come to campaign for *Jesus.*" Applause, for the faithful did not for a minute think the as yet unannounced candidacy of Jesus for president was a bar to his apostle Huckabee somehow making it onto the ticket. The preacher went on in his trademark lightly self-deprecating manner, a million miles from the sanctimonious rants expected of evangelicals in politics, to claim that politics and power weren't all they were cracked up to be, and yes, his heart was in this campaign, but whatever happened, it was always going to be more important to bring souls to the Lord. Sure. But such was the disarming charm, the mellifluousness of the voice, that for a moment you actually began to believe this, even while the Huckachopper was parked not far off by a golf course waiting to spring Slim to his next date with the voters.

When it came to Pastor Johnny's turn, you could see why he was such a hit with the suburbans. Dapper, he moved lightly on his hand-somely shod feet, a bobbing flash of pugilistic energy, like a sparring partner who was just teasing. For the denunciations against iniquity, Johnny could summon up the preacher's roar, a nod to the great Baptist yore, for he divined the craving for castigation among the BMW classes. But Woodstock is a sinner-friendly church, and the fulmination was carefully rationed, sharing sermon time with scenes from Johnny's personal odyssey, lost soul to glorious rebirth, that date being 17 January 1973. There were down-home stories of early benefactors. Praise be to Otis Scruggs, apostle in denim overalls, who bought him his first suit. "Johnny," said Otis, "I never missed anything I ever gave away." Lord be thanked for Jowls Watner, "a heavy man with a fist like a ham" who had written Johnny a letter every day for thirteen years just to make sure he knew he had a friend who would see he stayed on the straight and narrow. And rising to the climax of all Baptist services, little Johnny, big-time showman for the Lord, opened his arms wide to all those strangers to the Word who wanted, wanted NOW to come and be received . . . And, from the host in the stalls there arose just such men and women (but mostly men), who stepped or stumbled forward into the embrace of Pastor Johnny (or one of his numerous deacons stationed along the many aisles), some of them ending pros-trate at the front, heads bowed on the lowest step, gently raised by the pastor, and on came the choir of multitudes and the strings and woodwinds, fortissimo, and out came that anthem again like a dollop of molasses on morning corn mush: "He CAME, he DIED, he ROSE again on HIGH . . ."

And you would think that all this would make the pastor ambitious for power, as well as glory. But you would be wrong. Afterward, mopped down and freshened up, Johnny said, "You know, Simon, I don't think the answer lies in the White House." I had asked him why, Huckabee aside, all the predictions of a Christian vote deciding elections for the Republicans as the conventional wisdom said it did in 2004, seemed not to be repeating four years later. Surprisingly, Johnny replied that being too closely tied to the perceived purposes of the

Bush administration and the pre-2006 Republican Congress may have hurt rather than helped the cause. Would he like to see someone like Mike Huckabee, hell, Huckabee himself, in the White House? Yes, he would. Did the moral future of America depend on that outcome? Not a bit. This line significantly departed from the Moral Majority and Christian Coalition position, that the United States stood on the brink of perdition, unless prayer was introduced into schools, abortion outlawed, a constitutional amendment adopted that defined marriage as a sacrament between one man and one woman (at a time, a rider would presumably have to add). Notoriously, Jerry Falwell had been of a like mind with Mohamed Atta in seeing the massacre of 9/11 as a chastisement meted out by God on a sin-stained America. But modern pastors like Johnny Hunt were not of a vengeful temper, nor did they think whatever ailed the moral condition of the country could be remedied by legislation or the fiat of the Supreme Court. "The answer," said Johnny, "lies at home," and by home he meant the individual houses of his 21,000 congregants, but he also meant their shared home, the community he himself had fashioned. His pastorate was just that: a shepherding, and he was more interested in those who had strayed from the flock than the fat and fleecy. Anyone less like a beetle-browed theocrat it was hard to imagine. Once there had been Baptist brimstone. Now there was cool marketing, Christian rock music, and the high-school mission to Argentina.

Unblushingly theatrical demonstrations of faith, then, are not to be confused with a campaign for an American-Christian theocracy. It is certainly true that America, even before the revolution, has been fertile ground for self-appointed prophets, crusaders, and messiahs. There is nothing like a wide-open continent (save the original inhabitants) for postponing disenchantment, for if a prophecy fails to pan out it can always be relocated, preferably somewhere toward the west. There are still a host of Americans reading Tim La Haye's *Left Behind* books (fifty million at the last count) and if they believe what they read, are impatiently waiting for the Rapture, the Last Days' battles with the Antichrist somewhere in the vicinity of, say, Fallujah, to be followed by the inauguration of the Thousand-Year Rule of Christ. The deputy undersecretary for intelligence at Donald Rumsfeld's Department of Defense, Lieutenant General William Boykin, has become notorious for insisting that the Holy Spirit, and sometimes God in person, makes regular visitations to instruct him on strategy. As recently as last April the retired but unrepentant Boykin told a gathering in Israel that when (not if) the time came for him to be admitted to heaven he wanted to arrive on all fours, "with blood on my knees and elbows . . . standing [not kneeling] with a ragged breastplate of righteousness. And with a spear in my hand. And I want to say, 'Look at me, Jesus, I've been fighting for you.'"

It's safe to say that for General Boykin's regiment of holy warriors, the preservation of democracy, much less toleration, plays second fiddle to the execution of God's Ultimate Plan. But it was the wisdom of the Founding Fathers to ensure that while such visionaries are free to shout their dreams from the mountaintops, they are not at liberty to impose them on their fellow citizens. The First Amendment to the Constitution, which states that "Congress shall make no law respecting an establishment of religion, or prohibiting the free exercise thereof," opens an unlimited space for worship precisely in order never to be ruled by it. Which is exactly what makes the United States a different republic from, say, Iran.

This is not always well understood by habitually secular, skeptical Europeans, some of whom equate two fanaticisms without noticing that America's institutions are designed to protect citizens from religious coercion rather than enable it. A well-meaning lady to my left at a country lunch a year ago in Britain, marveling that I should choose to spend so much time among the Americans, asked, "Tell me why *are* they all so religious?" "They got it from us," I parried (dancing clumsily around the question), donning the professorial hat and pointing out that, Mormons aside, there was nothing so extravagant in American religious life that couldn't be found in England during the Civil War and the Interregnum. Americans may rant at each other and at the profanities of the modern world, but we killed each other and King Charles for just such matters. If there was, as my lunch partner implied, something hysterical and deluded about American religiosity, it came by it honestly. After all, Cambridge, Massachusetts, was not named arbitrarily. "Oh, but that was so *long* ago," my lunch partner replied, at which I countered with the Victorians who were not So Long Ago, a time when British churches were packed with piety. The really interesting question perhaps was not why Americans were believers, for most of the world outside Europe and perhaps east Asia remained believers too, but why the British had stopped believing. A sudden mass conversion to reasoned atheism around 1920? The bitter education of twentieth-century history (my preferred explanation)? Or an established church which saddled Christian theology with the baggage of an exhausted official institution?

In any event, the Founding Fathers by and large shared what Alexis de Tocqueville nicely called "commonly held opinions" in respect of the existence of a Creator, but differed a great deal on the intensity of their convictions. Jefferson, for example, marveled at the credulousness of those who believed that Jesus was the son of God, born to a virgin, while John Adams was mostly Unitarian. The main thing, though, was to make a place for profession of whatever sort, so that the pious or the impious would never feel obliged to kill each other on behalf of the victory of their convictions. That, they could (and did) fairly point out, was what Europeans had done since Christianity began. A bet was made with posterity that, by keeping the church from directing the state, or the state from compromising theology, religion might actually flourish rather than wither, since it could depend only on its own intrinsic persuasiveness.

Much of American history has been the vindication of that original gamble. The implications of the First Amendment have inadvertently, or not, backed America into the great question on which the peace of the whole world, not just the United States, will turn. And it is a question that secular Europe with its donnish bafflement that any properly, rationally wired human being could ever believe this guff, disqualifies itself from addressing if it invariably talks of the religious as though they were all visitors from Planet Loopy. A double standard not infrequently operates here, partly generated by British romanticism about Islam. American evangelicals, who—so far—are obstructed from imposing law, are madmen, but the ayatollahs who are not are merely misunderstood traditionalists. Sometimes liberal secularism does itself a disservice by deferring to intolerance, rather than debating how those claiming a monopoly of wisdom can be prevented from imposing it on others. The First Amendment makes avoidance of that debate impossible, even when it's something as inadvertently comical as, say, the state of South Carolina offering car license plates with a stained-glass cross and the motto "I Believe." (In a God who will overlook moving violations, for instance?) It's this unavoidable dialogue between faith and freedom, conviction and toleration, that has always been at the heart of American history and which is only crudely characterized as a "church–state separation debate." The unmistakable indifference of the American electorate to evangelical dogmatics in this election year, the clear sense—shared by both Johnny Hunt and Barack Obama—that evangelical politics has had its day, only comes as a surprise to those beyond America who imagined it would go on and on, eating away at democratic toleration. It's elsewhere in the world that dogma chokes on pluralism—the coexistence of conflicting versions of the best way to redemption—and uses state power to wipe it out. In the United States the Founding Fathers believed instead that religious truth would best be served by keeping the state out of the business of its propagation; that the power of religious engagement would not just survive freedom of conscience but be its noblest consequence. It was a daring bet: that faith and freedom were mutually nourishing. But it paid off and it has made America uniquely qualified to fight the only battle that matters, not General Boykin's quixotic reenactment of the true god against the false idol, but the war of toleration against conformity; the war of a faith that commands obedience against a faith that promises liberty. That, actually, turns out to be the big American story.

15. Raven, Virginia, 2008

It was when the men started chanting that I found myself slipping down a wormhole of time, emerging somewhere in the mid-seventeenth century. It was a sound I had never heard before in any church: a low tribal drone, diphthonged, nasal, as if exuded from human bagpipes; a sound that might, I thought, have been overheard by mud-caked sheep in some wetly ancient British valley, an adenoidal chant that had been overtaken by the more tunefully gracious hymns of Isaac Watts. Where in God's name *were* we?

In the Macedonia chapel of the Primitive Baptist Universalist Church, halfway up a mountain in far southwest Virginia near the small town of Raven. I was right about the British antiquity of what I was seeing and hearing, but wrong to guess that the human bagpipes must have been Scots-Irish. In fact, the droners were the descendants of Welsh

Tract Baptists who had settled around Newark, Delaware, sometime
in the early eighteenth century. Finding the East Coast too peopled
for their liking, and too patrolled by the elders of the church for the
free practice of their particular kind of Christianity, they moved on in
search of sheep pastures and coal mines new. They took their bricks
with them to build sturdy little houses amid the timber-frame cabins
of the hill country. But this church—not much bigger than a garage—
was built from stone, whitewashed against the hillside. Inside there
was no ornament at all, unless you counted the faded print of the
Last Supper on the back wall. Even the cross hanging just above the
doorway had been reduced to the most rudimentary wooden form,
as if made by schoolchildren (which it probably had been). Most of
the worshippers, men in long-sleeved white shirts and dark trousers,
big-hipped women in bright print dresses, were either very young or
very old. The "No Hellers" (for they refuse to believe in an eternal
inferno, only the punishing tribulations of this life) were unlikely to
turn into a megachurch anytime soon. The population of this corner
of the Appalachians had always been poor and had stayed that way;
a declaration the No Hellers made to themselves when they wanted
to profess a simple connection to the life of the Savior. "We are just
common people," one of the Brothers said in the middle of his Words,
not something one could imagine hearing at Woodstock. There was
the occasional mother with small children seated on benches parallel to
the end wall (for there was nothing remotely like an altar, but rather a
rough table on which, mysteriously, a blue picnic cooler had been set).
Most of the fifty or so worshippers crowded the benches toward the far
end of the chapel or sat behind and to each side of the reading desk,
at right angles to the wall. The service was free-form. Brother Farley
had warned me that all that was arranged was a meeting time. Other
than that, they had no idea exactly when they might start and even
less notion when they would finish. It all depended on the inclination
of the Spirit to show up and when He decided the visitation had run
its course. "It ain't a bus timetable," he said, adding, with a twinkle,
"Hope you like huggin'," and I thought, Who doesn't?

   At the end of the church there was neighborly chatting and greeting,
from which, without any warning, suddenly arose the skirling: "Day
and night, the lambs are crying . . . come, good shepherd, feed thy

sheep." This went on, over and over for a good ten minutes, the voices seldom rising or falling much but keeping to their hypnotic drone, words alone sounding the emphasis. Apparently, so one of the Brothers told me, they had had a visit from a Catholic woman who, on hearing this same chant, asked, "Are you Jewish? The last time I heard this was at the Wailing Wall." "Well," said the Elder, "we claim to be the Spiritual Jews, so maybe she had something." This wasn't as improbable as it sounds, for I, too, had been put in mind of chants I had heard in remote synagogues, far away from Ashkenazi operatics; songlines crossing time and space in unexpected dissonant weaves of music.

A visitor had come from a sister church: a nervous young man in the regulation long-sleeved white shirt, Brother Craig, who out of courtesy was offered the first chance to speak and did so with becoming diffidence, without any of the vocal confidence that heralds a powerful sermon. "God smote me in the year 2000," he said, looking neither happy nor unhappy with the fact, but merely acknowledging somehow his draft call. Like everyone else who spoke, Brother Craig was gently anxious about whether he was, in fact, worthy enough to be the instrument of the Lord's will and whether he would in time be saved from his "vile body." Poor thing, he had all the worry and none of the displaced urge to acquire trophies to assuage the anxiety. He was Max Weber's thesis about Calvinism minus the capitalism. Someone understood how Brother Craig, or possibly God, was feeling about this uncertainty for at one particularly sorrowful moment a wild cry went up from a dainty silver-haired woman in her seventies sitting a few feet from the reading desk. "GLORY BE!" she wailed, "GLORY BE TO GOD," her voice breaking into a possessed ululation, at which point the women around her delivered the soothing hug and gentled her back to silence.

These mountain people touched each other, and us. A lot. Literally. In midsentence, mid-spate, a Brother would suddenly extend a hand and give the handshake of Christian fellowship to anyone he felt needed it, or who might not need but would welcome it all the same. At other moments, the whole service would simply break up to allow the congregation to wander about the chapel offering hugs and neck kisses to all and sundry within reach, and if anyone was not within reach they would venture up the aisle until they were, pushing the benches aside.

"Good MORNING," they would say as they reached for a shake or a hug. "Good MORNING," I responded, dimly remembering a description of precisely the same practices by English Baptists and Quakers in the seventeenth century. There was constant body motion among the No Hellers, walking, singing, embracing, chatting. This is how it must have been, I thought, before Protestantism turned, irrecoverably, into an expression of the social order; the hierarchy of pews, the imposition of decorum; silence until bidden to sing, stand, pray, kneel, leave. The No Hellers, on the other hand, were living relics of radical Protestantism in its earliest purity; all tender sweetness, and nervous, neighborly joy; the kind I had only read about in books by Christopher Hill. Compared to them the Methodists with their Wesleyan Love Feasts were vulgar upstarts.

A series of mindful speakers pronounced, but everyone was waiting for the man who seemed most dependably to be taken by the Spirit and who had the Voice: Farley Beavers. Farley was a slight, bony soul, angular and awkward, in his late sixties or early seventies (the humble age fast in coal-stricken Appalachia). But Farley had the Gift. When Farley uttered, it began as it did with all the brethren, with a low and quiet pitch, lamenting the passing of brothers Willard, Curtis, and Melvin who "will stand at the right hand of Jesus," but then Farley got into his stride, intensifying pace, passion, and volume, cantering through a recitation of unworthiness and affirmation: HE formed the peace in the darkness, WHOAH, there was no hell only trib-u-lation down here, and why how could there be seeing that the Lord was a kind and tender lord and, WHOAH (sounding this as a rhythmic, punctuating moan), He would not wish any terror or dismay on his people, WHOAH no for He was full always of loving-kindness (wail from the silver-haired lady, Glory BE!), and now Farley Beavers was galloping along, full tilt, auctioneer speed, so fast I wasn't sure whether he was proclaiming the coming of a day in which "I" (the Lord) "will smite every horse with astonishment" or whether it was whores would be smitten if they were not already, but anyway, WHOAH, He wanted us all to think well of each other (handshake handshake, walkabout, handshake) and if there were bad things among men in power, well, WHOAH, we had the good power to change all that with God's Blessing and, WHOAAAH, peace be upon all of us . . . for are we all not brothers and sisters and are

sent to care and love each other . . . ? And eventually Farley Beavers
climbed to the summit and his bony little head looked out at all the
No Hellers, and slowed right down, a sign of the Spirit's so-long val-
ediction this particular Sunday morning in the Appalachians, and so
it was time, of course, for another walkabout and mass hug.

All of this was a small miracle; not the kind Brother Farley might
have been apprehensively waiting for, but a miracle of survival against
the odds anyway. Whatever the opposite of a full-service megachurch was (a microflock?), this was it, and the brethren and sisters
of Macedonia, Raven, were happy to leave it that way. No Church
on the Street for them. Toward the end Farley had said something,
on the face of it bizarre, flying on the wings of his free associations.
The dead brothers, Willard, Melvin, and Curtis, would all, he said,
"meet the Lord" along with the rest of the No Hellers "in the air."
Which led him, right away, to think about those who took to the
air in pursuit of power and money and the vain glories of the world,
the creatures of false doctrine . . . Not for us, said Farley,"'we don't
go out and deceive people, buy a big ay-ro-plane . . . but that's good
'cause then they leave us alone." That's the real miracle, I thought,
that they can indeed be left alone; at liberty to say whatever the Spirit
prompts, and that thanks to the peculiarly American bargain between
faith and freedom, the No Hellers could wail and drone unmolested
on the dark Virginia hillside.

16. Providence

Roger Williams sat on the frozen dirt in the winter of 1633 amid the
Pokanoket Indians trying his best to make out what they were saying.
"God was pleased," he later wrote, "to give me a painful, patient spirit
to lodge with them in their filthy, smoky holes even while I lived at
Plymouth and Salem, to gain their tongue." The accommodation may
have been poor, but Williams was not scornful of the Pokanokets
and the Narragansetts. "My soul's desire was to do the natives good,"
he wrote, and it could not have hurt his ambition to win their trust
that, unlike almost everyone else in the Massachusetts Bay Colony, he
believed the Indians were the legitimate proprietors of the land, and
claims made by the Crown and its charters to freely dispose of it were

patently false. Only those contracts made directly between the Indians and newcomers (such as he himself would draft) could properly transfer that land. The chief reason for learning their language was of course to lead the Indians out of pagan barbarism (as he saw it). But Williams already also knew that no church, certainly not his, could prescribe the right way to Christ. That, the natives would have to seek on their own. He could but lead them to the opening in the trees.

It was for thinking such things and, much worse, not keeping them to himself, that Roger Williams had got into bad odor with Governor John Winthrop in Boston and the Great and General Court that had care of bodies and souls in the Massachusetts Bay Colony. Sometimes their indignation puzzled him. He was, he thought, no agitator and certainly no Anabaptist rejecting the sway of all earthly princes and powers. Had he not always granted to the magistrates their power to rule in matters of "bodies and goods"? It was the remainder of a man's life (the part, he would have conceded, if pressed, that most mattered) over which no prince on earth could have jurisdiction. And no church either, for, whatever their claims, they were all unregenerate, contaminated by worldly governance, and would remain so until Christ's second coming, in expectation of which Williams had the liveliest hopes. In the meantime the best a true Christian could do was to separate himself from those false churches, and what he called "soul liberty," with his utmost strength.

There was little in Roger Williams's upbringing to make him purer than the Puritans. His father was a London merchant tailor, but must have moved in powerful circles, for the precocious Roger became adept at shorthand transcriptions of sermons and speeches for Sir Edward Coke, chief justice of the King's Bench and the sharpest thorn in the side of King James's assertions of divine-right sovereignty. Coke was impressed enough to become Williams's patron, sending him to school at Sutton's Hospital and then on, through the school's bursaries, to Pembroke College, Cambridge. But Coke's resistance to Stuart absolutism was legal rather than theological, based on the "immemorial constitution" vested in the common law. Williams was to go altogether another way. The history that spoke to him was not Magna Carta, but what had happened to the "visible church" when it became entangled with, and corrupted by, earthly power. The date that church historians routinely celebrated as a triumph—AD 313, the edict of Milan promulgated by the convert Emperor Constantine, making Rome a Christian empire—was, for Williams, a calamitous fall from grace. "Then began the great Mysterie of the Church's sleepe," he wrote, more than a millennium later: a usurpation of God's provision for history, and still more heinous, the coercion of souls, the "sword of stele" that Christ had expressly rejected. Had not Jesus said "my kingdom is not of this world?" But he had been disregarded by those who claimed to be his apostolic heirs, who had erected a government. That the Roman Church should seek to enforce its authority was no surprise; what distressed Williams was that Protestant churches, including the one into whose service he was supposed to be ordained, had, since King Henry VIII, claimed similar powers of conformity and brutally punished those who resisted it. Williams was unmoved by the argument that such measures were needful to stop apostasy or heresy in its tracks, for no man could find his way to Jesus except through his own free will, and to usurp God's own authority was worse than subjection under Rome. A great disentanglement was needed if true Christians were ever to find their way to salvation. Holiness, which he compared to a garden, needed "a hedge or wall of separation" enclosing it off from worldly matters if it was to hold sway in the hearts of men.

Williams must have been on the edge of these convictions rather than over it when he graduated from Cambridge in 1627 for, as planned, he was ordained into the Church of England, and accepted a chaplaincy with the Puritan Member of Parliament, Sir William Masham. What might have been the beginning of a settled life turned into the opposite. A rejected courtship sent the young chaplain into a sickly fever from which he was nursed by a member of Masham's household, Mary Barnard, whom he married in 1629. But the church whose ministry he was supposed to profess, was falling into the hands of Archbishop Laud, whose reforms were, for the Puritans, tantamount to Catholic Counter-reformation. Lord Chief Justice Coke happened to have an interest as a venturer-investor in the American colonies, so it might well have struck him that the best place for his free-speaking protégé might be the other side of the Atlantic. In December 1630, Roger and Mary sailed on *The Lyon*, arriving at Nantasket Harbor just south of Boston in the first week of February 1631.

Perhaps it was on the long sea voyage that Williams made his own journey of revelation, for he wasted no time in making trouble for himself. Welcomed in Boston, Williams protested that he could not serve a church that had insufficiently renounced and separated itself from the impious Church of England. For the moment Governor Winthrop was prepared to countenance the young man's eccentricities and sent him north to Salem to serve as teacher and preacher with an older minister. But what Williams began to teach and preach was intolerable. Oaths administered to the "unregenerate" in court, or routinely as an act of allegiance, he said, were blasphemous, since no earthly authority could invoke God's name; he claimed that the Great and General Court could regulate matters threatening civil peace but in no circumstances could prosecute those deemed heretic, much less punish them with flogging as was the prescribed penalty. Representations from Winthrop, who had initially greeted Williams as a likely "godly minister," failed to have any effect. By 1633 Williams had joined a godly community who called themselves Seekers and who believed that since all churches were corrupt, membership must always be voluntary. That suited Williams, who went south in search of just such a loosely organized gathering at Plymouth. It was there that he took himself off to the Indians, and by the time he returned to Salem the following year as its minister, his conviction that no part of "soul liberty" should ever be surrendered to those who had usurped Christ's own lordship had only hardened.

For Winthrop (who still professed to like him) the impossibly pure Williams had become a threat, a sower of discord. He was duly arraigned before the General Court on 1 October 1635. The gravest accusations were Williams's claim that the government of the colony had no right to punish infractions of the first four of the Ten Commandments and that oaths sworn on the Bible and in the name of God were blasphemous. He freely confessed that he believed no man ought to be obliged to maintain a church establishment whose beliefs he did not share. The sentence was banishment, and constables were sent to his house in Salem to enforce the writ and escort Williams to a sloop that would take him back to England. But he had already fled.

"I was sorely tossed for fourteen weeks in a bitter winter season," he recalled, "not knowing what bread or bed" he could expect. It was almost certainly his familiarity with Indian languages that saved him, for it was the natives who provided food and shelter when Williams most desperately needed it. Paddling a canoe up and down the Seekonk and Moshassuck rivers, Williams gradually emerged into the New England spring and in June 1636 did what he bid others, by entering into a direct agreement with the sachems, in this case of the Narragansett tribe, for the purchase of land "upon the fresh rivers of the Mooshausic [Moshassuck] and Wannasquatasket [Woonasquatucket]." It was there that Williams established Providence Plantation, out of the sight and jurisdiction of Massachusetts. From the beginning Providence was to refrain from any acts of forced conformity, nor was it ever to impose tests for the holding of office. It was not just the first American settlement to embrace such freedom of conscience, it was the first in the Western world. Massachusetts would in fact retain some of the moral and religious laws on its statute books until well into the twentieth century. But it had long since ceased to matter. It was the renegade Williams whose views had—eventually—come to prevail.

For some years, Williams spent his time "day and night, at home and on the water, at hoe and oar for bread," his children given names like Mercy and Providence. A small community of the persecuted clustered in and around the settlement and at Newport, where the minister John Clarke had opened his doors in the same fashion. But both of them realized their colony of conscience would not survive the hostility coming from Massachusetts unless they could get authority for it from England. As luck—or as Williams certainly assumed, providence—had it, Clarke and he arrived there just after the outbreak of the war between Parliament and the king, the quarrel not least being over grand matters of religious coercion. With the authority of the Church of England crumbling, Williams went to see one of the parliamentary leaders, Sir Henry Vane, whom he had met in Boston in 1635 and who, notwithstanding a more orthodox view, was himself a believer in toleration and the disestablishment of a national church. Vane saw no reason not to assent to Providence becoming a place where freedom of beliefs could be absolutely protected. Given the ordeal that England was undergoing, he might well have agreed with Williams's assertion in *The Bloudy Tenet of Persecution for Cause of Conscience*, that "God requireth not a uniformity of Religion to be inacted and inforced in any civil state, which inforced uniformity (sooner or later) is the greatest occasion of civill warre, ravishing of conscience, persecution of Christ Jesus in his servants and of hypocrisy and destruction of millions of souls." In 1644 Vane persuaded Parliament to authorize "an absolute charter [of liberty] for those parts of his [Williams's] abode." In 1651 Williams returned to England to have that charter confirmed, and stayed with Vane at his grand country house and was introduced to the Latin secretary to Cromwell's Council, John Milton, to whom he gave Dutch lessons. Vane had become one of the powers in the land, commissioner for the army and navy, and an intransigent opponent of Cromwell's attempts to browbeat and purge Parliament. That resistance would see him imprisoned, but not before he had managed to extend the life of Williams's colony, the only government Vane could have found ideal.

Four years before, in May 1647, a meeting of delegates from several towns in what would become Rhode Island formally declared that "it is agreed that the forme of Government established [there] is DEMOCRATICALL, that is to say a Government held by ye free and voluntary consent of the greater part of the inhabitants." Providence Plantation would be governed by an appointed president but would be accountable to an elected assembly of representatives from the townships. Thus, as Roger Williams wished, political and religious liberty were coupled and an American future made before there was a United States of America.

When the monarchy was restored in 1660, Williams must have feared that his little republic of free conscience would also be terminated. Sir Henry Vane had come to a bad end, imprisoned first by Cromwell, then by the Restoration government. Although Vane had disapproved of the execution of Charles I, he was tarred with the brush of the Parliament that had judged the king. In June Vane was tried in Parliament, where he defended to the end its sovereign authority, and was beheaded on Tower Hill nine days later after delivering on the scaffold a characteristically long speech. But in the following year, on 15 July 1663, Charles II signed a renewed charter giving the blessing of the Crown to the "livelie experiment" at Providence Plantation. This act and the document that proclaimed it are so astonishing, in that they follow almost to the letter what Williams would have himself sought, that they are almost unaccountable for a government then very much in the business of reestablishing tests of conformity. But since the charter begins by addressing "our trustie and well-beloved subject

John Clarke," it may be that Williams's old friend, who had stayed in England, had something to do with what then followed. The document bears quotation at length for it unmistakably makes the very un-puritanical figure of King Charles II rather than, say, William Penn or Thomas Jefferson, the establisher of free conscience in America!

John Clarke and the rest of those who had petitioned on behalf of Providence Plantation, the document proclaims, "professing with peaceable and loyall minds their sober, serious and religious intention of . . . edifyeing themselves in the holie Christian ffaith and worshipp as they were persuaded, together with the gaineing and conversion of the poore ignorant Indian natives . . . have not only byn preserved to admiration but have increased and prospered . . . whereas in their humble address they have ffreely declared that it is their hearts (if they may bee permitted) to hold forth a livelie experiment that a most flourishing civill state may stand and best bee maintained and that among our English subjects with a full libertie in religious concernments and that true pietye and religion grounded upon gospell principles will give the best and greatest security to sovereignty and will lay in the hearts of men the strongest obligations to true loyaltie . . . Now know that wee beinge willinge to encourage the hopefull undertaking off the sayd loyall and loveinge subjects, to preserve unto them that libertye in the true Christian ffaith . . . and worshipp of God which they have sought with so much travaill and with peaceable myndes have . . . thought fit and doe hereby publish, graunt, ordeyne and declare, that our royal will and pleasure is that noe person in the said colonye at any tyme and hereafter shall bee any wyse molested, punished, disquieted or called into question for any differences in matters of religion and doe not actually disturb the civill peace . . . but that all and every persons may from tyme to tyme and at all herafter freelye and fully have and enjoy his and their own judgment and conscience in matters of religious concern-ment . . . they behaving themselves peaceablie and quietlie and not using this libertie to lycentiousness and profaneness nor to the civil injurye and outward disturbeance of others."

That last reservation, which Clarke and Williams certainly shared, is important for the future history of what they began in Rhode Island. To breach the separation of church and state in the interest of keeping the civil peace, the wall-breachers have had to make the case that the particular transgressions they seek to have the state outlaw—alcohol or abortion, say—are indeed matters that threaten social peace. Were he around today, would Williams concur that the former was the handmaid of mayhem, or the latter a form of murder, a crime against "person?" Maybe, although drink and dead babies were in plentiful supply in his world. But there can be no doubt that he would have regarded any suggestion that taxes ought to be used for the support of either ministers, schools, or any "faith-based" institutions unacceptable, given that there would be taxpayers not sharing the beliefs of those whom their money maintained. Or would he have thought that, for instance, gay marriage was a manifest threat to the social order, or would he have reserved that judgment for the Almighty?

Did Charles II, Defender of the Faith according to the Church of England, actually read the revolutionary charter issued in his name, a document more radical than anything Cromwell had ever endorsed? At precisely the time he signed it, his bishops were busy hounding from their livings any minister who showed the slightest dissent. But if Charles hadn't taken seriously the implications of what was now established in Rhode Island, others who needed freedom of conscience certainly had. In 1658, a ship carrying fifteen families of Sephardi Jews had sailed into Newport Harbor looking for better days. In all likelihood they were a remnant of the communities of Recife who had thrived in Brazil before its recapture by the Portuguese in 1654. Their own origins were in Spain and Portugal, and their continued covert existence there after the expulsions of the 1490s as pseudo-Christian Marranos and "conversos" had given them a network of language and culture that in both the Mediterranean and Atlantic worlds was a priceless asset for the Dutch. Pragmatic toleration had been extended to them, communities had settled; synagogues had been built; printing presses had turned out literature in Hebrew and the vernacular language of the Sephardim, Ladino. The fall of Brazil had made them wanderers again. That same year, 1654, a group of twenty-four families had arrived in New Amsterdam on the island of Manhattan, looking for some semblance of the existence they already enjoyed in the other Amsterdam. But the governor, Pieter Stuyvesant, thought their religion an "abomination" before God and felt that if he granted rights of private worship, the next thing would be a demand to build synagogues. A direct application to the Heeren of the West India Company back in Holland produced permission to meet for prayer in their own homes, but it was short of the rights they had expected.

Some of them, at least, wanted more. Was England and its American empire an answer? After a meeting with the Amsterdam sage Menasseh ben Israel, Oliver Cromwell's hope that the conversion of the Jews might usher in the Last Days and the reign of Christ, as well as his canny sense of opportunity to steal an asset from the Dutch, led him to permit the de facto residence of the community in the country that had expelled them in the thirteenth century. Evidently, though, Rhode Island promised more, perhaps even the full enjoyment of civil rights alongside Christians. Hence their optimistic voyage to Newport.

The promise was only partially fulfilled. As in New York, worship was in private houses; there was no thought that Jews might hold public office or vote, and it was not until 1677 that a plot of land on the edge of town was acquired for a cemetery. But by the mid-eighteenth century, exactly a century after the first Jews had arrived, the community was strong and wealthy enough to call for a rabbi— the novice, Isaac Touro—to come to Newport from Amsterdam. A year later, in 1759, the foundation stone was laid for the synagogue that miraculously still survives on Spring Street. Commissioned from Peter Harrison, the most distinguished architect in New England, it's the most elegant as well as the oldest synagogue in the United States: two stories, arched leaded windows, a balustraded gallery for the women congregants. Inside the ark of Jeshuat Israel, the Salvation of Israel, is a deerskin *sefer Torah*, thought to have been carried from Spain in the year of the expulsion, to have sojourned in Amsterdam and then been brought to the Atlantic world by the first immigrants, who came seeking Roger Williams's "soul liberty." The Torah is a work of great beauty, startling in the clarity of its Hebrew calligraphy, the hide enriched by time to a deep tawny hue.

When not in use, the scroll is kept open to the chapter of Exodus describing the crossing of the Red Sea; the passage from bondage to freedom, the birth of a nation. But for the Jews of Jeshuat Israel true redemption would come only with the revolution.

17. "Whereas Almighty God hath created the mind free . . ."

In the dog days of August 1790, President George Washington paid a courtesy visit to Newport. The purpose of his journey was partly emblematic. The first Congress of the United States, following the adoption of the Constitution, had adjourned for the summer recess, and Washington was minded to show the People the face of their president. The morning walkabout (with Washington apparently setting a clip that fatigued those trying to keep up with him "fortified by wine and punch" at four different houses) was especially meaningful to Newport, which had suffered heavy losses of material, building fabric, and population during the Revolutionary War. In the autumn of 1776 the British had occupied the port to preempt it becoming an American base from which an attack on New York, their strategic jewel and hostage, could be mounted. Repeated attempts by American forces to dislodge them failed, with the town and port in 1778 turning into a battle site assaulted by sea and land. When the British finally evacuated Newport in 1781 and Washington arrived for a meeting with the French admiral Rochambeau, the place was a shell of its wealthy mercantile former self. Half of its prewar population of 9,000 had gone, dispersed elsewhere in New England and the mid-Atlantic states, never to return. The least he could do, the new president figured, was to offer in his person some encouragement for its restoration.

But there was another reason for Washington to go to Rhode Island and that was to gin up the state's ratification of the Bill of Rights, the first ten amendments to the constitution. Though minute in territory, or perhaps because of it, Rhode Island, as Washington knew, had mixed feelings about the Union. Its citizens were notoriously protective of their idiosyncrasies and quick to suspect any attempt by the rest of America to compromise them. Excluded from the New England Confederation of states during the colonial centuries for their excessive generosity in matters of conscience, not to mention the unaccountable attentiveness to Indian land titles, the citizens of Providence and Newport knew they were joked about as "Rogue Islanders" elsewhere on the eastern seaboard, especially in neighboring Massachusetts. Though their merchants and seamen had been the first to take violent resistance to the British, firing on their ships as early as 1772 and again in 1774, and had also been the first to make a formal break from allegiance to the Crown, Rhode Island was the last of the thirteen states to ratify the Constitution, refusing to send delegates to the Convention in Philadelphia. Only the threat of being treated as a foreign nation, and made subject to customs duties, overcame their pesky reluctance at being integrated into the new Union.

So the president was paying a call on the dog-in-the-manger of the United States, and he was not taking anything for granted. Rather, he was doing what all successful presidents have done ever since: making his presence felt in American cities that had gone through hardship, glad-handing the people, drinking with them (very important), promising a better future, and diplomatically giving the prickly Rhode Islanders a sense that they were being personally consulted on the amendments to the Constitution; that though they might be the most modest state in the Union, in his own balance they were as weighty as any New York or Virginia. Washington was a huge success: his peculiar combination of rugged simplicity and noble bearing working magic as it almost always did. But there was one section of the Newport community especially eager to pay homage: the Jews of the Kahal Jeshuat Israel. Many of them had departed with their fellow citizens at the time of the British occupation, leaving only a few like the *parnas* (warden) and banker Moses Seixas to protect the deerskin Torah and the fabric of the Touro Synagogue from harm, notwithstanding the latter's appropriation for storage of arms and ammunition, making it a prime target for enemy guns.

The Seixases were a little Jewish empire all to themselves. Originally from Lisbon, they had dispersed during the revolution to Connecticut, New York, and Philadelphia, where Moses's pious brother Gershon was *haham* (rabbi), *hazzan* (cantor), *mohel* (circumciser) and *shochet* (ritual slaughterer), a full-service minister to the community. Benjamin, another brother, had been an officer in the New York militia. You didn't get any more Judeo-patriotic than the Seixases. So it was natural for Moses, who, while never swerving in his loyalty to the American cause, had stayed behind in Newport during the occupation, to seize the moment of Washington's visit to clear up one or two matters concerning the First Amendment which promised that "Congress shall make no law respecting an establishment of religion, or prohibiting the free exercise thereof." On the face of it this sounded like unequivocal good news, the redemption, at last, of Roger Williams's promise of "soul liberty." No church would be the national church of the United States, and evidently the Jews would be free to "exercise" their right to free worship. But they had had that allowance already for more than a century. What they wanted to know was whether under the Constitution they would finally secure what had been long denied: equal rights as citizens, including the right to hold public office and, most precious of all, the right to vote. Seixas forbore from quoting Williams in support of his case, but he might have for the Seeker's view, explicitly expressed in *The Bloudy Tenet Yet More Bloudy*, was that since religious profession could be no criterion for the exercise of public office, but rather "a [more general] morall vertue" such as "morall fidelitie, abilitie and honestie," this could as well be found in "other men (beside Church-members) [who] are by good nature and education, by good Lawes and good examples nourished and trained up." All that Williams really wanted was for principled men to do what was right by their own lights. "I commend that man, whether Jew or Turk or Papist or whoever steers no otherwise than his conscience does." But the notion that any of the above might actually exercise positions of public trust in the United States was, for the vast majority of Americans, still an abhorrent notion.

Seixas hoped that, since the implication of the First Amendment was to separate entirely religion and government, it might go otherwise in Rhode Island. So the letter he penned on 17 August for presentation to the president the following day was a nosegay of praise to the Father of the Nation (and not before time, since the Jews of the several American congregations had been tardy in offering congratulations on Washington's inauguration earlier that year—but then getting a handful of *kehillot*, community leaders, to sign on the same page of anything counts as a miracle). Between the lines Seixas was also seeking clarification. Had the great day finally arrived when Jews would be treated as all other citizens? Could they now be magistrates, councillors, constables? Above all could they now *vote*?

Being "the stock of Abraham," Seixas took an ornamentally Hebraic tone with the general, reflecting on "those days of difficulty and danger when the God of Israel, who delivered David from the peril of the sword shielded Your head in the day of battle . . . and we rejoice to think that the same Spirit who rested in the Bosom of the greatly beloved Daniel enabling him to preside over the provinces of the Babylonish

Empire, rests and will ever rest upon you." That must have softened
the old boy up—visions of David at Yorktown, President Daniel. Then
the nub of the matter: "Deprived as we heretofore have been of the
invaluable rights of free Citizens, we now with a deep sense of gratitude
to the Almighty disposer of all events behold a Government erected
by the Majesty of the People, a Government which to bigotry gives
no sanction, to persecution no assistance—but generously affording
to all Liberty of conscience and immunities of Citizenship—deeming
every one, of whatever Nation, tongue or language equal parts of the
great governmental Machine—This so ample and extensive Federal
Union whose basis is Philanthropy, Mutual Confidence and Public
Virtue [nice touch that, putting *tzedakah*, righteous charity, first in the
Masonic trio], we cannot but acknowledge to be the work of the Great
God who ruleth in the Armies of Heaven." Cleverly Seixas wasn't
asking. He was in this manner merely describing what he took to be
self-evident, leaving Washington to demur if he must. "For all these
Blessings of civil and religious liberty which we enjoy under an equal
benign administration, we desire to send up our thanks to the Ancient
of Days." (God, not the president.) May he like Joshua when gathered
to his Fathers be admitted into "Paradise to partake of the water of
life and the tree of immortality."

Washington loved this kind of thing. The next day, after an all-out
dinner in the Old State House, he responded to Moses Seixas in a
way designed to make Jeshuat Israel happy. "The Citizens of the
United States of America have a right to applaud themselves for
having given to mankind examples of an enlarged and liberal policy:
a policy worthy of imitation. All possess alike liberty of conscience
and immunities of citizenship." And then came Washington's endorse-
ment of the presumption that active citizenship for all Americans
was indeed what was understood in the Constitution and the Bill
of Rights. America was the republic in which toleration was not
bestowed as an "indulgence of one class of people" to another but the
"exercise of their inherent national gifts." Then, in what one hopes
Seixas took as a compliment to his prose, Washington simply lifted
the Jew's lovely characterization of a nation "which gives to bigotry
no sanction, to persecution no assistance" and then rather grandly
failed to acknowledge (or perhaps notice) that he had taken it from
Moses Seixas's letter. Not until the poet Emma Lazarus came along

would a Jew manage to supply so perfectly felicitous phrasing for
what the United States was supposed to stand for. Rest assured, the
president concluded, in this benign state of affairs, every one of the
"Stock of Abraham" "shall sit in safety under his own vine and fig
tree and there shall be none to make him afraid." A neat touch this,
straight from the prayer book of the psalmist. Well, Washington had
just been compared to David.

Now vines and fig trees were all very nice, especially if you lived
oceanside in Rhode Island, but did this mean that Jews could, after all,
be eligible to be *magistrates*, have the *vote*? It seemed indeed that it did.
And this could not have been more important, because it can hardly
have escaped the Jews of Newport that this was emphatically not the
case elsewhere. The Jews of Baltimore, for example, had to wait until
the 1820s for the Maryland "Jew Bill" which cleared matters up.

There was someone else on hand in Newport on 18 August, for
whom this little exchange was of more than casual interest: the secretary of state, Thomas Jefferson. Jefferson knew better than to steal the
president's thunder and diligently played second fiddle to Washington's
stentorian brass. But this particular turn in the proceedings had a special
significance for him, as the principal author of the Virginia Statute of
Religious Freedom, enacted in his and Washington's home state four
years before in 1786. The fight to keep matters religious and matters
of state apart, to institute toleration and equal rights for those of all
beliefs or none, was not, for Jefferson, nor for his friend James Madison,
a revolutionary afterthought. It *was* the revolution just as much as the
institution of democracy itself. In 1776 what was it that he described as
"the severest contest in which I have ever been engaged"? The battle
against the British in Massachusetts? No, the overthrow of a church
establishment. What was the first political campaign Madison fought?
The defense of dissenters in Culpeper County. If the two of them were
around today and needed a flag to wave at the zealots who slaugh-
tered New Yorkers on 9/11, the Statute of Religious Freedom would
replace the Stars and Stripes. Read this, they would say, and you will
read America. Jefferson's authorship of the bill (and the much better
known Declaration of Independence and his creation of the University
of Virginia) were the achievements he wanted inscribed on his tombstone.

Jefferson knew that not everyone in America felt quite the same way, especially not one of his personal bugbears: John Adams. The constitution of Massachusetts, presented and ratified by the General Court in 1780, drafted by Adams, and most usually remembered as a "mild and equitable" treatment of religion, was, in fact, nothing of the sort. But it does represent—to this day—precisely the other side of the American dialogue between the Williams–Jefferson tradition of a clean cut between public power and private conscience, on the one hand, and the Winthrop–Massachusetts side of insisting on the indispensability of Christian-grounded moral regulation for the good order of society. This argument never goes away. As I write this, the junior senator from Oklahoma, Tom Coburn, whose Web site declares him to be a member of the Muskogee New Community Church, is holding up the passage of an AIDS-assistance bill through Congress on the grounds that it includes provision for health education that pays insufficient attention to abstinence. This is purest John Adams in his Massachusetts 1780 mode, decreeing any thought of political action uncoupled from religious morality to be a reprehensible abandonment of civic responsibility.

To those for whom Adams, a Unitarian, albeit with a Calvinist cast of temper, represents a beacon of New England liberalism, it may come as a surprise to find him so adamantly on the side of Christian public politics. But the importance he assigned to this issue is apparent from the fact that it takes up articles II and III of the Massachusetts constitution, right at the front of the document, preceded only by the ritual recycling from the Declaration of Independence (on which he had collaborated with Jefferson) that "All men are born free and equal and have certain natural, essential and unalienable rights." Absolute freedom to exercise their conscience to the point of opting out from supporting the clergy, however, much less leading an irreligious life, was not among those natural rights. Article II states that "It is the right *as well as the duty* [my italics] of all men in society, *publicly and at stated seasons*, to worship the Supreme Being, the great Creator and Preserver of the Universe." In other words, no American could consider himself a right citizen unless he had fulfilled that duty of public worship. Then came the sweetener, drawn from Jefferson's rejected draft for the Virginia statute in 1778–79, and ultimately from Roger Williams and the Rhode Island charter of 1663, that "no subject" (are there still subjects in the republic?) "shall be hurt, molested or restrained in his person, liberty or estate, for worshipping God in the manner and season most agreeable to the dictates of his own conscience . . . provided he doth not disturb the public peace."

It was in article III, however, that Adams got down to serious business. His premise was, and it is expressed as if unarguable (though we have seen the impassioned Christian Roger Williams would have contested it), that "the happiness of a people and the good order and preservation of civil government, essentially depend upon piety, religion and morality." Since "these cannot be generally diffused through a community but by the institution of the public worship of God and of public instructions in piety, religion and morality . . . to promote . . . and secure the good order and preservation of government, the people of this commonwealth have a right to invest their legislature with the power to authorise and *require* the several towns, parishes, precincts and other bodies politic or religious societies [as if they were interchangeable], to make suitable provision at their own expense, for the institution of the public worship of God and for the support and maintenance of public Protestant teachers of piety, religion and morality, in all such cases where such provision shall not be made voluntarily." Notice the *Protestant*. Catholic worshippers and schoolteachers expecting public funding, much less Jews or "Mahometans," could not expect to be provided for. Notice also the element of compulsion Adams has smuggled in. The good people of the commonwealth could volunteer to finance churches and religious schools, but should they wish to opt out, they would be taxed for that purpose anyway.

It's safe to say that what the Right likes to call the People's Republic of Massachusetts, or "Taxachusetts," is not the favorite state of, say, Pat Robertson or the Christian Coalition. But there is not a word of articles II and III of the constitution of 1780 with which they could possibly find fault. The document, the first of the state constitutions and meant as a template for the others, rides roughshod over Williams's ideas and means to make moral regulation a habitually indispensable feature of American public life. Under the Adams constitution, Sunday church attendance was compulsory, the law being repealed only seven years after his death, in 1833. Blasphemy could be punished by a year's imprisonment, a public whipping, time in the pillory, or "standing on the gallows, a rope about the neck." That law stayed on the books in

Massachusetts for sixty years after the adoption of the state constitution. And—in what would become one of the most Catholic states in the Union—a test affirming Protestantism was required for most public offices. A law defining sodomy as an "Unnatural and Lascivious Act" is still on the statute book of the commonwealth, carrying a penalty of twenty years' jail time, although since 1974 it is deemed not to apply in cases of private consensual acts. Without noticing that the second president was the founding patriarch of its cause, the evangelical crusade in American politics was fueled by the ambition to re-create John Adams's commonwealth of Christian virtue until this election season, when it finally ran out of gas.

Why? Because there has always been an alternative American tradition in competition with the Massachusetts model, that of the other state that likes to call itself a commonwealth: Thomas Jefferson's and James Madison's Virginia of 1786. This may seem upside down. Is not the address of the Moral Majority Coalition, Lynchburg, *Virginia*? Is not Massachusetts gay-marriage friendly? But matters were differently assorted at the founding, and it was the hypocritical equality-mouthing slave-owning philosophical gentry who put on their statute book one of the most eloquent documents of cultural liberty ever penned.

It was, to be sure, a precious moment of philosophical clarity and moral courage, sandwiched, rather tightly, between that other authentically American phenomenon: the outpourings of Christian instinct known as Great Awakenings, the first in the 1740s and midcentury, the second coming hard on the heels of Jefferson's election to the presidency in 1801, which was taken by his enemies as the elevation of a shameless atheist to the highest office in the land. But in a way the Great Awakening, with its spectacular manifestations of itinerant sermonizing by the likes of George Whitefield, Jonathan Edwards, and John Wesley, helped prepare the way for Jefferson and the First Amendment. When the Christian wildness, the appeal to passion over doctrine, began, between 60 and 80 percent of Americans belonged to the established churches, either Anglican or Congregationalist. But the hot-gospellers of the Awakening ignored parish boundaries, church decorum, and the obligations of hierarchy, and many of the most viscerally extravagant preachers won followings on the frontier, taking the physical ministrations of Christian revelation up rivers, into the wilderness, taking Lord Jesus to the mountains.

By the eve of the revolution the numbers of those attending established churches had fallen below 50 percent, and in Virginia it was more like a third, with the backcountry farmers overwhelmingly Baptists or Presbyterians. And naturally the dissenting churches had a strong interest in making sure that what had been the favored establishments of British-dominated ecclesiastical order disappeared with the revolution, along with everything else about imperial rule.

And, eventually, it was the passion, not of high-minded freethinkers like Jefferson, but of Baptist enthusiasts that would make the difference between the failure of Jefferson's draft for toleration and its eventual successful enactment, steered by the less glamorous but politically astute James Madison through the Virginia Assembly. That connection between Christian enthusiasm and freedom of conscience, the authentically American bet that religion would flourish best if left as a matter of purely private choice, as distinct from, say, Taliban coercion, is what prevailed in that part of the United States that did not hew to the Adams version. For Europeans it may be hard to get their heads around this paradox; for America it's second nature. And if the world wants to find a way to confront theocratic fanaticism with more than expressions of ridicule, the American way may offer a more persuasive strategy of cultural disarmament.

Jefferson was not an atheist. In fact he thought that the observable universe, being as intricate and harmoniously engineered as it was, must presuppose some Designer, the Enlightenment's watchmaker-deity who, once the machine was completed, allowed it to run itself, with perhaps occasional checkups for reducing (or aggravating) the friction of parts. This made Jefferson a deist, incredulous of those who conceived of the world as arbitrarily arranged physical matter. But this being so, Thomas Jefferson could not possibly have hoped to run in the election of 2008 and have any chance of winning. Though Jefferson held Jesus in high esteem as perhaps the greatest of history's moral teachers, he thought it absurd, if not offensive, to compromise that standing by fairy tales declaring him Son of God, born of a virgin, a water-walking corpse resuscitator, and such foolishness. Anything worthwhile in the teachings of religion ought to withstand rational scrutiny if it were to be upheld. In 1787 he counseled his nephew Peter Carr to be a man, philosophically, and "shake off all of the fears and servile prejudices under which weak minds are servilely crouched. Fix reason firmly in her seat and call to her tribunal every fact, every opinion. Question with boldness even the existence of God because, if there be one, he must more approve of the homage of reason than that of blindfolded fear. Read the Bible then as you would read Livy or Tacitus." If biblical events like the sun standing still for Joshua should provoke doubt, that skepticism ought not to be shaken by being told that the story had been divinely revealed. Nothing was to be accepted merely as a matter of blind faith: "examine, therefore, candidly, what evidence there is of his being divinely inspired. The pretension is entitled to your inquiry because millions believe it."

This was extraordinary counsel from the benevolent Uncle Thomas, but Jefferson, in common with the Enlightenment *philosophes*, believed that adhesion to unexamined and irrational beliefs had been the greatest cause of contention and slaughter in the world, for there could be no arguing with those who asserted from revelation alone. Nothing about our own epoch would be likely to shake Jefferson from that view, though doubtless he would be dismayed that the human race had somehow failed to shake off its thraldom to myth. Dispose of those myths, he argued, and you would neutralize the carnage. If only mankind could somehow be persuaded to hold only those beliefs that could withstand the empirical scrutiny of reason, there was a chance for some sort of universal consensus on the characteristics of the divine that did, or did not, make sense. Then men might at last forbear from imposing their particular monopoly of revealed truth on others. Neither Taliban nor televangelist would make him feel better about the remoteness of this eventuality. In common with Roger Williams, Jefferson held that nothing, however, could justify criminalizing religious or irreligious opinion. In his *Notes on the State of Virginia*, Jefferson described as intolerable the situation in his own state inherited from earlier laws by which anyone denying the Trinity, or questioning the divine authority of scripture, was disqualified from holding office. A second offense along these lines would disable the offender from any right to sue, and could lead on conviction to a prison sentence and the removal of his children from parental custody. "This is a summary view of that religious slavery under which a people have been willing to remain who have lavished their lives and fortunes for the establishment of their civil freedom."

Governments, Jefferson went on, may only have rights over qualities submitted to them in the first place, but rights of conscience have never been so submitted. "We are answerable for them to our God. The legitimate powers of government extend to such acts only as are injurious to others. But it does me no injury for my neighbour to say there are twenty gods or no god. It neither picks my pocket nor breaks my leg." This position of the deist was, in fact, remarkably close to Williams the Baptist, although Jefferson's immediate source was much more likely to have been John Locke's *Letters on Toleration*, which was far better known in the eighteenth century. He might also have read the work of the Scottish schoolteacher James Burgh, written in the 1760s, making much the same case. But there was time for him to have gotten to know Roger Williams before the visit to Newport in 1790, as the Massachusetts Baptist Isaac Backus, whom Jefferson knew very well, published an edition of Williams's work in the 1770s.

Some of Jefferson's sardonic militancy at this time no doubt came from keeping like company in pre-Revolutionary Paris, where intelligent sniggering at the follies of the benighted was de rigueur in the salons. His conviction that any religion worth its salt ought to be accessed through the mind, rather than through metaphysical mystery, was pure Locke, even though unlike Locke, Jefferson denied the divinity of Jesus. And some of Jefferson's passion was a product of his frustration at the inability to get the Bill on Religious Freedom adopted by the Virginia Assembly when it had come before them in 1779, largely due to the vocal opposition of Patrick Henry. Just what was the liberty Henry had been thinking of? Jefferson might have asked himself when he had postured rhetorically "give me liberty or give me death." Jefferson doubtless took some comfort from the fact that the assembly had also denied Henry his motion to support religious teachers from public funds: the creation, in effect, of multiple Protestant establishments. Both motions were shelved for the duration of the war. No doubt the vexed Jefferson was arrogant, remote from understanding the human craving for the myths he found so puerile. But the merest look at his draft for the statute—arguably the greatest and bravest thing he ever wrote—is to forgive him.

"Whereas Almighty God hath created the mind free" runs the first sentence in his revised version, and with that one plangent phrase an oxymoron becomes an American truism, "all attempts to influence it by temporal punishments or burthens or by civil incapacitations tend only to beget habits of hypocrisy and meanness and are a departure from the plan of the holy author of our religion who, being Lord of both body and mind yet chose not to propagate it by coercions on either." Thus, Jefferson continues in a high Williamsite vein, it is only the presumptuous impiety of weak men and rulers to usurp the Almighty's sovereign power and presume to do what he refrained from. "To compel a man to furnish contributions of money for the propagation of opinions in which he disbelieves [as Henry was arguing should be the case in Virginia, and Adams would insert into the Massachusetts constitution] is sinful and tyrannical; that even the forcing him to support this or that teacher of his own religious persuasion is depriving him of the comfortable liberty of giving his contributions to the particular passions he feels most persuasive to righteousnes." The last sentence describes exactly the American pattern of philanthropy, as good in its instinctive impulse as forced support for the clergy was bad in its moral cowardice.

As Jefferson warms to his task, the modern reader can feel the indignation and contempt rising in him for all those who needed to support their views, religious or otherwise, with anything other than the pure force of their truth and wisdom. And suddenly, or so it seemed to me in the Virginia State archive, as I held the version that would finally be enacted six years later, but that kept the ringing eloquence—Jefferson was addressing something more than the cramped and timorous prejudices of the day. He was steaming ahead into dark modernity with a coda that was imperishably connected to what America stood for over the long haul of history.

Sentences like Jefferson's great coda are what should be the text that schoolchildren throughout the American republic ought to recite each day instead of the numb and, since the 1950s, mindlessly reverent Pledge of Allegiance. Then they would understand, right away, the proper meaning of their nation's existence. "Truth is great," wrote the man who could be hypocrite, egotist, utopian, beady-eyed stratagem-maker, all in the same week; yet if he had written only the following, he would have still warranted the gratitude of posterity. It is the unflinching answer to moral and immoral bullying (whether by Americans or others), to the sweaty insecurity of the fanatics, to the secret policemen and thug-triumvirs. It is why it is never sensible to give up on America. "Truth is great and will prevail if left to herself . . . she is the proper and sufficient antagonist to error, and has nothing to fear from the conflict unless by human interposition disarmed of her natural weapons, free argument and debate, errors ceasing to be dangerous when it is permitted freely to contradict them."

But truth did not prevail, left to itself, at least not immediately. It would take Patrick Henry, pushing his luck by obstinately returning to his scheme for a "general assessment," for James Madison to realize that the statute might yet have another chance of enactment. Dissenters were now a majority in Virginia, and more than a hundred petitions and addresses, bearing 11,000 signatures against Henry's proposal, poured into the assembly toward the end of 1785. Many of the most adamant against Henry's proposal were from backcountry areas like Cumberland County where Baptists were especially strong. Before Madison introduced his eloquent reiteration of Jefferson's arguments, titled "Memorial and Remonstrance Against Religious Assessments," he made sure that Patrick Henry was got out of the way by being elected governor of Virginia. Once that was accomplished, he was free to go on the attack, describing the appropriateness of public alarm at what was "the first experiment [i.e., assault] on our liberties." It was an ominous precedent, he went on, for "who does not see that the same authority which can establish Christianity in exclusion of all other religions may establish with the same ease any particular sect of Christians in exclusion of all other sects." To deny to others the liberty to profess anything they believed would be an offense against God, not against man. To introduce this kind of preference, Madison argued, would destroy American harmony. "Torrents of blood have been spilt in the old world by vain attempts of the secular arm to extinguish religious discord, by proscribing difference in Religious opinion." The American "theater" proved that if quarrels could not be eradicated, at least, in warranty of equal liberty, they could be defanged, their "malignant influence" drained away.

For Madison and Jefferson, toleration and religious pluralism were America's greatest blessing, a freedom arising "from the multitude of sects" that, absent government interference, would naturally flourish and multiply. The variety of faiths was not, of course, a hallmark of Madison and Jefferson's own time, but it would certainly become America's distinction and was in Madison's words "the best and only security for religious belief in any society, for where there is such a variety of sects, there cannot be a majority of any one sect to persecute and oppress the rest." For them both, moreover, that "variety" extended beyond Christians. In his autobiography Jefferson made it clear when referring to those who had wanted to insert before the words "author of our holy religion" the qualifier "Jesus Christ," that they were outvoted precisely because the protection offered by the statute "was meant to comprehend . . . the Jew, the Gentile, the Christian, the Mahometan, the Hindu and infidels of every denomination." Remarkably, this pluralism was reaffirmed during the administration of John Adams, when in the treaty made with the bey of Tripoli in November 1796 concluding hostilities, article XI, written by Jefferson's friend, the poet-diplomat Joel Barlow, declared that "as the United States is not in any sense founded on the Christian religion it has no character of enmity against the laws, religion and tranquillity of Mussulmen." A pity, then, that apparently the translation into Arabic failed to convey the forthrightness of that profession, which certainly would have come as news to the Maghrebi rulers (and still would, today). But the treaty in its entirety passed muster in Congress with no votes against it, and the religiously inclined President Adams signed it in 1797.

Three years later, however, Adams was happy enough to run for reelection with the help of a smear campaign designed to represent Jefferson as a Jacobinical atheist. "GOD or JEFFERSON AND NO GOD" ran the flyers, and Federalists like John Mitchell Mason said it would be "a crime never to be forgiven for the American people to confer the office of chief magistrate upon an open enemy of religion." The result would be the enthronement of the "morality of devils, which would break in an instant every link in the chain of human friendship and transform the globe into one equal scene of desolation and horror where fiends would prowl for plunder and blood." Jefferson won a three-way contest anyway after a protracted count in the electoral college. But what is often overlooked is that the forgotten third man in the election, Charles Cotesworth Pinckney, had himself steered one of the most tolerant statutes on religious liberty through the legislature of South Carolina, making that one of the few states where Jews could indeed hold public office, not an academic point given the presence of a lively community and hand-some synagogue in Charleston.

In the end, did the Jefferson–Madison view prevail across the United States, concentrated as it was in the First Amendment, the crystal residue in the alembic of so much fiery debate? Not exactly. The "establishment clause" merely bound the federal government, leaving states like Massachusetts to create a government that was aggressively invested in the patrol of religion and public morals. Notwithstanding Madison's sponsorship of the amendment in late 1789 and his handling of the revisions, Virginia was actually one of the few states not to ratify the First Amendment on the grounds that it offered only "inadequate" protection against the dominance of a single sect. But throughout the nineteenth century, those still excluded from public office—espe-cially Jews and Catholics—could sue under the terms of the First Amendment and often won.

As for President Jefferson, he was happily unrepentant, knowing that the Virginia statute in particular gave encouragement to those elsewhere in the country who would now campaign for their states to follow its example. He was especially happy to receive, on New Year's Day morning 1802, from the Massachusetts Baptist preacher John Leland, a gift of a 1,200-pound bright red Cheshire cheese, made by the grateful farmers of Cheshire, Massachusetts, from the milk of 900 local cows, every one of them, Leland promised the president, good "Republican cows." The "Mammoth Cheese" was cut open later that morning, and in the afternoon, the happy Jefferson penned a letter to the Baptists of Danbury, Connecticut, also engaged in bringing the spirit and letter of the Virginia statute to their state. "Believing with you that religion is a matter which lies solely between Man & his God," the president wrote that he contemplated the First Amendment with "sovereign reverence," establishing as it did "a wall of separation between church and state." In his first draft of the letter (for Jefferson seldom dispatched anything in a single draft) he had written "eternal" before "wall." But Jefferson knew full well that even, or especially, in the United States nothing was eternal.

18. National Sin

So there was Thomas Jefferson, president no longer, comfortably ensconced in his study, patented rotating reading stand on the desk, still working to make America rational. Good luck; it never hurts to try. And he was tenacious in this enterprise, sensing the moment might never come again; that out on the frontier where he imagined sturdy yeoman farmers to be building the nation, men and women were being told by circuit-riding preachers and in thunderous camp revivals to let go of their reason for the Lord, to open themselves to his Light, to shake and shiver when it pierces the quick. But at Monticello there was no quick-piercing, rather the cool labor of the mind, at which there was no more resolute toiler than Jefferson, who devoted himself to the capstone of the project of American enlightenment: the University of Virginia. In that sanctum, he prescribed, there shall be no school of divinity, no Sunday services, no chaplain and no chapel. Instead: a rotunda, with an oculus at the top, as in the Roman pantheon, so that the rays of reason may mantle the undergraduates as they go about their studies. Not everyone in the college was delighted by Jefferson's instructions. Accusations of a nest of atheism being introduced into a Christian commonwealth had been made, and the great founder was politely asked if it were not possible after all for those who might wish to enjoy the blessings of worship, whether such assemblies might not be permissible if voluntarily funded? Jefferson relented a mite, only to the point of finding it acceptable that students could pray howsoever and wheresoever they wished, provided their solemnities were conducted beyond the boundaries of the university. Then he returned to his provisions for their scientific instruction.

What is wrong with this picture? Its frame is too narrow. It takes in the prospect from Jefferson's window: the kitchen garden, the botanical pharmacopoeia, the trim fields. But it does not take in those who worked them: slaves, nor the slave quarters, hidden from the house as if entrenched in a Virginian ha-ha. It does not seem to pay much attention to the expressions of faith among the slaves, which were passionate rather than reasonable. For them, Jesus was most certainly the Son of God; the Bible was His Word. They knew exactly what was meant by the sufferings of Christ, endured to save all men and to grant them the hope of salvation, which for them was a matter of body as well as soul. They did not wish to be told that Jesus was but a teacher, not least because in 1819, Mr. Jefferson's Commonwealth of Virginia, that pillar of separation between church and state, made the instruction of slaves, by black or white, illegal, punishable by imprisonment or twenty lashes or both. They heard from slaves in other plantations that sometimes, white "missionaries" (as they called themselves) would come and preach that if the hands respected the master they might in return expect kind treatment. But the slaves knew their Bible well enough to recite from the Gospel of Matthew, which says, as every Christian knew, do unto others as you would be done by for that it is the law and the prophet, and they didn't see that much "do as you would be done by" around those parts, not even in Monticello. So when they could they stole away to Jesus, at night, where they heard tell of the Old Israelites taken from bondage by the Lord Almighty, and they sang (way down low lest they be found out) after their own manner "I 'lieve I'm a chile of God, and this ain't my home, cos heaven's my aim . . ."

Much farther away, in a country of white ash and standing corn, at the Oneida Institute in northwest New York, its upright president, Reverend Beriah Green, the author of *The Bible Against Slavery* and *The Chattel Principle: The Abhorrence of Jesus Christ and the Apostles or No Refuge for American Slavery in the New Testament*, singled out Jefferson for castigation by simply quoting him, chapter and verse. In his *Notes on the State of Virginia*, Green reminded his readers, Jefferson had described slavery as "the most unremitting despotism on the one part and degrading submissions on the other . . . I tremble for my country when I reflect that God is just: that his justice cannot sleep for ever . . . the Almighty has no attribute which can take side with us in such a contest—But is it possible to be temperate and to pursue this subject through the various considerations of policy, of morals, of history natural and civil?"

No, it is not, Beriah Green declared, not if one were a true Christian. And so thought all the seers and prophets of the Second

Great Awakening, then burning its way through New England, the
Ohio Valley, western Pennsylvania, the Adirondacks and Appalachians
down through Kentucky and Tennessee. The awakening was from the
torpor of formal church decorum and doctrine; the sense of church
as a building with doors that opened and shut at appointed times.
The awakeners wanted twenty-four-hour Christians. They wanted
men—and especially women, who they felt had not been brought
fully within the power of the gospel—to be riven and shriven. And
the notion that public matters were off limits to the religious was
cowardice, an indolence of the soul. There could be no possibility of
moderation on what the most eloquent of them all, Charles Grandison
Finney, called "the national sin": slavery. Merely for church people
and its ministers to stay silent on such matters, on the grounds that
it was not a spiritual matter, was to betray Christ's teaching and any
possibility of America's redemption from the damning iniquity. "Let no
man say," wrote Finney in one of his lectures for the *Oberlin Evangelist*,
in the town where he was professor of divinity and had opened the
doors of the college to blacks, "that ministers are out of their place in
exposing and reproving the sins of the nation. The fact is that ministers and all other men not only have a right but are bound to expose
and rebuke the national sins. We are all aboard the same ship. As a
nation our very existence depends upon the correct moral conduct of
our rulers . . . shall ministers be told, shall any man be told, that he
is meddling with other men's matters when he reproves and rebukes
the abominations of slavery?"

Finney would have had no truck with being told by people like Joel
Barlow that the United States was not a Christian nation. Except that
he agreed that a country so steeped in blood could not yet qualify, he
hoped that his entire life would be devoted to making it so. Finney,
an elongated wizard of a man, six feet three, with seemingly elastic
arms that he would extend out from his cuffs up to the heavens, or
imploringly over the wide-eyed tear-stained faces of reprobates electri-
fied by his lightning bolts of rhetorical ferocity; Finney with the startling periwinkle blue, slightly exophthalmic eyes, with pinprick pupils,
swiveled to great effect like an evangelical chameleon; Finney of the
chiseled cheekbones, high-domed cranium, and a diapason of a voice
that seemed more potent than the most melodious pipe organ; Finney
could make women and men from Cincinnati to Memphis faint and

tremble with the desire to be disembodied and reborn, right there and then in the dewy Ohio fields.

He was not the first of the great evangelical thunderers. George Whitefield, John Wesley, and Jonathan Edwards had been artists of sacred despair and joy two generations before. But Finney had something they did not: a touch for democracy. He was the Andrew Jackson of the soul's ardor. He wanted crowds; he loved crowds; and he gave them the spectacle of terror and the thrill of mercy. Finney had an almost tribal instinct for the excitement of sacrifice: a show of sinners who would be brought to redemption thumbs down, thumbs up, just like at the Colosseum. And in between, props decorated his theater of doom and rescue: an "anxiety bench" where those in trepidation of losing their soul could sweat it out, eyes tight shut while the congregation looked on for telltale tics of redemption. Finney was horribly good at this precisely because he was no Elmer Gantry, no charlatan, but someone who genuinely believed he was a liberator: the emancipator of ordinary men and women who would otherwise be doomed to the "old" Calvinist view of subjection to a preordained fate. There was, he thought, something wretchedly un-American about such passivity. In its place he would supply something more natural to the Yankee: bootstrap salvation. Americans already had the will to make money; he would give them something even more precious: the irresistible urge to stand up and be saved.

A self-educated farm boy from Connecticut, Finney had a notion he would be an attorney. But it did not take him long to realize that such a life would be too narrow a cage for zeal such as his. A traveling Presbyterian laid hands on him, and he came quickly to his understanding that frontier Americans, accustomed to believe they were makers of their own destiny, would respond warmly to the same prospect of determining, through acts of individual will, the fate of their souls. To effect this, though, could not be the work of a mere morning's church service. Camp revivals expanded sacred time: the holy assembly lasting days, improvised sermons that could be fleet minutes, or hours on end; great moving walls of hymn and jubilation; outbreaks amid the crowds of joy and trembling as the living spirit broke forth from the dry carnal husk; the errant shaking with glee as they were brought to the fold; and the long arms of Charles Finney gathered them to the shepherd. Ah, he loved to see stumbling sinners draw nigh to the enveloping merciful love of Jesus. It drenched him with sweet gratification.

But once on this road to redemption, there could be no compromise, no shilly-shallying with what Finney called "hindrances to revival" of which the heaviest was the National Sin. That anyone could possibly remain silent on this odium, who could imagine it was but a private matter, was incomprehensible to Finney, who in Lecture XV of his *Lectures on Revivals* called slavery "pre-eminently the *sin of the church.*" Finney made it known that he would refuse Communion to any slave owners and demanded that the church "*take right ground in regard to politics.*" This did not mean, he explained, forming a Christian political party but making sure that only honest men, men who would not be silent in the face of the abomination, would be supported. "Christians have been exceedingly guilty in this matter," he wrote. "But the time has come when they must act differently or God will curse the nation and withdraw his spirit."

On the other hand, Finney the professor (and later president of Oberlin) was nervous about committing students and flock to militant organization themselves. Others in the church were not so selectively demure. Often, their moment on the road to Damascus took place on the shifting borderland into which the slave economy was moving, only to collide with the Christian furies in the North.

In the late summer of 1822, the Reverend James Dickey was returning from a family excursion through the Kentucky Barrens to his home near Paris. The jaunty sound of fiddling came to them over the prairie grass, and the Dickeys assumed they were about to meet up with some sort of festive parade or a "military fair." Instead what they saw was a procession of about forty manacled black men and women, "the foremost couple furnished with violins" and another forced to raise, from handcuffed wrists, the Stars and Stripes, which waved above their bowed heads; glee gone mad. The minister learned that this public ordeal was a collective punishment meted out to slaves as a result of one of them, a woman, having physically resisted being shipped off (almost certainly taken from her husband and children) and who had had the temerity to raise a hand against her purchaser. "My soul was sick" at the spectacle, Dickey wrote. "As a man I sympathized with suffering humanity. As a Christian I mourned over the transgressions of God's Holy Land and as a republican I felt indignant to see the flag of my country thus insulted. I could not forbear from addressing the driver: 'Heaven will curse that man who engages in such traffic.'"

We know about Dickey's confrontation with the grotesque procession near Paris, Kentucky, because it was included in the book that lit thousands of fires up and down the United States: John Rankin's *Letters on Slavery*, now one of the least read but most trailblazing of all the early abolitionist works. Rankin, a Presbyterian, who established himself, after many run-ins with slaveholders and mobs, in Ripley, Ohio, at the top of a hill where he could light a beacon to guide fugitives on the Underground Railroad toward his asylum, had made the painful discovery that his own brother in Kentucky had become an owner of slaves. The *Letters* were written in an effort to persuade him to reject the iniquity but also to set out all the reasons why slavery, which "hangs like the mantle of night over our republic and shrouds its rising glory," was an offense against God, "an unhallowed thing . . . fraught with the tears and sweat and groans and blood of hapless millions of innocent and unoffending people."

Everything proper to a Christian nation—the sanctity of the family, instruction in Scripture, the nobility of free labor—was defiled by slavery. Sunday schools had been attacked in Kentucky, the teachers and pupils stoned, and in some places slaves were barred from worship lest they Get Ideas. "I have seen the Preacher and Elder bow their knees around the family altar," wrote Rankin, "while their poor slaves remained without as if like mere animal herds they had no interest in the morning and evening sacrifices." Still worse the ubiquitousness of mulattos testified to the depravity that whites forced onto defenseless slave women. Thus slavery "is the very sink of filthiness and the source of every hateful abomination. It seems to me astonishing that any government, much more that of the United States, should sanction such a source of monstrous crime."

Christian quietism for ministers like John Rankin was unthinkable even if activism came with risks. One night, a local mob came to attack his house on the hill, and only Rankin's six sons, all armed with guns, saved the day and their father from being torn to pieces or lynched. It was these kinds of scenes that frightened Charles Finney, who preferred to keep his denunciations to the pulpit and the lecture hall. But his own protégé, Theodore Weld, pushed such timidity aside, marrying up the revival fervor of the camp revivals, held where the formal church dared not go—in fields and woods—to the spiritual soldiering of abolitionism. Weld's band of "the 70" were not cloistered men of the cloth but hard-bitten, zeal-driven circuit riders, the shock troops of the new crusade, ranging far and wide on the western frontier, men with the Bible in saddlebag next to the shotgun. It was men and women (for Weld married the impassioned and determined Angelina Grimké) who, by equating a Christian life with the attack on slavery, made sure that as America moved west, slavery did not automatically move with it; that at the very least there would be a battle over bodies and souls, although no one yet remotely imagined that the eventual battle would cost more than half a million American lives, white and black.

The fervor of the abolitionist evangelicals complicates the way we might feel about the "wall of separation" erected by the Virginia statute and the First Amendment between morality and politics. Of course it was entirely possible to arrive at an abhorrence for slavery from rationally derived ethics; the degradation of man to commodity, the violation of natural right to sovereignty over person, and so on. Historically, though, both in the early nineteenth century, and again in the 1960s, the force of shame directed at slaveholders and segregationists was religious. Realistically, it is unlikely that the propagation of Enlightenment views of humanity would have swayed millions of nineteenth-century white Americans against slavery. After all, such moral principles convinced Jefferson and Patrick Henry of the infamy of the institution, but still failed to move them to liberate their own slaves, so what hope was there of persuading less high-minded southerners to make sacrifice of their property, or what Henry described as "inconveniencing" himself? Both in the 1830s and 1840s, and then again in the 1960s, it was the determination of the Rankins and Finneys, and Fannie Lou Hamers, to cross the line between religion and politics and appeal to the country's Christian conscience that brought white Americans into brotherhood with persecuted blacks. For secular humanists (like this writer), this is an awkward historical truth to acknowledge, accustomed as they are to equating evangelical fervor with illiberal reaction. The abolitionist argument that some enormities were so vicious that they had to be made accountable to the principles of the gospel, even if that meant breaching the establishment clause of the First Amendment in the interests of a higher good, is not altogether different from the way Right to Life evangelicals argue today. History sets such snares to make us think harder.

If appeals to his Christianity had any effect on Thomas Rankin, the slave-owning brother of John, no record survives of a repentance. So the likelihood is that Thomas took the moral drubbing no more warmly than Henry Meigs did from his brother Montgomery. If Thomas Rankin's dander was up, he would have been more in tune with much of the plantation South in the 1830s, which felt itself under heavy siege from abolitionist righteousness. In July 1835 a shipload of thousands of abolitionist tracts and treatises was unloaded from the steamer *Columbia* onto the docks at Charleston Harbor, a cargo of fervor dispatched by the New York evangelical Lewis Tappan. The following night, a bonfire was made of them. And the sense of resentment that the gospel was being manipulated by the "fanatics" (as slave society supposed) was particularly fierce because at least two slave rebellions (one, Denmark Vesey in 1822, nipped in the bud; the other, Nat Turner in 1831, brutally successful) involved leaders who claimed the inspiration of holiness. Vesey had been one of the founders of the African Episcopal Methodist Church in Charleston, and his chief coconspirator was "Gullah" Jack Pritchard, who also preached in the church and was said to have married up African conjuring to Christian liturgy. At least as important for the jumpy guardians of order in South Carolina as hanging the ringleaders and thirty-three others was the closure and destruction of the Methodist church, now demonized as a cover for insurrection. Nat Turner, who killed fifty-seven whites in Virginia before being captured, and was known as "'the Prophet" to his slave followers, took authority for his revolt in personal dreams of an impending struggle between Christ and Antichrist and was famous for the intensity of his piety and prayers.

At least as frightening to the white South as evidence of actual rebellions were incendiary calls for liberation coming from the North. The most eloquent of those was the 1829 revolutionary tract of the free black tailor David Walker, *One Universal Cry*. A ferocious attack on the institutionalized hypocrisy of the United States and its canonical founding texts—the Declaration of Independence and the Constitution—the gravamen of Walker's charge was that by tolerating and profiting from slavery, America had revealed itself a godless, unchristian nation. This had been the message of Rankin, Finney, and Weld, too, but coming from the pen of a black, it had much greater power to enrage and terrify. Thus was born in the mind of the paranoid South an ultimately unholy alliance of violent black rage and naive white "fanaticism." Much the same would be said of the civil-rights alliance of "interfering Jews and priests" and "uppity niggers" in the 1960s. It became suddenly urgent to foreclose opportunities for slaves and free blacks to read seditious literature coming from the North. An Alabama Baptist, William Jenkin, who went on to become a minister, was one of those who devoured illicit abolitionist writings but was terrified of being discovered. "I had rather been caught with a hog than a newspaper," he later said, "because for the hog I was likely to get a whipping but for the newspaper I might get a hanging." Even the places where slaves might get the rudiments of an education were now suspect. Sunday schools became a special hate target of organized mobs; in Kentucky teachers and students were beaten, the buildings wrecked or burned.

But some on the pained receiving end of the barrage of antislavery agitation—southern Baptists, Methodists, and Congregationalists—thought they couldn't burn, hang, or flog their way out of trouble (though it helped). A two-pronged counterattack was needed. In the first instance they pulled out their Bible, as they had been regularly doing since the rise of abolitionism in the 1770s, to demonstrate that if the Hebrews had bondsmen and women, then the Almighty in his wisdom must have condoned if not actually designed the practice. Then came the argument that the slaves were from such a savage culture that they were far better off in the rice fields of Savannah than the African savanna.

But beside dueling with the white evangelical fanatics, a number of churchmen came to believe that it was futile to prevent blacks from going to Christianity, for that seemed to encourage the rise of Denmark Veseys and Nat Turners, uncontrollable priests and prophets who would turn the gospel into a license to murder their masters. Better perhaps to take charge of the Christianization and education of the slaves and use the gospel to instill precepts of obedience, respect, and humility, in return for which, benevolent treatment—not to mention the blessings of salvation—would be ministered to them. A plantation "mission" movement was thus born.

Owner of rice and cotton plantations in Midway, Georgia, not far from Savannah, Charles Colcock Jones felt especially called to this mission to create a mutually benign, God-obliging slave world. Jones felt this way because he had been educated in the North, at Phillips Academy and then at Princeton seminary (where in fact both defenders and attackers of slavery could be found). The young Jones passionately believed in the evil of the system on which his own wealth was based, and that, eventually, it would be wiped out. But correspondence with his cousin Mary, whom he ended up marrying, and his own sense of a local tragedy for black and white alike, should a head-on collision be accelerated, persuaded him of the need for a middle road. Returning to Midway, and the pretty church on the green where he had made his first confession in 1821 and whose pulpit he now occupied, only confirmed him in this course. The slaves, degraded by their oppression as they were, were not yet ready for liberty but needed an apprenticeship in moral and religious "amelioration" before this could be safely granted them. Charles Colcock Jones and his fellow plantation missionaries would devote their lives to bringing this about: seeking out and berating masters who were inhumane and brutal; and providing the slaves with education, medicine when needed, for the body and soul. Thus they would be elevated for—eventual—freedom. It comes as no surprise to learn, then, that the sister of one of Jones's best friends in the North was Harriet Beecher Stowe. And if you go to Midway, Georgia, opposite the great magnolia shading Colcock Jones's lichen-covered tomb, you will find two doors offering entrance to the white church: a grand one for white worshippers and a mean one, off to the side, for all those slaves to whom Jones was ministering. For many, South and North, slave and free, this indignity was at odds with the precept of universal admissibility to the merciful grace of God. They didn't want to be let into heaven through a stooping door. They wanted a church of their own.

19. Jarena Lee

I was born 11 February 1783 at Cape May, state of New Jersey. At the age of seven years I was parted from my parents and went to live as a servant maid with a Mr. Sharp at the distance of about sixty miles from the place of my birth.

My parents being wholly ignorant of the knowledge of God had not therefore instructed me in any degree in this great matter. Not long after the commencement of my attendance on this lady she had bid me do something respecting my work which in a little while after she asked me if I had done, when I replied—Yes—but this was not true.

At this awful point in my early history the Spirit of God moved in power through my conscience and told me I was a wretched sinner . . .

In the year 1804 it so happened I went with others to hear a missionary of the Presbyterian order preach. It was an afternoon meeting but few were there, the place was a schoolroom; but the preacher was solemn . . . at the reading of the Psalms a ray of conviction darted into my soul. These were the words composing the first verse of the Psalms for the service: "Lord I am vile conceived in sin / Born unholy and unclean . . ."

This description of my condition struck me in the heart and made me feel in some measure the weight of my sins . . . but not knowing how to run immediately to the Lord for help I was driven to Satan . . . and tempted to destroy myself. There was a brook about a quarter of a mile from the house in which there was a deep hole where water whirled about around the rocks; to this place it was suggested I must go and drown myself. At the time I had a book in my hand; it was on a Sabbath morning about ten o'clock; to this place I resorted where on coming to the water I sat down on the bank and on my looking into it, it was suggested it would be an easy death. It seemed that someone was speaking to me saying put your head under, it will not distress you. But by some means of which I can give no account my thoughts were taken entirely from this purpose when I went from this place to the house again. It was the unseen arm of God that saved me from self-murder.

1809

I went to the city of Philadelphia and commenced going to the English church the pastor of which was an Englishman named Pilmore . . . But while sitting under the ministration of this man which was about three months . . . it appeared there was a wall between me and the people which was higher than I could ever see over and seemed to make this impression on my mind, this is not the people for you . . . But on returning home I inquired of the head cook of the house of the rules respecting the Methodists as I knew she belonged to that society . . . on which account I told her I should not be able to abide such strict rules not even for one year—however I told her I would go with her and hear what they had to say.

The man who was to speak in the afternoon was the Reverend Richard Allen of the African Episcopal Methodists in America. During the labors of that afternoon I had come to the conclusion that this is the people to which my heart unites and it so happened that as soon as the service closed he invited such as felt the desire to flee the wrath to come, to unite with them and I embraced the opportunity. Three weeks from that day my soul was gloriously converted to God under preaching. The text was barely pronounced which was "I perceive thy heart is not right in the sight of God" when there appeared to view in the center of my heart one sin and this was malice against one particular individual who had strove deeply to injure me which I resented. At this discovery I said LORD I forgive every creature. That instant it appeared to me that a garment which had entirely enveloped my whole person even to my fingers' ends split at the crown of my head and was stripped away from me passing like a shadow from my sight—when the glory of God seemed to cover me in its stead.

That moment, though hundreds were present, I did leap to my feet and declare that God, for Christ's sake, had pardoned the sins of my soul. Great was the ecstasy of my mind for I felt not only that the sin of malice was pardoned but that all other sins were swept away together. That day was the first my heart believed and my tongue made confession unto salvation—the first words uttered, a part of that song which shall fill eternity with its sound was glory to God. For a few moments I had the power to exhort sinners and to tell of the wonders and of the goodness of Him who had clothed me in salvation. During this the minister was silent until I felt the duty of my soul had been performed.

1814

Between four and five years after my sanctification, on a certain time an impressive silence fell upon me and I stood as if someone was about to speak to me yet I had no such thought in my heart. But to my utter surprise there seemed to sound a voice . . . which said to me, "Go preach the gospel!" I immediately replied, "No one will believe me." Again I listened and again the voice seemed to say, "Preach the gospel; I will put words in your mouth and will turn your enemies to become your friends." At first I supposed that Satan had spoken to me for I had read that he could transform himself into an angel of light for the purpose of deception. Immediately I went into a secret place and called upon the Lord to know if he had called me to preach and whether I was deceived or not; when there appeared to my view the form and figure of a pulpit with a Bible lying thereon which was presented to me plainly as if it had been literal fact.

In consequence of this my mind became so exercised that during the night following I took a text and preached in my sleep. I thought that there stood before me a great multitude while I expounded to them the things of religion. So violent were my exertions and so loud my exclamations that I awoke from the sound of my own voice which also awoke the family of the house where I resided. Two days after I went to see the preacher in charge of the African Society who was the Reverend Richard Allen . . . to tell him that I felt it was my duty to preach the gospel . . . But as I drew near to the street where his house was which was in the city of Philadelphia my courage began to fail me so terrible did the cross appear it seemed that I should not be able to bear it . . . several times on my way I turned back again but often I felt my strength renewed . . .

I told him that the Lord had revealed to me that I must preach the gospel . . . But as to women preaching he said that our Discipline knew nothing at all about it—that it did not call for women preachers. This I was glad to hear because it removed the fear of the cross from me—but no sooner did this feeling cross my mind than I found that the love of souls had in a measure departed from me, that holy energy which burned within me as in a fire began to be smothered . . .

If a man may preach because the Savior died for him why not the woman seeing he died for her also? Is he not a whole Savior instead of a half one as those who hold it wrong for a woman to preach would seem to make it appear? If to preach the Gospel is the gift of heaven, comes by inspiration solely, is God straitened? Must he take the man exclusively? May he not, did he not and can he not, inspire a female to preach the simple story of the birth, life, death and resurrection of our Lord and accompany it too with power to the sinner's heart? . . .

In my wanderings up and down among men, preaching according to my ability, I have frequently found families who told me that they had for several years been to a meeting and yet, while listening to hear what God would say *by his poor female instrument, have believed with trembling—tears rolling down their cheeks, the sign of contrition and repentance towards God.*

1821

It was now eight years since I had made application to be permitted to preach the gospel during which time I had only been allowed to exhort . . . the subject now was renewed afresh in my mind; it was as a fire shut up in my bones. During this time I had solicited of the Reverend Richard Allen who . . . had become Bishop of the African Episcopal Methodists in America, to be permitted the liberty of holding prayer meetings in my own hired house and of exhorting as I found liberty, which was granted me . . .

Soon after this the Reverend Richard Williams was to preach at Bethel Church—where I with others was assembled. He entered the pulpit, gave out the hymn which was sung and then addressed the throne of grace . . . the text he took is Jonah, 2nd chapter, 9th verse "Salvation is of the lord." But as he proceeded to explain, he seemed to have lost the spirit; when in the same instant I sprang, as by altogether supernatural impulse, to my feet to give an exhortation on the very text which my brother Williams had taken . . .

I told them I was like Jonah; for then it had been nearly eight years since the Lord had called me to preach but that I had lingered like him and delayed to go to the bidding of the Lord . . .

During the exhortation God made manifest his power in a manner sufficient to show the world that I was called to labor according to my ability and the grace given unto me . . .

At the first meeting . . . at my uncle's house there was with others who had come from curiosity to hear the woman preacher, an old man who was a deist and who said he did not believe the colored people had any souls— he was sure they had none. He took a seat very near where I was standing and boldly tried to look me out of countenance. But as I labored on in the best manner I was able, looking to God all the while it seemed to me I had but little liberty, yet there was an arrow from the bent bow of the gospel and fastened in his till then obdurate heart. After I had done speaking he went out and called the people round him, said that my preaching might seem a small thing yet he believed I had the worth of souls at heart . . . he now seemed to admit that colored people had souls whose good I had in view . . . He now came into the house and in the most friendly manner shook hands with me saying he hoped God had spared him to some good purpose. This man was a great slave-holder and had been very cruel, thinking nothing of knocking down a slave with a fence stake or whatever might come to hand. From this time it was said of him he became greatly altered in his ways for the better . . .

The Lord was with me, glory be to his holy name. I next went six miles and held a meeting in a colored friend's house . . . and preached to a well-behaved congregation of both colored and white. After service I again walked back which was all twelve miles in the same day.

1822

I returned to Philadelphia and attended meetings in and out of the city . . . I felt a greater love for the people than ever.

In July I spoke in a schoolhouse to a large congregation . . . here we had a sweet foretaste of heaven—full measure and running over—shouting and rejoicing—while the poor errand bearer of a free gospel was assisted from on high. I wish my reader had been there to share with us the joyous heavenly feast . . .

I was sent for by the servant of a white gentleman to hold a meeting in his house in the evening. He invited the neighbors, colored and white, when I spoke according to the ability God gave me. It was pleasant to my poor soul to be there—Jesus was in our midst . . .

I next attended and preached several times at a camp meeting which continued five days. We had Pentecostal showers—sinners were pricked to the heart and cried mightily to God for succor from impending judgment and I verily believe the Lord was well pleased at our weak endeavors to serve him in the tented grove.

1823

In the month of June 1823 I went on from Philadelphia to New York with Bishop Allen and several elders to attend the New York Annual Conference of our denomination where I spent three months of my time . . . On 4 June I spoke in the Asbury Church from Psalms chapter 33. I think I never witnessed such a shouting and rejoicing time . . . The spirit of God came upon me I spoke without fear of man . . . the preachers shouted and prayed and it was a time long to be remembered.

1824

In company with a good sister who took a gig and horse I travelled about 300 miles and labored in different places. Went to Denton African Church and on the first Sabbath gave two sermons. The church was in a thriving, prosperous condition and the Lord blessed our word to our comfort . . . by request I also spoke in the Old Methodist Church in Denton which was full to overflowing. It was a happy meeting. My tongue was loosened and my heart warm with the love of God.

I have travelled in four years 1,600 miles and of that I walked 211 and preached the kingdom of God to the falling sons and daughters of Adam counting it all for the sake of Jesus . . .

In Milford . . . at night the people came in their carriages from the country but were disappointed for I spoke in a colored church. The doors and windows were open on account of the heat, but were crowded with people; pride and prejudice were buried. We had a powerful time. I was quite taken out of myself; the meeting held till daybreak, but I returned to my home. They told me sinners were converted, backsliders reclaimed, mourners comforted . . . Then they wished us to stay until next night to preach again but I thought it best to leave them hungry.

I made an appointment at a place called Hole in the Wall, it was a little settlement of colored people but we had no church but used a dwelling house and had a large congregation. I had no help but an old man, one hundred and odd years of age, he prayed and his prayers made us feel awful, he died in the year 1825 and has gone to reap the reward of his labor . . .

Although in a slave state we had every thing in order, good preaching, a solemn time long to be remembered. Some of the poor slaves came happy in the Lord, walked twenty to thirty and from that to seventy miles to worship God. Although through hardships they counted it all for joy.

1827

I went to Baltimore with the bishop and enjoyed great preaching. We had a good time rejoicing in the Lord. I left them for Albany . . . Glory to God . . . the people in Niagara seemed to me to be a kind and Christian-like people. The white inhabitants united with us and ladies of great renown. The slaves that came felt their freedom, began to see the necessity of education . . . I . . . crossed the lake from Buffalo to Fort George and spoke about eight miles from there; it was cold and snowed very fast—it was four o'clock in the afternoon—the congregation had been there and gone. We were on a sleigh and the driver got lost; we were all brought up in a swamp among fallen tree tops but we turned round and found a house and lodged all night . . . after I spoke to the people I left them and made an appointment for the Indians; two of the chiefs called at where I stopped to see me. I asked them to pray for us,

they complied but in their own tongue. I felt the power of God in my heart.
That year I travelled 2,325 miles and preached 178 sermons.

On and on went the inexhaustible road warrior, Jarena Lee, walking
and riding, going by gig and wagon, by steamboat and railroad, sleigh
and mule train; lugging her portable lectern with her, through Ohio
and Illinois, New York and Delaware, Massachusetts and across the
border in Ontario, through every corner of Pennsylvania, and into
the slave state of Maryland and the half-slave city of Washington;
teaching in schools, exhorting in field and forest, in camp revivals and
Love Feasts, comforting the dying, of which there were a good many
in the terrifying cholera year of 1831 when in New York 160 perished
horribly every day; clocking a record 692 sermons in 1835, the year of
the great abolitionist push in the South. In her prodigious diary—one
of the great unread black narratives—can be heard the exultation of
those meetings and services: the shouting and clapping, the sobbing
and singing, in city chapels and backcountry churches. She had begun
as a little slave girl, struck by guilt when she told her mistress a
fib, saved from drowning by a sense of God's help, and become
an authentic American phenomenon, preaching to overflowing
congregations, the first, in her way, of the great black orators. She
was in her own person what W. E. B. DuBois in *The Souls of Black
Folk* identified as the first kind of true black leader: the Preacher
and the Teacher. And the astonishing thing is, that although revolutionary in her way, Jarena was by no means alone. By the time
that she hung up her lectern in the 1840s, there was a whole black
sisterhood of traveling preachers, defying mobs and magistrates, male
prejudice and skeptical indifference: Amanda Smith, "Elizabeth, a
Colored Minister of the Gospel," Mary McCray in Kentucky, and
Bethany Veney, "Aunt Betty."

Though they were very much mistresses of their own vocation,
all these women needed some help from the Protestant black church
which was becoming powerfully entrenched in the cities, as Jarena
and Bethany Veney were going on their travels. Jarena's most significant conversion was the famous Richard Allen, the black Methodist
bishop of the Mother Bethel Church in Philadelphia. Once he was
persuaded that Jarena was a force to be reckoned with, Allen took her
along on his own preaching tours to New York and Baltimore, both

slave cities, where to be a black preacher at all was to invite assault and sometimes not just the vocal kind.

But the work of Allen and his counterparts in Savannah—Andrew Bryan and later Andrew Marshall—was to create an entire black dominion of the saved, out of reach of both the slaveholders and the patronizing plantation missionaries. In towns like Savannah, the local citizenry often had plenty of second thoughts as to whether it was such a good idea to have crowded black churches instructing their flocks. And even though Bryan in particular always claimed he was no threat to the institution of slavery, he was viciously beaten up on the streets of the town until his master Jonathan Bryan found him a disused rice barn in which he could hold services. The history of these black Baptists of the South had begun as a flight to freedom, when thousands of them had taken advantage of the British offer, made during the Revolutionary War, to give freedom to escaped slaves from rebel plantations who would serve the king. When the British cause was lost, many left with the royal army, north to Nova Scotia or to the Caribbean. But those who stayed found mutual support in worship and somehow hung on in their own churches amid intimidation and poverty. Though those first heroic generations of Savannah Baptists are buried (needless to say) in a separate lot in Laurel Grove cemetery, they were nonetheless the origin of the freedom church in the South, the first encampment of an army that would have to fight its way through war and a century of Jim Crow segregation before it got anywhere near the Promised Land.

In the histories I read at school decades ago, American slaves before the Civil War were never capable of shaking off their chains, mental as well as physical, except by flight. Their imprisonment in the system of degradation was so total that the best they could do, other than become a fugitive, was to wait for deliverance at the hands of white evangelical abolitionists like John Rankin, Charles Finney, and Theodore Weld. The notion that from North to South, New York to Baltimore to Savannah, there existed a free black church, numerous (1,400 alone at Bethel Church, Charleston, where Denmark Vesey's sect had been uprooted, maybe 400,000 through the prewar South according to W. E. B. DuBois), a church that was vigorous, well disciplined, restlessly active at saving both souls and bodies, with dauntless outriders like Jarena Lee before whom whites and blacks quailed, rejoiced, and celebrated—all this would have seemed far-fetched. But a look at the rich trove of memoirs and spiritual autobiographies of that period, and in anthologies collected later in the century like the *Cyclopedia of the Colored Baptists of Alabama*, reveals an entire world of daring self-determination.

And even where it was impossible to organize in the manner of the city black churches of Baptists and Methodists, slave religion found a way to shake off the yoke. The historian Albert Raboteau has collected evidence of a religious counterculture existing under the noses of the plantation overseers and missionaries. Their real worship, not the permitted decorum of churches like Colcock Jones's at Midway, but one inflected with the body movements, chants, and storytelling of Africa. When they could do so more or less openly in fellow slaves' cabins, the brothers and sisters worshipped as they wanted, as Jarena Lee often heard, with the power of the ring shout, call and response, handclapping, and the exclaimed A-MEN! But where such worship was suspected of instigating some sort of quasi rebellion, the slaves resorted to "hush harbors," in woods, ravines, and gullies, where the slaves would "steal away to Jesus." The ex-slave Washington Wilson explained that "when de niggers go round singin 'Steal away to Jesus' dat mean dere gwine be a religious meetin dat night. De masters . . . didn't like dem religious meetins so us natcherly slips off at night down in de bottoms or somewhere. Sometimes us sing and pray all night." In Prince George County, Virginia, Peter Randolph described the boughs of trees bent over and held in place to fashion a natural chapel, slaves openly recounting what they endured the past week and—just as in Raven with the Primitive Baptists—breaking off to give each other the handshakes of Christian fellowship. When they had a visit from a traveling black preacher, probably more down at heel and ragged than Jarena Lee, he would also deliver a touch of African American poetry when recounting the Exodus or the sufferings of Christ at the Passion, stories that meant something to slaves that no white oppressor or for that matter benefactor could understand in quite the same way: "I see the sun when she turned herself black, I see the stars a-fallin from the sky and them old Herods coming out . . . and then I knew 'twas the Lord of Glory."

The temptation to raise the voice in praise and hope was so strong that wet quilts or iron pots turned upside down were used as mutes (how painful that must have been for a world that lived to set the voice free). Another slave, Anderson Edwards, reported "we didn't have no song books, and the Lord done give us our song." Sometimes, though, the music could be sung in the open, in the summer after work and supper at bonfires lit "to keep the mosquitoes away and listen to our preachers preach half the night." There would be singing and testifying and shouting. And then, the "Frenzy" that DuBois describes, "the silent rapt countenance or the low murmur and moan to the mad abandon of physical fervor—the stamping shrieking and shouting, the rushing to and fro, and wild waving of arms, the weeping and laughing, the vision and the trance." Often others from neighboring plantations would come and join in at the fire. And sometimes the blacks in the midst of their song would turn and notice, at the edge of the circle, white faces lit by the burning wood.

20. The sovereignty of the voice

Thomas Wentworth Higginson, Unitarian minister turned Union army colonel, walked toward the bonfire as was his habit after supper; heading for the shout. In other regiments across the theater of war, taps would have already been sounded; men would be stretching out for the night in their tents. Not here, though, at Beaufort on the sea island of Port Royal off the Carolina coast in 1861. Higginson and his commander and comrade in arms, General Rufus Saxton, had been of one mind. Let the 1st South Carolina Volunteers sing. Had not Cromwell's men sung as they girded themselves for the morning fight? And Higginson was sure that not since the New Model Army had there been such a spirit of religion among soldiers as among his freed slaves. He and brother-abolitionists in the North had long spoken of a "gospel army," but they had meant it metaphorically, Fighting the Good Fight, Christian Soldiers, and so forth. After a while Higginson had found, somewhat to his surprise, the figure of speech disingenuous, shaming, a sign that fighting against slavery would be done *merely* with words and prayers. Higginson, Massachusetts Brahmin, Harvard man, a fixture of the literary world, had become exasperated with the rhetoric. He hungered to smite the despotic enemy hip and thigh, to lay about them with a mighty whack. His sermons at Newburyport turned intemperate, so complained the local citizenry, and it did not help matters that he used the pulpit to castigate the way they saw fit to run their cotton textile businesses, calling on them to feel shame for buying from the South. He was politely warned to moderate. He declined to do so, and Newburyport bid him farewell.

At Worcester Free Church, the people were more inclined to allow Higginson to agitate hard for the Anti-Slavery Association, although his appetite for action could still try tempers. The Fugitive Slave Act of 1850 requiring the return of runaways triggered the creation of a Vigilance Committee organized to thwart seizure. Higginson was one of their most ardent militants, participating in an attempt to storm the Boston Courthouse in order to liberate the ex-slave Anthony Burns. Even this was too remote a gesture for the handsome, tirelessly immoderate Reverend Higginson. He had run for election to Congress as a Free Soil candidate, so the division of republic into free and slave worlds was, for him, the ultimate national question. In Kansas, on the frontier between those versions of America, Higginson swore to be the scourge of the slave hunters, the refuge of the escaping slaves. Known from Wichita to Lawrence as a dangerous traveling preacher man, Thomas kept the outward sign of the cloth but rode with a revolver at his belt; running guns to the blacks and their helpers, getting food to the fugitives, laying on shelter and his own hands (as much in aid of homeopathic cures as blessings), still the smooth white hands of the Harvard divinity student. One night in a saloon (for even God's ministers needed a little warmth against the prairie bitterness) he heard men talking about someone they were going to have to deal with, tar and feather, maybe, but run out of town for sure: a preacher man. "First he has his text," the ringleader looking for like-minded men explained, "and he preaches religion, then he drops that, pitches into politics, then he drops that too and begins about the suffering niggers. Well boys we've got to take care of that." Higginson was unafraid. God and Samuel Colt would shield him from harm; the Lord's work had still to be done. The true fight had scarcely begun. There were those among his bien-pensant friends in the North who frowned on those such as John Brown as a raver, a mischief-maker. But he thought Brown, with his silvery mane and his high exhortations, another such as had come at the time of Oliver: a holy warrior. And Higginson raised secret funds for Brown's raid on the arsenal at Harpers Ferry and when it went awry, raised more money for his hero's defense. After Brown was hanged, Higginson went to see his widow and children at New Elba and offered what poor comfort he could.

Perhaps it was the sense that some outlet must be found for the fighting minister before some serious harm befell him that moved General Rufus Saxton—who was himself a transcendentalist-feminist-abolitionist West Point graduate—to offer Higginson a military posting. And the post was not as chaplain but regimental colonel! The notion was less outlandish than it seemed for after the execution of John Brown, Higginson had removed his cloth and put himself through a second ordination as soldier, teaching himself from manuals of arms, drill, firing, maneuvers. Word had got around, for the circles of the righteous were in communication with each other, and the perfect opportunity seemed to arise in 1861, when a Union expedition to the sea islands off the South Carolina and Georgia coast had liberated tens of thousands of slaves on rice and cotton plantations. Land had been distributed to the ex-slaves in what thus became the first "experiment' in free black farming. Those who wished were offered service in the Union army. With such a force of devoted soldiers, fighting, as Higginson would write, "with a rope about their necks," those islands, Port Royal in particular, could be held as strategic posts between the Carolinas and the Georgia–Florida coast. Saxton was set in command of almost a thousand blacks, transformed from slavery to soldiering almost overnight: the first ex-slave regiment.

So there was Higginson, now thirty-seven years old, still wearing his hair long in the manner of a Boston poet, approaching a cabin roofed with palm and palmetto fronds: the tabernacle of worship, around which a line of men was circling, chanting their songs, clapping their hands, bodies swaying, as they moved. Somewhere a drum was beating, and it wasn't to marching time. Listening to this, Higginson felt lifted into a state of grace. He was beholding the perfect innocence of the church primitive in all its ancient purity.

On the morning of New Year's Day 1863, the soldiers of the 1st South Carolina and the men, women, and children of the islands were assembled by General Saxton to hear President Lincoln's proclamation of their freedom. Saxton had sent steamers to the neighboring islands to fetch the people, and most of them were women, fine-looking, with brilliant African kerchiefs on their head, many carrying children in their arms or holding hands as a crocodile line made its way to an improvised parade ground, beneath overarching live oaks. A little platform had been erected for the speakers and officers, and to Higginson's surprise, some white visitors had arrived in gigs and carriages, parked under the trees in which they sat listening attentively to the proceedings.

After prayers, colors were presented to the black regiments and a South Carolinian doctor who had long since freed his own slaves read "Pres Linkum's" words. "The very moment the speaker had ceased and just as I took and waved the flag which now for the first time meant anything to these poor people, a strong male voice (but rather cracked and elderly) into which two women's voices blended, singing as if by an impulse that could no more be repressed than the morning note of the song sparrow

> My country 'tis of thee
> Sweet land of liberty
> Of thee I sing . . .

"People looked at each other and then at us on the platform to see when this interruption came not set down in the bills. Firmly and irrepressibly the quavering voices sang verse after verse; others of the colored people joined in; some whites on the platform began but I motioned them to silence. I never saw anything so electric. It made all other words cheap. It seemed the choked voice of a race at last unloosed."

Colonel Higginson had found his moment. A good officer to his men, he led them on gunboat raids up the Southern rivers, liberating slaves as they went, getting the proclamation read and understood. On the Edisto in South Carolina, his soldiers saw a water meadow all at once "come alive with human heads . . . a straggling file of men and women, all on a run for the riverside . . . old women trotting on the narrow paths, would kneel to pray a little prayer, still balancing the bundle, and then would suddenly spring up, urged by the accumulating procession behind." "De brack sojer so presumptuous," said one of the freed slaves, "dey come right ashore, hold up dere head. Fus' ting I know dere was a barn, 10,000 bushel rough rice, all in a blaze, den mas'rs great house all cracklin up de roof. Didn't I keer for see em blaze. Lor mas'r, didn't care nothin at all. *I was gwine to de boat.*"

Every moment when he wasn't commanding the liberator boats, or seeing to the orderly welfare of the camp, Higginson devoted to the African American music, already known as "Spirituals"; the "Sorrow Songs" that DuBois described, rightly, as "the most original and beautiful expression of human life and longing yet born on American soil."

He understood just what the songs had meant in slavery; knew that some of them had been jailed in Georgetown, South Carolina, at the start of the war, just for singing:

> We'll soon be free
> We'll soon be free
> We'll soon be free
> When de Lord will call us home

"Dey tink *de Lord* mean for say *de Yankees*," a drummer boy told him, smiling, though they did mean the Lord. But when they sang "Many Thousand Go," there was no ambiguity:

> No more driver's lash for me
> No more, no more
> No more driver's lash for me
> Many thousand go.

Whenever, wherever they could, they sang. One morning it was raining, the last drenching after a night's evil storm. Higginson was concerned for the pickets out in exposed country and walked to the edge of the camp looking for their return like a fretful mother. Then he heard the sound of voices:

> O dey call me Hangman Johnny
> O ho o ho

And there were the soldiers, water streaming from their black rubber blanket-coats, marching into camp, broad smiles and big baritones.

> But I never hang nobody
> Hang boys hang

Transcribing the songs, "the vocal expression of the simplicity of their faith and the sublimity of their long resignation," now became Higginson's obsession, his personal battle against the oblivion he thought could swallow up folk culture. The notebooks tucked inside his coats grew thicker like quilting. "Often," wrote Higginson, "in the starlit evening, I have returned from some lonely ride by the swift river or on the plover-haunted barrens, and entering the camp, have silently approached some glimmering fire, round which the dusky figures moved in the rhythmical barbaric dance the Negroes call a 'shout,' chanting, often harshly, but always in the most perfect time. Writing down in the darkness as I best could with my hand in the safe covert of my pocket—the words of the song. I have afterwards carried it to my tent, like some captured bird or insect, and then, after examination, put it by." The gap between listening and recording meant that, often enough, some words slipped away; others made no sense to Higginson. But he had help filling in those gaps from his black corporal Robert Sutton, "whose iron memory held all the details of a song as if it were a ford or a forest." The melodies he could "only retain by ear" and few can have survived the way the black soldiers sang them. But his collection—some of it published in the *Atlantic Monthly* in 1869—is still a precious document; the first anthology of the Sorrow Songs.

> My army cross over
> My army cross over
> O Pharaoh's army drownded
> My army cross over

Queen Victoria quite liked this, thought it charming; so did Mr. Gladstone, one of the few judgments they had in common. Having listened to the Fisk Singers in the spring of 1873, the queen noted in her diary, "they are *real* Negroes; they come from America and have been slaves." And with some amazement, "they sing extremely well together." Like most everybody else, the queen was more accustomed to hearing fake Negroes: the blackface whites doing doo-dah minstrelsy; the music that DuBois in particular singled out as a degradation of the true soul-liberty music. But everyone who heard Ella Sheppard, Fred Loudin, and the other five knew they were hearing something raw and unsettling in its passion. Only the irreproachable devotional intensity saved it from indecorousness. But the voices conquered just as surely as had the black soldiers. The year before their English tour, the Fisk Singers had performed in the White House at the invitation of President Grant before an audience of congressmen and senators, the diplomatic corps and cabinet members. But they had still been expelled from their Washington hotel for defying the conventions of segregation.

Their manager-mentor George White was pitched into a fury about this, but recognized there was little he could do about it, for all the fame of his ex-slave singers. They must needs take an event at a time, and he had pledged his life to the concerts, which raised funds for Fisk University in Nashville where all this had started. White was a blacksmith's son from rural New York who had seen the worst at Gettysburg, Chancellorsville, and Chattanooga, where he had seen Montgomery Meigs bringing in wagons of provisions for the beleaguered Union soldiers. Like Higginson, White had listened to the sorrow songs he had heard at the campfire, had begun to take notes; and when, after the war, he had offered his services to the Fisk Free School for Negroes in Nashville, it was music (as well as penmanship) that he taught.

The school, hard up, even in the heyday of Reconstruction, recognized in White someone committed to build a new life of the ex-slaves; someone who understood that the Teacher alongside the Preacher was to be the way ahead. He was appointed school treasurer, which kept him awake at nights worrying about its future. When he heard one of the students, Ella Sheppard, who had spent most of her sixteen years as a slave, sing more sweetly than anything he had ever heard, White conceived of a choir that might give public performances of the spirituals to raise money for Fisk. This was mostly frowned on even by those sympathetic to the Negro colleges, like the American Missionary Association. Taking young blacks on the road with no support but what a collection plate might yield, and in an atmosphere of, at best, mixed sentiments, even in the North, was a provocation.

But White was resolved. He would have his young men and women "sing the money out of the hearts and pockets of the people." The choir of seven, with no proper cold-weather coats or clothing, opened their tour at Oberlin in October 1871, where they were warmly applauded, but the clapping failed to translate into strong receipts. The same was true in Cincinnati: many huzzahs but just fifty dollars. In New York everything changed as, often, everything does. The Reverend Henry Ward Beecher brought the Fisk Singers to the Plymouth Church in Brooklyn, and overnight they became a sensation. Verbal brickbats were thrown at them but so were dollars, enough to pay their expenses and ensure the survival and even flourishing of Fisk. There was a price to be paid for the relentless touring: Mabel Lewis harmed her larynx; White himself, Maggie Porter, and Fred Loudin all became ill with bronchitis and pneumonia. White's tubercular cough became chronic, incurable; stifled somehow during the recitals. They were playing now to audiences of 10,000 and more.

> Steal away
> Steal away
> Steal away to Jesus

White pleaded with the president of Fisk, Erastus Cravath, for a break, but the college needed funds and the choir went on a second European tour to England (again), Holland, and Germany in 1875. Many of the singers were now on the verge of collapse. White had left, feeling he had created a monster; a form of show-business slavery. After the choir disintegrated in sickness and argument, he re-formed them in 1878, as the Fisk Jubilee Singers, the Jubilee of course being the biblical moment when all bondsmen were liberated. When TB prevented White from being with them, Fred Loudin took charge for tours that took the singers through Asia to Australia and back to the West Coast of the United States. In 1895 White's own lungs finally gave out. At his funeral the choir sang "Steal Away," their voices lifting the Nashville chapel roof off.

The Fisk Jubilee Singers had become an institution, exactly at the moment of a racial counterrevolution that put into question the victory of the Civil War. Outside the United States (and frequently inside), it is often forgotten that during the Reconstruction decade from 1865 to 1875, an extraordinary flowering took place. Protected by the Fourteenth and Fifteenth amendments to the Constitution conferring full citizenship and voting rights on the former slaves, and abolishing segregation in public places (other than schools and cemeteries!), African Americans flocked to the polls. Federal troops, still stationed in the South, were there to uphold their right to vote should that become necessary. The first African American governor was elected in Louisiana; congressmen and senators followed. Turnout was on the order of 70 percent; a level which Barack Obama would be happy to achieve almost a century and a half later. Among the defeated white population of the South, with some honorable exceptions, all of this was viewed as tragic farce; the enforcement of an occupation; the blacks as puppets of sinister northern carpetbaggers who had battened on their ruined country like ravening coyotes. The sentimental literature of defeat now actually hymned the virtues of the slaves: honest, toiling, decent in their simple way; while the monsters unloosed by the Freedmen's Bureau were promiscuous, idle, and empty-headed. The only hope of overturning such an unnatural order of things and restoring God's proper racial hierarchy lay with the Democratic Party. So when their candidate Samuel Tilden won a plurality of the popular vote in 1876, the bargain struck with the Republican, Rutherford B. Hayes, included the withdrawal of federal troops from the Southern states. And that was that: the Civil Rights Act set at naught; poll taxes and literacy tests put in the way of the vote; segregation everywhere triumphant; violence and intimidation unleashed against any blacks presuming differently. Thomas Wentworth Higginson resumed a literary life as Emily Dickinson's mentor, protector, and posthumous editor while Jim Crow reigned in the South.

Except in two places: black colleges like Fisk, Howard, and Atlanta (later Morehouse, where Martin Luther King studied) and the black church. In both those institutions, a battle was fought, at the end of the nineteenth century, for the "souls of black folk," between Booker T. Washington's practical gradualism, bought at the expense of political self-determination, and DuBois's call for church and school to produce a liberation vanguard. DuBois was a light-skinned son of Great Barrington, Massachusetts, which was about as far away from the sharecropper South as you could get (though he would write movingly and accurately of the impoverished rural counties of Georgia). Fisk had given him an undergraduate education, but he had been a graduate student at Harvard, studying with William James and George Santayana, as well as the Friedrich Wilhelm University in Berlin. Before he came to Atlanta University he taught at another black college, Wilberforce in Ohio, and the University of Pennsylvania. What DuBois came to want from black education is pretty much exactly what America has in the Democratic nominee in 2008: someone not only unapologetic about the empowerment of learning in an age of mass democracy, but who could also convincingly project knowledge as a tool of liberation. DuBois pinned his hopes for an educated black future on the "talented tenth" of black America—that would be embarrassingly elitist now—but when he turned away from an intellectual vanguard toward the mass of his people he looked in exactly the same direction as Obama: toward the church.

For although the Harvard-educated pragmatist himself became more skeptical as he aged, he always knew that the black church functioned as more than a house of worship. The 24,000 black churches were also "the social center of Negro life in the United States"; a communal government that, because it had roots deep in the antebellum world of the Richard Allens and Jarena Lees, functioned far more effectively as a government than anything the more thinly attached politics of Reconstruction had managed. After the liquidation of Reconstruction the church "reproduced in microcosm, all that great world from which the Negro is cut off by color-prejudice and social condition." Allen's own Mother Bethel Methodist church in Philadelphia, at the turn of the twentieth century, had more than 1,100 members, "an edifice seating 1,500 persons and valued at $100,000, an annual budget of $5,000, and a government consisting of a pastor with several assisting local preachers, an executive and legislative board, financial boards and tax collectors, general church meetings for making law . . . a company of militia and twenty-four auxiliary societies." DuBois could also have mentioned schools, sickness insurance, and burial societies. In its cradle-to-grave inclusiveness Mother Bethel was just one of the black megachurches, indistinguishable, except in pure numbers, from their black and white counterparts of today.

And they were not limited to the old abolitionist North. The most cursory look at the prodigious archive of memoirs and local histories provided by Charles Octavius Boothe's *Cyclopedia of the Colored Baptists of Alabama* reveals a world of astonishing cohesiveness and richness; the matrix from which eventually the civil-rights generation would spring.

It would be possible to read the accounts of the temperance clubs, city missions, sewing schools, the hundreds of Sunday schools operating in Birmingham alone, as evidence of the formation of a Booker T. Washington world of practical, politically self-effacing "Negroes." But look a little further and you find the embryo of vigorous self-determination: a Colored Baptist Convention in Montgomery, 1888; the Colored Deaf and Dumb Asylum; an entire network of state school inspectors; a University in Selma, the offspring of the St. Phillip Street Church founded by Samuel Phillips, an ex-slave who had been freed for serving as a soldier in the Mexican War; post offices; fire stations; ambulance services; men like Addison Wimbs of Greensboro, who was (so he boasted) the first black in Alabama to use a typewriter, and then an Edison phonograph—for the white governor of the state at the turn of the century. It was in those places that the early history of the "freedom church" and the self-emancipating world within slave culture was preserved and passed on to the next generation.

In his book Boothe reviewed the distance he thought his people had come since slavery, going out of his way to comprehend the bitter rage of the white South, but making sure that the history of self-making against the odds in the years after the war, in a landless, sharecropping world, was put on record. "With homeless mothers and fathers, with homeless wives and children, and with oppression on every side—with all these burdens and much more which cannot be told upon us—we bravely undertook the work of building the walls of Zion. The writer knows a minister who (between 1886 and 1875 especially between '66–'77 during the reign of the 'K. K. Klan' when the people could not in many places be induced to open their doors after dark for fear of being shot) has endured some of the severest privations and performed some of the hardest toils known to the ministry at his own charges. This case is only one in hundreds." At the end of the book, the Alabama Publishing Company of Montgomery that printed it advertised, along with the Reverend Pettiford's manuals on *Divinity in Wedlock*, the autobiography of Frederick Douglass, which promised liberation rather than sewing schools.

21. Easter Sunday, 2008, Ebenezer Baptist Church, Atlanta

Hanging on the vestibule wall were fading photos of past pastors of Ebenezer, stretching back all the way to Reconstruction Atlanta: a gallery of nobly chiseled faces, composed in attitudes of dignified sobriety; some embellished with the luxuriant whiskers of the prophets, all austerely dressed; faces that looked as though they knew they belonged in an ecclesiastical genealogy even before they actually did. Inside the church, perfectly costumed, impossibly beautiful children were giving everything they had to the Easter play, the odd shouted line "Oh MARY!" the only sign of nerves. Watching them, faces wreathed in smiles, were the parents, the grandparents, friends, the schoolteachers, all of them dressed in elegant suits and print dresses. Slopes of white lilies rose sheer behind the pulpit and in front of the choir, which if not quite on the scale of First Woodstock, out in the piny suburbs, was still a goodly hundred voices. At Pastor Johnny's megachurch there had been but two black faces in the choral ensemble; in Ebenezer, one white woman, back row of the sopranos, presumably with a set of power lungs on her.

This Ebenezer was just the latest reincarnation of many predecessors, all in the same neighborhood of Atlanta, a block or two from Martin Luther King's old brick temple, now canonized as a national historic site. The interior of the new church represented the history of its congregation: African teak and stained Georgian oak, married together in decorative designs, inset into the columns supporting the gallery. As wide as it was deep, the church seemed at once shelter and opening, which was, I suppose, more or less what its original founder had in mind two millennia ago.

I was assuming holy fireworks. A few days earlier, the former pastor of Barack Obama's Chicago church and a longtime family mentor, Jeremiah Wright, had been denounced on right-wing talk radio as an America-hating fanatic. A video clip of Wright, his voice rising in hoarse rage to proclaim "God *bless* America? God *damn* America" for its manifold sins of racism, had been aired in an endless loop on cable television stations, gleeful at the ratings gift that had suddenly come their way. No one was sure whether or not the current incumbent of Ebenezer, Raphael Warnock (Harvard Divinity School), would grasp the nettle; or whether he'd be saying anything political at all on Easter Sunday, but in any case, we had brought our own protagonists. On one side of me sat Angela and Fred Gross, business professionals in elegant early middle age; on the other, Mark Anthony Green, Morehouse College political science student, male-model looks and shoes to match. Fred and Angela were on fire for Hillary; Mark Anthony had come not to bury Obama but to praise him, to the skies if that's what it took.

We had first met Fred and Angela a month earlier on Super Tuesday, at their house in an upscale Atlanta suburban cul-de-sac, manicured swathes of lawn shaded by tall firs and cedars. Leading off the center hallway were plumply cushioned living rooms, a formal dining room and the obligatory gourmet kitchen. Swimming against the tide, Angela had invited friends round to run a phone canvass for Hillary. Theirs was a classic story of baby-boomer black prosperity. Fred had come out of the air force and made money in the catering business; Angela had been teacher, lawyer, executive. Like the other black women who arrived at the house, she was offended by the presumption that she was bound to fall in line behind the African American candidate. Gender mattered more to Angela than race. "See," she said, leaning on the table, fixing me directly in the eyes, "men mouth off a whole lot, women get to clean up the mess." Fred (also a Hillary supporter) turned and did something noisy with the ice bucket.

We talked about the civil-rights movement and the part the churches had played in it. This was, after all, Martin Luther King's and W. E. B. DuBois's Atlanta. "Look," said Angela's friend Lisa, "we're all church-going here; every Sunday rain or shine. Sure, faith mattered back then; how could it not? The church was the only place our people felt safe, bound together. But things are different now. Our religion is just our own private business. It's a hard world out there and it needs hard-headed people to cope with it. Hillary knows what's what. She's not just hot air." "How do we know Obama has a clue how to fix what's broken?" Angela chimed in. "She's proven; she's been tested." And the look on her face made me think Angela wasn't just talking about the Senate.

Earlier that day, I'd sat down with Mark Anthony on a low wall at the Morehouse campus. The statue of Martin Luther King was around the corner. How important was faith to the Obama campaign? "It means a lot to me," the young man said, "maybe everything. I just know that after all that we have been through, this is our moment."

Later on the doorstep of a black neighborhood in the city he would ask the lady who opened the door, "Do you believe God wants Barack to be president?" "I surely do," she said, smiling. So Obama was the prophet and Mark Anthony and countless other kids up and down the country were his evangelists. They wouldn't be surprised at his announcement that he plans to extend the faith-based initiatives begun by George Bush in the delivery of schooling and social services. But unlike the outgoing president and in keeping with his reading of the establishment clause of the First Amendment, Obama would open access to those services irrespective of denomination.

That evening in the Democratic Party watering hole, the two camps coexisted uneasily beneath the television screens registering primary tallies from around the country. By ten in the evening, amid the debris of hot dogs and beer bottles, it was already apparent that Hillary and Barack were going to split the vote, but that he had run away with Georgia in a landslide. Remarkably, the deeper south the primaries went, the better he was doing. Even more astoundingly, he had won more than 40 percent of the white male vote in the state. Brought together in a corner of the bar, the women were in a feisty mood. Jabbing a finger in Mark Anthony's direction, Lisa said, "If he weren't black would you want him to be president?" Answering right back, Mark Anthony moved his beautiful face closer to his rhetorical assailant and said, "If her name weren't Clinton and she weren't a woman would you want her as president?" "How DARE you?" Lisa yelled back over the tavern din. "How DARE you?" She and Angela might have been lecturing their wayward teenage son. "You think you have all the answers, and I thought so too when I was your age, a head just full of dreams and fancy notions, but let me tell you that's not the way the world works and you'd better wise up fast because nothing your fine preacher man says is going to give folks who don't have health care the drugs and the treatment they need, nor get kids who drop out to stay at school, or keep the world from getting messed up a whole lot worse than it is already, and if you don't believe it you are going to find out the HARD way."

And smiling under the storm of fury he had set off, Mark Anthony stood there, holding his ground, keeping the faith; saying there were just times when the old rules didn't apply, when America needed to turn the page. This was the time, his time.

Now the protagonists were gathered in the same row in Ebenezer.

Fred and Angela were there with their son, daughter-in-law, and the cute three-year-old granddaughter who kept climbing over the back of the pew in front to widen her green eyes at me and make faces. I made faces back; I'm good at that. The peace was kept. The Ebenezer choir, to my disappointment, had stuck to a grandiose version of the hymnal. Handel was much in evidence. But then they broke into one of the old spirituals, and in an instant, there was the swaying, clapping, the Music and the Frenzy that DuBois loved to hear, and the walls of the church shook with the jubilation of it.

Pastor Raphael Warnock, resplendent in a white gown with a high collar, trimmed in scarlet thread, the purity and the blood sacrifice turned into ecclesiastical chic, began the Easter sermon. He was loose and powerful at the same time, familiar and august; joking about his own childhood nerves at having to do the Easter play, and then warming to the lesson at hand, he made it clear from the get-go that he would not shrink from the controversy. "For some days now, the black church, America's freedom church, has been under attack from the press." Jeremiah Wright, who may or may not have been guilty of intemperate statements, was only the proximate target. It was the church itself that was a thorn in the side of the right-wing rabble-rousers. They wonder why we are *angry* the young pastor asked, his voice rising sardonically. "Three hundred years of slavery and segregation and they wonder why we are ANGRY?" (The last word a great roar of fury itself.) Then followed a brilliant disquisition on the selective conscience of white America; the dishonesty with which it trumpeted the virtues of democracy without confessing the sins of perpetuated inequality. And I thought of all those who had come before Warnock, before King; of David Walker in 1829, of Frederick Douglass in 1851 who asked rhetorically what could the Fourth of July and the Declaration of Independence possibly mean to *him* while slavery persisted in the republic? And Warnock built and built the music of his sermon, stepping from the pulpit out into the congregation, bidding them stand up, to rise up, for that was the message of the Easter Passion and resurrection, stand up for salvation, and the whole congregation did, shouting and singing and acclaiming, and at that moment in that church they were all there again, Andrew Bryan and Richard Allen and Jarena Lee and Fannie Lou in one great communion of purpose, and on cue the choir burst into voice and you thought for a moment the roof was going to be raised and we would be opened to the blue Atlanta heavens.

22. Great white hopes?

Now here's a peculiar thing: the Republican candidate, on a hilltop in North Carolina, come to seek the blessing of the evangelical Grahams, father and son, Billy and Franklin, is caught in a photo op looking as though he would rather be anywhere else, the smile glued on, his whole demeanor a portrait of a man in extreme discomfort. Meanwhile the Democratic candidate goes to Zanesville, Ohio, the kind of small town where nineteenth-century circuit riders shook up the faithful and is seen amid tow-haired students at a Christian community college saying that support for faith-based community services would be "at the moral center" of his campaign.

Obama has always pitched his appeal ecumenically, wanting Americans divided by the culture wars to come together in the broadest of tents. Peeling off white believers from their allegiance to the Republican Party seems, at first sight, a stretch. The history of white Protestantism after the Civil War seemed to take it as far as possible from that of the black church. In the South, its ministers were determined to sanctify the defeated heroes—Lee, Davis, and above all General Stonewall Jackson, whose life was canonized as a model of selfless Christian gentility. Every time another Confederate general died, the funeral was used by the church to project an image of the South as a shot-through citadel of Christian virtues, holding the fort against an oncoming tide (especially in the New South) of commercial debasement, sexual depravity, and liquor-drenched stupor: the sins of the metropolis marching on the corn-fed encampment of the righteous. While it was appreciated that the black churches were all that stood between this bastion of "civilized" America and the terrifying specter of Blacks on the Loose, extreme vigilance was needed if what remained of the right order of things was to survive the assault of modern vice.

It was in that paranoid atmosphere that, during Reconstruction, the founders of the Ku Klux Klan presented themselves as an order of modern crusaders. With their headquarters in Nashville, cheek by jowl with Fisk and its singers, and their imperial wizard the Confederate

Heroic full-length of the quartermaster general Montgomery C. Meigs, by Mathew Brady, in his Roman pose of tragic meditation. Meigs himself was an accomplished, enthusiastic photographer.

Gilbert Stuart's 1805 portrait of Thomas Jefferson, the presidential founder of West Point.

John Trumbull's 1806 portrait of Alexander Hamilton—ex-soldier, secretary of the treasury, who had much more ambitious plans for a peacetime army and a military college.

The inauguration of Abraham Lincoln, 4 March 1861, beneath Meigs's unfinished Capitol Dome.

Montgomery Meigs in command of the force mobilized to defend Washington from a Confederate offensive, 1864.

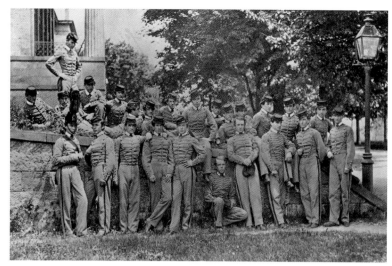

The West Point Class of 1863. John Rodgers Meigs is sitting against the wall, front row, center. Photograph taken probably in 1862.

John Rodgers Meigs in the uniform of a lieutenant of the Engineers when he was attached to Sheridan's army, 1864.

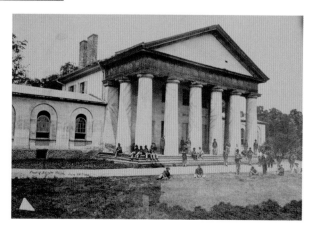

The beginning of Arlington National Cemetery: the occupation, on Meigs's orders, of Robert E. Lee's house and grounds on Arlington Hill, 28 June 1864.

Mass-produced pin, worn by those who wanted to advertise their patriotic indignation at the sinking of the USS *Maine* and their enthusiasm for the Spanish-American War.

Theodore Roosevelt in full cry, 1905.

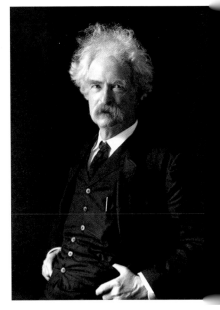

Mark Twain, middle-aged gadfly of imperialists.

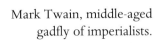

Fannie Lou Hamer, speaking to the Credentials Committee on behalf of the Mississippi Freedom Democratic Party at the Democratic Party Convention, Atlantic City, August 1964.

The campaign to register black voters, Virginia, 1960.

A BILL *for establishing* RELIGIOUS FREEDOM, *printed for the consideration of the* PEOPLE.

WELL aware that the opinions and belief of men depend not on their own will, but follow involunta-rily the evidence propofed to their minds, that Almighty God hath created the mind free, and manifefted his Supreme will that free it fhall remain, by making it altogether infufceptible of reftraint: That all attempts to influence it by temporal punifhments or burthens, or by civil inca-pacitations, tend only to beget habits of hypocrify and meannefs, and are a departure from the plan of the holy author of our religion, who being Lord both of body and mind, yet chofe not to propagate it by coercions on either, as was in his Almighty power to do, but to extend it by its influence on reafon alone: That the impious prefumption of legiflators and rulers, civil as well as ecclefiaftical, who, being themfelves but fallible and unin-fpired men, have affumed dominion over the faith of others, fetting up their own opinions and modes of think-ing, as the only true and infallible, and as fuch, endeavouring to impofe them on others, hath eftablifhed and maintained falfe religions over the greateft part of the world, and through all time: That to compel a man to furnifh contributions of money for the propagation of opinions which he difbelieves and abhors, is finful and tyrannical: That even the forcing him to fupport this or that teacher of his own religious perfuafion, is depriving him of the comfortable liberty of giving his contributions to the particular paftor whofe morals he would make his pattern, and whofe powers he feels moft perfuafive to righteoufnefs, and is withdrawing from the Miniftry thofe temporal rewards which, proceeding from an approbation of their perfonal conduct, are an additional incitement to earneft and unremitting labour for the inftruction of mankind: That our civil rights have no dependance on our religious opinions, any more than on our opinions in phyficks or geometry: That therefore the porfcribing any citizen as unworthy the publick confidence, by laying upon him an incapacity of being called to offices of truft and emolument, unlefs he profefs or renounce this or that religious opinion, is depriving him injurioufly of thofe privileges and advantages to which, in common with his fellow citizens he has a natural right: That it tends alfo to corrupt the principles of that very religion it is meant to encourage, by bribing with a monopoly of wordly honours and emoluments, thofe who will externally profefs and conform to it: That though indeed thefe are criminal who do not withftand fuch temptation, yet neither are thofe innocent who lay the bait in their way: That the opinions of men are not the object of civil government, nor under its jurifdiction: That to fuffer the civil Magiftrate to intrude his powers into the field of opinion, and to reftrain the profeffion or propagation of principles on fuppofition of their ill tendency, is a dangerous fallacy, which at once deftroys all religious liberty; becaufe he being of courfe Judge of that tendency will make his own opinions the rule of judgment, and approve or condemn the fentiments of others only as they fhall fquare with, or differ from his own: That it is time enough for the rightful purpofes of civil government for its officers to interfere when principles break out into overtacts againft peace and good order: And finally, that truth is great and will prevail if left to herfelf; that fhe is the proper and fufficient antagonift to errour, and has nothing to fear from the conflict, unlefs by human interpofition, difarmed of her natural weapons, free argument and debate; errours ceafing to be dangerous when it is permitted freely to contradict them.

WE the General Affembly of *Virginia* do enact, that no man fhall be compelled to frequent or fupport any religous Worfhip place or Miniftry whatfoever, nor fhall be enforced, reftrained, molefted, or burthened in his body or goods, nor fhall otherwife fuffer on account of his religious opinions or belief, but that all men fhall be free to profefs, and by argument to maintain their opinions in matters of religion, and that the fame fhall in no wife diminifh, enlarge, or affect their civil capacities.

AND though we well know that this Affembly, elected by the people for the ordinary purpofes of legiflation only, have no power to reftrain the acts of fucceeding Affemblies, conftituted with powers equal to our own, and that therefore to declare this act irrevocable would be of no effect in law; yet we are free to declare, and do declare, that the rights hereby afferted are of the natural rights of mankind, and that if any act fhall be hereafter paffed to repeal the prefent, or to narrow its operation, fuch act will be an infringement of natural right.

Printed version of Jefferson's original draft for the Virginia Statute of Religious Freedom, 1778–79, sections of which were preserved in the version James Madison steered through the Virginia Assembly in 1786.

Charles Grandison Finney, the abolitionist thunderer of Oberlin, with the eyes that made sinners and slavers cringe.

Jarena Lee, itinerant black preacher in the Atlantic states; frontispiece from her autobiography.

African American schoolchildren in North Carolina around the time of emancipation, 1863.

Very early photograph of First African Baptist Church, probably Savannah.

Thomas Wentworth Higginson, preacher, literary editor, colonel of the 1st South Carolina (black) Volunteers and the first anthologist of black spirituals.

The Fisk Jubilee Singers as they would have looked at the time of their recital before Queen Victoria, 1875.

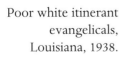

Poor white itinerant evangelicals, Louisiana, 1938.

Frontispiece portrait of J. Hector St. John de Crèvecoeur from the first edition of *Letters from an American Farmer*, 1782.

The beauteous Citizen Know-Nothing, imagined as the frontispiece to sheet music for the party at the height of its political power, 1854.

Anti-Mexican newspaper cartoon at the time of the Mexican War, complete with lustful, dagger-wielding Latinos.

Chinese railroad construction workers,
Montana, 1870s.

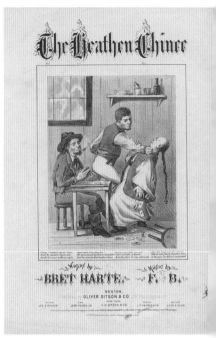

Illustration for a song sheet,
the musical version of Bret Harte's
"The Heathen Chinee."

Colonel Fred Bee, Chinese consul, in his familiar plug hat, with Chinese and
U.S. Army investigators into the Rock Springs, Wyoming, massacre.

"Where the Blame Lies": anti-immigration cartoon, 1891, the complete gallery of ethnic caricatures of the "inferior races."

Graduation Day, 1916, at Henry Ford's English School in Dearborn, Michigan. *Pluribus Unum* is the name of the steamer bringing immigrants to the "Melting Pot," in which the graduates stand, holding their Stars and Stripes.

Grace Abbott, 1925, two
decades after founding
the Immigrants Protective
League in Chicago.

Immigrant matrons from eastern Europe, doing their bit for wartime America, 19

John Gast's *American Progress*, 1872, a much-reproduced image from the years leading up to the Centennial, embodying the inevitability of western manifest destiny: the telegraph and railroad completing what covered wagons and stagecoaches had begun.

Franklinia alatamaha, native to western Georgia, from William Bartram's *Travels*, 1791.

Seal of the Cherokee
Nation, 1839.

Colonel Return Jonathan Meigs Sr., at the time of
his residence among the Cherokee.

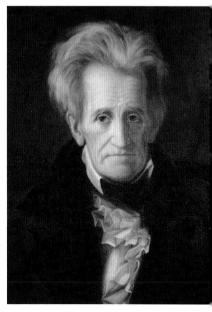

John Ross, chief of the Cherokee Nation,
Jackson's antagonist, during the years of
their independence and mass deportation
over the Mississippi.

George Healey's portrait of Preside
Andrew Jackson, author of the Indi
Removal Act of 1830.

The last great land run, Cherokee Strip, Oklahoma Territory, 16 September 1893.

Cherokee family, photographed in the 1930s in North Carolina, in a rural setting much like that from which they had been expelled a century earlier.

Major John Wesley Powell, one-armed explorer, geologist, and ecological prophet, with Tau-gu, chief of the Paiutes, Colorado Basin, 1873.

Dorothea Lange's great portrait of a migrant mother from the Dust Bowl, 1936.

"Bath rings" and boat trapped in the dried bed of the drought-depleted Lake Mead reservoir, 2007.

ex-general Nathan Bedford Forrest, the Klan rapidly developed into a vigilante organization, bent on terrorizing anyone opposing the perpetuation of white supremacy or supporting the institutions of Reconstruction. In the minds of the Klansmen they were resisting an alien Northern occupation, but their violence was so out of control that Congress outlawed the order in 1871. Some thirty years later, however, a North Carolina Baptist minister, Thomas Dixon Jr., gave it a second life. His novel *The Clansman* featured a Presbyterian preacher to the Klan giving his blessing to their battle to protect "Christian white civilization" from the depravity of black scalawags and their Yankee carpetbagging puppeteers. Filmed by D. W. Griffith as *The Birth of a Nation*, the movie fired thousands to rejoin a Klan reborn after the ex-Methodist minister William Simmons led a procession up Stone Mountain outside Atlanta. Beneath relief carvings of the Confederate heroes Lee, Davis, and Jackson, an altar was erected and the first Klan cross burned. By 1921, there were more than 100,000 members, organized in cells from the Midwest right through the old South, and many of them heard sermons and lectures from Protestant clergy who endorsed their claim to be protecting southern Christianity from the bestiality of the lower races—Jews as well as blacks.

The Klan's perversion of faith was extreme but it was a character-istic response to the experience of defeat and dispossession, clinging to a nostalgia for a form of Christian gentility that had only ever existed on the backs of slaves. Both black and white churches in the century after the Civil War were life rafts for people living in poverty and alienation. But while the black churches were construc-tive and forward-looking, emphasizing education and economic skills, planting in country and city the germs of self-determination, the white churches that ministered to the hard-up were most often defensive, fighting a furious rearguard action against the assembled forces of modernity: secularism, alcohol, Charles Darwin, and the latest satanic monster, showbiz. The answer to the last threat, of course, was to fight fire with fire, and launch what was in effect the Third Great Awakening: using the techniques of mass entertainment to steer the weak and sinful away from Babylon. It was a priceless gift that Billy Sunday, the battler against liquor, had been a professional baseball player (his name helped too). Movies of his revival sermons show Sunday using his slugger's swing to maximum effect when demonstrating what he would do with demon drink and those who profited from it.

White Pentecostalism, the Holiness movement, all made their deepest inroads, though, in the regions of the United States most vulnerable to economic distress: the prairies where corporate agriculture was uprooting the ancient American ideal of the family farm; the Great Plains where the dust storms blew away the hopes of the small farmer along with his topsoil; the Appalachians where coal miners were paid by "tickets" redeemable only through the company store, and were unprotected from the brutal emphysema and lesions that erupted in their bodies after years of toil at the coal face. Small mining towns like Pocahontas, West Virginia, where immigrant labor came from Hungary, Poland, and Ireland, became the theater of an all-out struggle between vice and salvation; the number of churches struggling to keep up with the saloons and whorehouses that competed for the miners' Sundays. The mining and railroad companies that wanted to protect their investment often funded the building of churches and the expenses of the minister. But the appeal of religion to all the Americans living hand to mouth in good times and disasters did not need the sanction of the bosses to make it compelling. Very often—whether in big-city tenements or broken-down coal towns—it was the church that ministered to the sick; looked in when pantries were empty; made sure the kids were in school. And, to the dismay of management, the unorthodox fringe churches would often help out strikers when times got really rough. Near Birmingham, Alabama, the pastor of the Mount Hebron Baptist Church, where the congregation were mill workers and small tenant farmers, was Fred Maxey, who made no apology for showing up at union rallies, going out of his way to minister to black and white alike. Needless to say a Klan cross was burned outside his church.

The twentieth-century church of the white poor, then, was not monopolized by reaction. There had been, in fact, a change in party identification. The great Christian revival orator and Democrat populist William Jennings Bryan had mobilized a huge rural constituency behind his attacks on the bastions of capitalism: Wall Street, the banks that, in the name of the gold standard, made credit tough for the little man. Conversely, after Theodore Roosevelt left the White House, the Republicans cozied up to business. By the time the roof fell in on American capitalism in 1929, the natural alliance between Irish Democrats and the Catholic Church had been augmented by another between white rural Protestants in the South and Midwest. In the harder-hit areas of the industrial South, joining an evangelical revival church was a way to cut loose from management-endorsed Sunday worship and establish an independent sense of community. One disgruntled textile mill owner complained that holy rollers "tear up a village, keep folks at meetings till all hours at night so they are not to fit to work the next morning."

A remarkable Alabama woman, Myrtle Lawrence, illiterate white sharecropper, stalwart of the Taylor Springs Baptist Church and the Southern Tenant Farmers' Union, came to New York in 1937 for National Sharecroppers' Week, organized by New Deal government agencies. What the united ranks of union organizers and government progressives would have liked, as the historian Wayne Flynt observes, was someone beauteous in poverty, straight out of a Dorothea Lange photograph. What they got instead was toothless Myrtle Lawrence, snorting snuff, chewing quids, and hurtling gobs of juice into her pink-wrapped spit box before plunging into prayers and hymns with raw Old Time gusto. But on one principle, Myrtle could not be faulted by the high-minded New Dealers: she thought blacks fellow Americans, welcomed in the love of Jesus. At the YWCA Summer School in North Carolina, she told an English class, "They eat the same kind of food we eat; they live in the same kind of shacks that we live in; they work for the same boss men that we work for; they hoe beside us in the fields; they drink out of the same bucket that we drink out of; ignorance is killing them just the same as it's killing us. Why shouldn't they belong to the same union that we belong to?" Myrtle and Fannie Lou would surely have got along.

23. Ruleville, Mississippi, 31 August 1962

She knew why she had to get on that bus, and it wasn't because no out-of-state kids who'd come to Mississippi had told her to. No, sir, it was what she'd heard in the little chapel in Ruleville from Reverend Story that Sunday; God speaking directly to her, though the words came out of the mouths of all those young men with their fine hopes. Fannie

Lou was sick and tired of being sick and tired. She had had enough of breaking her back every day in the fields so she could put bread on the table; enough of being frightened she might be put out of her house if she so much as went near the folks organizing in Greenwood for the vote; talking about sitting in on lunch counters; in the front of the bus. Wasn't it enough that she couldn't have children of her own; that they had taken her and done something to her that made sure of that without so much as asking her first? And now what was she supposed to do with all that welling up of feeling; knowing that it was more than high time that promises made should be promises kept; that she and people like her could have the decencies of life, starting with the vote? And if she got scared at the thought of what she was going to do, she just did what she couldn't but help: she sang. She would lean over the cot and sing to her little Vergie. "This little light of mine, I'm gonna let it shine," she would sing, and her adopted girl would smile and sleep.

They were going to Indianola, twenty-six miles down the road, the road that went straight through the cotton fields past W. D. Marlowe's plantation where she sweated for her share or kept time for the croppers. In Indianola, Mr. Charles McLaurin had said, they would all register to vote as was their born right and something would get started in the Delta. The yellow bus rolled up by the water tower, and on they all got, mostly the folks of the church and a few of the organizers to see everything went as it should. But there was a pit in their stomachs and they sat there, low and quiet like the children who would usually be in the seats. The drive seemed endless. Every time a pickup went past, they could see pink and red faces looking hard at the bus, some of them scowling, the trucks making a swerve to bother them. Those twenty-six miles seemed the longest ride of their lives.

When they got to the city offices by the bayou, out they climbed. Word must surely have got out as there was a whole bunch of state troopers and highway patrolmen there, looking at them hard as if they were taking photographs. And they made it plain to McLaurin that there would be just two of them let in for the registration that day and if they didn't care for that, well, that was just too bad. Now Fannie Lou had worked her hardest at preparing herself for all the complicated questions she knew would be put in her way like a briar patch, but they had sure done a good job in making it hard and she was beat if she knew when this and that clause of the constitution of the state of Mississippi had become law and others had not. There was a good deal of head and pencil scratching; hours of it, the August sweat made worse by her nerves, before she handed in the papers to the hard-eyed white man at his desk. Ushered outside to wait, she already knew she wasn't going to be given what she had come for, not that day anywise, so it was no surprise when Fannie Lou was told she had failed the test the state needed to verify her credentials as a voter. One day soon Fannie Lou knew it would be different, so when they climbed aboard the bus she could not take it as defeat.

The bus pulled out of Indianola, but as the evening closed in, so did the bad feeling. They all sat there in the bus as quietly as they had come; but then the traffic outside on the road picked up, and much of it was men in trucks who had taken the shotguns off their racks and were waving them in the air at the bus, hollerin, "Fuck you niggers, we'll *kill* you niggers before you EVER get the vote." And then came the police cars flashing their lights and cutting them off, and the officer got the driver out and told him he had no right taking passengers seeing as the bus was too yellow; that it was "impersonating a school bus." There was a fine to be paid; and there was a go-round the bus for the money and it got handed over, but still they took the driver away.

So how were they to get back home now? No one on the bus had ever driven anything, that's for sure. So there they sat, in the thick Delta night, a stew of fright and despair. Even McLaurin seemed to have lost his voice. No one knew what to do; what might become of them. They could feel their own sweat go cold. But then, from somewhere in the back of the bus, a sound started; a voice, pure and low, just easing out into the air; Fannie Lou's voice.

"This little light of mine," it sang,

> I'm gonna let it shine
> This little light of mine
> I'm gonna let it shine

And the voices joined in. They were sovereign; invincible. They had the Lord with them. They would overcome.

PART THREE

III: WHAT IS AN AMERICAN?

24. Twilight, Downing Street, June 2008

The evening was winding down. The British-and-damned-proud-of-it grub (minty pea soup, salmon, roast beef, berries) had been polished off. The last of the daylight was fading through the sash windows. Some fifty or so guests, faces shiny with satisfaction and wine, were pushing chairs back and slowly making their way into the adjoining room for coffee and a dram, though neither the prime minister nor the president would touch a drop. We were a mixed bunch: cabinet ministers in varying states of merriment or beleaguered solemnity; the World's Most Important Media Tycoon (saturnine, imperiously unfriendly); and an unlikely prattle of historians. Invitations had gone out indicating passionate interest in the subject on the part of both chief executives. This is certainly true of the prime minister, who has a PhD in history, but it was a stretch to imagine George Bush with a heavily thumbed edition of Gibbon on the bedside table, even though a chronicle of the fate awaiting overextended empires might not go amiss.

Some of the tribe of Clio were loyal enthusiasts of the president, whose popularity ratings at home were sinking even faster than the Dow Jones. Others among us had reconciled our uneasy consciences by telling ourselves that if we were in the history business, how could we possibly stay away? But what sort of business was that? In what, precisely, were we being implicated? The general idea of the evening seemed to be that if a gathering of historians was convened in sufficient density, the moment itself would, by some act of cultural osmosis, become Historic. We who had communed in our pages with Churchill and de Gaulle, would now sprinkle a little Significance around like air freshener in the parlor of exhausted power.

Into the drawing room—a Quinlan Terry extravaganza in terra-cotta, edged with gilt reliefs, the result of a decorating epiphany experienced by Mrs. Thatcher in Nancy Reagan's White House—strode the hurting titans. During the early stages of the dinner, Bush seemed oddly ill at ease, especially for someone getting a respite from the gloomy sibyls at home. A generously fraternal speech by Gordon Brown, lauding the bond between the two democracies and ending in a toast, was met with an aw-shucks response as POTUS got to his feet, evidently underbriefed about this particular moment in the proceedings. "Guess I'd better say something," he thought out loud before offering the communiqué that "relations between the United States and the United Kingdom are good . . . [long pause] in fact, *damned* good." And that was pretty much it for hands across the sea. He seemed oddly diminished, someone who, for reasons he couldn't quite fathom, had discovered the armchair he'd been sitting in for years was now too big and his legs no longer touched the floor; the incredible shrinking prez. Over the pea soup, the shoulder-slapping bonhomie withered to a dry little tic of humorous coughs, eyebrows suspended between hilarity and uncertainty, like a stand-up comic sweating through a tough house on a Monday night in Milwaukee.

For a minute or two after the photo op, George Bush was left to his own devices and came my way. (Improbable, but true.) He looked unmistakably like a man in need of a drink; but since he had sworn off ever having one, and since I had a glass of cognac in my hand, the least I could do was say a word that would get him through the rest of the evening, and that word was: "Texas." The pick-me-up worked an instant transformation: the officially synthetic grin replaced by the real thing. Our television crew was off there soon, I explained, to make a film about immigration history. Then I reeled off a string of place names that were country music to his ears: "El Paso," "Brownsville," "Laredo." Some of the earlier discomfort, I realized, was that of a man who traveled abroad only so that he could feel a little pop of joy when he got within honking distance of the Houston freeways. "Your policy on immigration," I said, the cognac making me cheeky, "is about the only policy I've agreed with; and that's because it put you way to the left of your party." It was true, if patronizing. The line in the Republican Party over what to do about the twelve million illegal immigrants in the United States, the overwhelming majority Hispanic, runs between those for whom fencing, discovery, and deportation are the beginning and end of policy, and those like Bush and McCain, who have long been in the world of the Hispanic Southwest, and want to find some way to get the illegals to citizenship. Even the lamest of lame-duck presidents didn't need a pat on the head from someone who was not going to be lining up anytime soon for the job of ghosting the memoirs. But Bush took the condescension in good part because, I suspect, his head was off somewhere far away from Whitehall, mooching around the chaparral, where the lights of Houston oil refineries were still benevolent twinkles on the unlimited horizon of Texan plenty.

Bringing him back from the sagebrush, I asked, "So why do you suppose immigration turned out not to be the hot-button issue for your party?" I mentioned the standard-bearer of anti-illegal hard-line enforcement among the candidates, the Colorado congressman Tom Tancredo, who, despite the ranting of hard-core radio demagogues prophesying the End of America, had gone nowhere in the primaries. "Tancredo," said Bush, incredulously, rolling his eyes and lightly snorting like one of his own horses, adding under his breath, "is an idiot." He then complained a bit how he could have got a benign immigration reform bill enacted, guaranteeing medical and social services to at least the children of illegals, had it not been for the Machiavellian obstructionism of the Democratic Senate Majority leader Harry Reid. Then he made a suggestion for our film shoot. "I knew the border real well, back when it was just about the worst, dirt-poor place in America; I could tell you stories of those days in the fifties; of how people who'd jumped the line survived, what they worked for day and night." He paused, brows knitted again, and then suggested we take a look at old stills of the way Brownsville and the pueblos the other side of the border were half a century ago—shacks and rats—and compare them to the way those towns are now; which is, for all the trouble and grief, materially better. There had been, he reminded me, a huge shift within Mexico itself; people from the country moving steadily, unstoppably north, to the world of the Norteamericanos.

Back in the States at that very moment, two sure things were going down. Someone at a gas station was jacking up the price to as close to five dollars a gallon as they could get it, and along the Rio Grande, and in the unforgiving desert, *coyotes* (people smugglers) were trying to get a hundred souls *every hour* to make that crossing. Homes were being repossessed every time you looked; banks were hitting the deck, taking the bankers with them. Suddenly there might be a lot fewer center-hall colonial McMansions to clean, or prairie-size lawns to cut. The job lines would rise, skyrocketing oil prices would hike transportation costs, and the cost of food would head north too. Though one had nothing to do with the other, the pressure on jobs would spur hard-core right-wing radio ravers, like the comically self-named Michael Savage, to yet more foaming waves of abuse. The immigrants were parasitic vermin, barely human, battening on the wasted flesh of the body politic, sucking taxes from law-abiding citizens, taking jobs away from lunch-pail Joe.

But none of this matters to the unstoppable stream. On it comes. Twelve million undocumented and around a million more each year for the foreseeable future; right now the biggest human migration on the face of the planet. And they come not just from Mexico and Central America; but to New York from Uzbekistan and Tajikistan, from Cambodia and Liberia and Senegal. If they could only get out of the prison their government has made for them, the entire population of Myanmar would fetch up in New York Harbor tomorrow. And why? Because of two dramatic utterances. The first was Tom Paine's pronouncement in *Common Sense* (1776) that the "New World has been the asylum for the persecuted lovers of civil and religious liberty." But that was to define America as the abode of free conscience and would have been a little lofty for those just wanting to climb out of a Pomeranian mud hole, let alone a corrugated-roof shack in Guatemala City, into something approximating a human existence. For them, another work was written, by a native-born Frenchman on his farm in Orange County, New York. The book, which seemed like a memoir but was at least as much romance, was called *Letters from an American Farmer* (1782), and it was the first great work of American literature. Chapter Two of *Letters* is titled "What is an American?" A whole century before the Sephardi Jewish poet Emma Lazarus declaimed in words graven into the pedestal of the Statue of Liberty that America would be decent enough to receive the "wretched refuse of the teeming shore," J. Hector St. John de Crèvecoeur had already lit the candle of hope that would reach across oceans and continents to millions. The emblems of American self-sufficiency, the log cabin and the homestead farm, were already declared by Crèvecoeur to be the social cells of democracy. And they were within the reach of anyone who had the gumption to choose to live a free life in an open country. In Pennsylvania, New York, Connecticut, the servile peasant and the human effluent of the cities would be reborn as John Citizen. (Jane was not discussed.) Released from having to defer to aristocrats and ecclesiastics, common humanity would be liberated to realize its true, boundless potential. Happiness was on the horizon. "We have no princes for whom we toil or bleed,"wrote Crèvecoeur. "We are the most perfect society now existing in the world. Here man is free as it ought to be." Now why, then, should a man remain in the social prisons of European tradition? Why would anyone with a grain of common sense or self-respect feel attachment to the accidental geography of his nativity, for "a country that had no bread for him, whose fields procured him no harvest, who met with nothing but the frowns of the rich, the severity of the laws, with jails, punishments, who owned not a single foot of the extensive surface of the planet?" To leave such vexation behind was to experience social rebirth. What is an American, then? "He is an American, who leaving behind him, all his ancient prejudices and manners, receives new ones from the new mode of life he has embraced, the government he obeys and the new rank he holds. He becomes an American by being received into the broad lap of Alma Mater. Here individuals of all nations are melted into a new race of men, whose labours and posterity will one day cause great changes in the world." Let such melting commence! And the misery-stricken of the world would no longer be the defenseless creatures of the mighty. In America they would be something else: self-made men; citizens.

25. Citizen Heartbreak: France, August 1794

How they haunted him, those words written twenty years before. "What, then, is the American, this new man?" he had asked. Not him, apparently, not according to the mirthless James Monroe, the new American minister to the French Republic. Since Crèvecoeur had been the first to ask that rhetorical question in print, the sour irony was not lost on him. He, who had bidden the unfortunate of the world to come to America, who had promised that any of them might become that new man, was being denied that opportunity. But at least he had given voice to a universal ideal, something the impoverished and the harassed wanted. They would want it in blighted Ireland and pogrom-terrorized Russia; in famine-struck south China and arid Mexico. Perhaps Crèvecoeur, for whom Americans were mostly English, German, Dutch, Scottish and Irish, had an inkling of this epic immigration to come? Certainly he supposed that being American was, in the first place, a state of mind—the rejection of geographic fatalism—as much as an occupancy of a place on the earth. But then what was *he* himself? A true American, in spite of Mr. Monroe's rejection? Or was he, in his marrow, still French: Michel-Guillaume-Jean de Crèvecoeur, heir to the estate of Pierrepont near Caen in Normandy (land that would be watered by American blood in 1940)? That old birthright he had left behind, more from circumstance than any act of will. Remote though he became from his *pays* and mother tongue, Crèvecoeur never forgot the loamy Norman soil, the high hedgerows walling the narrow lanes, the clannish rooks in the orchards. But two revolutions on either side of the ocean had turned him from *seigneur* to *cultivateur*, squire to farmer. It was imprudent in such an altered world to indulge in pastoral nostalgia. Those who did so usually betrayed themselves as having lived off the sweat of the peasantry. So many liberal ci-devants, former gentry, had lost their liberty and then their heads, for less. Poor Condorcet, the social philosopher, had been captured by the Terror when, offered an omelette by the farmer sheltering the fugitive, he politely asked for it to be made from a dozen fresh-laid eggs. Force of habit had exposed the marquis beneath the trappings of the republican citizen. So for the moment it was better that he be just plain Citizen Crèvecoeur.

Or did he dream in English? In his heart of hearts was he still the husbandman of Pine Hill, New York, husband to Mehitable Tippet; known in Chester and the townships of Orange County as Mr. Hector St. John? Or had that old honeyed existence of his, the bee swarm beneath the locust trees, been consumed in the flames? Was it his lot now to eke out his days as Citizen Heartbreak, eccentric survivor of wars and revolutions, dimly remembered as botanizer, occasional man of letters, sometime consul, a one-man chamber of commerce who had created the packet-boat run between L'Orient and New York; who had introduced alfalfa (*luzerne*) to America, imported along with cargoes of cognac and human hair?

There were days when he scarcely knew himself. But he did know that he yearned to follow the setting sun across the sea, just once more before he died, to revisit the country he had delivered to the world from his mind's eye. But in the summer of 1794 Crèvecoeur couldn't just ship off. The French Republic was at war with Britain even if the United States had stayed neutral. To be anglophone was itself a cause for suspicion now, and it was no longer as delightful to be an American in Paris as it had been in the days when court and city had been thrilled by the sight of Benjamin Franklin's beaver cap. Too many conspicuous Friends of America had fallen foul of the Jacobins and Robespierre's Terror. Quite apart from the tricky nature of the language, the government in the United States had signed a peace treaty with the British; had apparently forgotten its debt of gratitude. *Amérique perfide!* So how could he, Crèvecoeur, who had been a soldier for the monarchy, a squire in British America, who seemed to change nationalities and languages with his linen, be trusted to travel to the United States? Many of his friends had suffered for the same compli-cations of allegiance. The philosophical La Rochefoucauld had been stoned to death by a mob. Guillaume-Louis Otto (erstwhile Count Otto of Strasbourg), married to Crèvecoeur's daughter America-Francès, had been imprisoned. Had he somehow been responsible? At the height of the Terror, fearful for the rest of his family, Crèvecoeur had tried to use the goodwill of the American agent, Joel Barlow, to get them out of the country, but neither Barlow nor the irascible minister, Gouverneur Morris, had any goodwill to spare. Now, at the end of August 1794, matters should have been different. Maximilien Robespierre had been forced to stick his head through "the republican window" where so many had perished for the crime of insufficient patriotic ardor. Parisians, including those with suspicious American connections, now breathed a little more easily.

And there was a new American minister to the French Republic: the Virginian lawyer and senator, James Monroe, who, naturally, like others before him (John Adams came readily to mind) spoke not a word of French and was dependent on the kindness and translations of Crèvecoeur to get his bearings. It was easy for Crèvecoeur to sympathize. When he himself had first returned to France in 1781 after quarter of a century in America, he had all but forgotten his native tongue, and the process of reacquiring fluency had been unexpect-edly painful. He made himself write a French edition of his book, not just a translation, but a manuscript almost twice as long, and given a different tonal music from the original, with lashings of strategically anti-British sentiments. Monroe seemed appreciative of his editorial and translating help, and Crèvecoeur's hunt for a summer house suitable for the representative of the United States. So Crèvecoeur dropped hints about his past standing as an American citizen-farmer; of some mission perhaps he might usefully perform in the United States. When no forthright response came back, the hints became broader. In the summer of 1794, Crèvecoeur took it upon himself to supervise the making of three American flags. The largest was intended for the French legislative chamber, the National Convention, where it would hang beside the tricolor. A second was for the American Embassy, and the third for Monroe's diplomatic carriage. It was the new flag, the revolution's circle of thirteen stars increased by two more representing the newly admitted states of Kentucky and Vermont, arranged in an irregular oval pattern while fifteen alternating bands of red and white made up the field. It was very beautiful, and Monroe was happy. But not happy enough to oblige Crèvecoeur, who, raising the matter once more, was met by so chilly a rejection that it would have to be the last. The man who thought old identities could be struck off like chains from the slave, that to wish to be an American was already a license to become one, was being disabused of his romance. He had been born a Frenchman and would die in, of all places, the rustic district of Sarcelles, today neither rustic nor peaceful but the scene of uneasy Jewish-Muslim coexistence.

The boy Michel-Guillaume had been no runaway. He had gone where his parents had pointed: first to the Jesuits in Caen, and then a place as un-Jesuitical as could be imagined, Salisbury, where the family had distant relatives and where he was to learn English. Since it is well known that the most expeditious road to fluency is driven by amorous energy, and since there were few more charming lures in Wiltshire than a young French beau attempting to say "thee," Michel-Guillaume caught the ear and eye of a local merchant's daughter, and in turn became smitten with her. They were each other's; they would be betrothed and have little milkmaids and shepherds. But before the romance could flower, the girl died, and for the first time the boy, thinking how it translated into Wiltshire English, was lanced by the irony of his name. The sensible thing, of course, would have been to go home to Pierrepont, to his parents and the imperturbable herds. But instead Crèvecoeur took ship for New France in 1755, arriving just in time for a war; the one which lost Quebec for Louis XV. The soldier in white smartly turned into Hector St. Jean de Crèvecoeur, who charged, as ordered by the irrationally gallant General Montcalm, directly at the British guns lined up unsportingly on the Plains of Abraham. The young soldier was badly hit but not killed. One of the hobbling wounded, he was now, whether he chose it or not, a Briton. If this mortified the survivor, he would not say, not then, nor ever after. But at many points in his life, Crèvecoeur the Franco-American would speak tenderly of the qualities of the English, the very ones Frenchmen despised: their taste for liberty, their habit of moderate manners and genteel decorum. Well, he never spent time in the bagnios of Covent Garden.

When his wounds allowed, Crèvecoeur left the ruins of Quebec and made his way south into deep British America, doing what pleased him most: sketching, mapping, botanizing; closely observing the fauna. Sometimes he walked, sometimes he rode through the granite Allegheny passes, down ancient Indian hunting trails choked with sumac and overhung with wild grapevines. Around the black reflecting lakes he paddled, listening to the loons at dusk, the rasp of a million segmented legs at night, squatting by the fire with Oneida Indians who named their adopted son Cahio-ha-ra. In the upper Hudson Valley, the mountain peaks were softened fringes of conifers; below, the forests had been cleared for sheep pasture. Church spires rose by the ponds; woodsmoke drifted over the valley, and the clink of a blacksmith's hammer carried over the Hudson Gorge. Dark-rigged sailing barges plied the stream. And there, the wounds of Crèvecoeur's heart and body, love, loss, and battle, finally became bearable. The possibility of a new American life intimated itself as delicately as a candle's glimmer.

Crèvecoeur became a tenant farmer in the uplands. He fished and kept cattle, poultry, and a prolific beehive. He was now someone else: J. Hector St. John Esq., "St. John" pronounced in the proper English fashion "Sinjun." He neither bared his wounds nor told campaign tales, but instead joined the company of the country at horse fairs, county assemblies, and taverns, and went to church in their Protestant way.

At some point his path crossed with the mercantile Tippets of Tippet's Neck, lower Yonkers, whose grand house welcomed him down its poplar-lined carriage drive. A daughter, Mehitable (a name favored by her family over many generations), received Crèvecoeur with particular tenderness, and they were married by a Huguenot pastor in September 1769. And in the way of now being a solid Anglo-American, St. John earned enough credit in the eyes of his neighbors and, more important, his in-laws, to buy himself and his wife an uncleared property of some 120 acres, thirty-five miles west of the Hudson, three miles from the village of Chester. Husband and wife began their farming life as every American was supposed to, in a self-built log cabin. But as the first crops came in, Mr. St. John wanted something more enduring and set a new house upon sturdy brick foundations, broad enough for five upper-story windows, which looked down on a little rustic "piazza," as he called it. The place was called Pine Hill, for behind the house there were both, lending a picturesque aspect to the property. A year after her marriage, Mehitable bore a girl, whom in recognition of his happy double identity Crèvecoeur called America-Francès (Fanny); and then, at biennial intervals, two boys, Guillaume-Alexandre (Ally) and Philippe-Louis. Mrs. and Mr. St. John prospered and had honeybee days beneath their locust trees undisturbed by intimations of rebellion.

To read the first fifty or so pages of *Letters from an American Farmer*, one would never know anything could ever be amiss with the idyll. Crèvecoeur insisted that he had not written a true memoir—that characters like "Andrew the Hebridean" were invented—but he knew very well that there is nothing much to separate him from his alter ego "James" other than the fiction of having inherited fortune and land from a benevolent American father. (Was Crèvecoeur projecting what still whispered to him from Normandy?) The voice yearns for what Thomas Jefferson would declare to be an American, indeed a universal, natural right: the pursuit of happiness. "Where is that station," Crèvecoeur wrote, "which can confer a more substantial system of felicity than that of an American farmer possessing freedom of action, freedom of thoughts, ruled by a mode of government which requires little from us?" Each year he kills 1,500 to 2,000 hundredweight of pork, 1,200 of beef, and "of fowls my wife has great stock; what can I wish more?" What indeed, although the fact that "my Negroes are tolerably faithful and healthy" may have helped maintain the arcadia.

"When I contemplate my wife by my fireside while she either spins, knits, darns or suckles our child, I cannot describe the various emotions of love, of gratitude, of conscious pride, which thrill in my heart and often overflow in involuntary tears. I feel the necessity, the sweet pleasure of acting my part, the part of a husband and father, with attention and propriety which may entitle me to my good fortune." When he plowed on low, even ground, the father set his little boy on a seat screwed to the beam of the plow, in front of the horses; and he prattles away while Crèvecoeur leans over the handle . . . "the father ploughing with his child to feed his family is inferior only to the emperor of China ploughing as an example to his kingdom." In the evening as he trots home he can see a myriad of insects "dancing in the beams of the setting sun."

In his pages, the landscape of freedom is animated by stupendous natural drama through which Crèvecoeur strides like Gulliver, brawny and amazed. Its profusion electrifies his pen. Passenger pigeons blot out the sun, so dense are their flocks in flight to the Great Lakes. The craw of a kingbird opened by the curious farmer is found to contain precisely 171 bees. Crèvecoeur lays their tiny bodies out on a blanket as if accommodated in an apiarian hospital. And fifty-four miraculously revive, "licked themselves clean and joyfully went back to the hive." Biting, stinging things that in less blessed lands might have been swatted or trodden out of existence are celebrated by Crèvecoeur as authentic American creatures, paragons of diligence and ingenuity, living in mutuality with humans. The hornets that nest in his parlor alight gently on the eyelids of his children to carry off the flies that have irritatingly settled there. Wasps form a multicellular "republic" and of course the bees are model citizens, swarming for his delight, their honey more richly fragrant than anything that had heretofore slipped down his gullet. "I bless God for all the good He has given me," Crèvecoeur writes at the end of this chapter, "I envy no man's prosperity and with no other portion of happiness than that I may live to teach the same philosophy to my children and give each of them a farm, show them how to cultivate it and be like their father, good, substantial, independent American farmers."

This is the Crèvecoeur most of his readers would have remembered, honored by critics from William Hazlitt to D. H. Lawrence: the artless husbandman dwelling in happy hollow, the anticipator of Thoreau, the man whose contentment bid multitudes to try their fortunes in America. But advance through the pages, beyond the scenes of Nantucket's whalers, beyond the story of "Andrew the Hebridean" from impoverished Barra who finds happiness in Pennsylvania, and get to the chapter where, suddenly, the American skies darken. The Aesopian tales become horrors; the bees make way for copperhead snakes, whose bite inflicts agony on the victim so that they die writhing, with their tongues stuck through their teeth in grotesque mimicry of their killer. Crèvecoeur's vision of the helpful hornet gives way to mortal combat between a water snake and a black viper, and before long it is Americans themselves who are locked in death struggles. "I can see the great and accumulated ruin as far as the theater of war has reached; I hear the groans of thousands of families now desolated . . . I cannot count the multitude of orphans this war has made nor ascertain the immensity of the blood we have lost." Into the idyll had burst the demon of politics.

Which made, again, for heartbreak, this time on a scale the happy husbandman could never have anticipated. Crèvecoeur had imagined the act of becoming American as a steadily transforming migration, not one of violently traumatic separation. But the particular way in which America the place became America the nation un-Americanized it for Crèvecoeur, sullied its innocence. He made no bones that it was America's first civil war; one that had set neighbors against each other, robbed children of their sweet dreams (his own came crying to him in terror of their nightmares), and put such men as he in an unbearable dilemma. Mehitable's family and many of his friends, like the banker and magistrate William Seton, were loyalists; those the Patriots called Tories. Crèvecoeur himself had felt warmly about the mildness of British rule in the colonies. Jefferson's assertions that Americans had been the victims of military tyranny would have struck him as self-serving, patently absurd hyperbole. But staying true to those moderate sentiments would make him an enemy to his country. "Must I be called a parricide, a traitor, a villain . . . be shunned like a rattlesnake?"

As it happened, the conflict along the New York–New Jersey rural and river frontier—his own and the Tippets' backyard—was one of the most bitterly unsparing of the entire war. Patriot militia and their houses were attacked by loyalist partisans, some of them escaped armed slaves who had worked their farms. Reprisals were merciless: summary hangings, farmhouses and stores burned where they stood. Back and forth the battle raged. The Tippets had thrown in their lot with the loyalists and would end up in Nova Scotia. Crèvecoeur wanted his family out of harm's way and seems to have toyed with taking his wife and children to the backcountry Indians—either the Oneida or the Seneca—whom he claimed to know well. But the tribes were themselves pulled into the conflict, enlisted by both sides as irregulars. After months of uncertain wanderings, Crèvecoeur, declaring openly for neither party, took the same painful decision that countless others in his predicament made: to divide the family. One part (Mehitable, their daughter, Fanny, and the younger son, Louis) would remain at Chester to protect the claim to their farm should the Patriots close in on it, while he decided to go to British-occupied New York with the six-year-old Ally.

But there, instead of finding safe refuge, Crèvecoeur was arrested. An anonymous letter sent to the British commandant General Pattison claimed that Crèvecoeur had made drawings of the British harbor defenses. All he knew that summer day in 1779 was that General Pattison requested an interview. Expecting assistance, he left Ally with acquaintances in Flushing on Long Island. The boy waited in vain for his father who had been thrown into a filthy basement cell from which he could hear the shrieks of men being flogged. Hours became days, days weeks, weeks months, and the father who had set his little boy on the plow seat went into mental shock, sobbing and gibbering, attempting to hurt himself. Somehow the promises of an old man who had shared his cell, that, once released, he would go to Flushing and find Ally, comforted Crèvecoeur. And before too long, William Seton put up a bond for his release. A slave from another friendly family, the Perrys, then arrived to tell him that Ally, though feverish, was staying with them. There was a reunion, no doubt of indescribable joy. The boy's fever cooled.

But Crèvecoeur's miseries were not over. It was impossible to get news of Mehitable or the children, much less travel safely to Orange County, still a fiercely embattled war zone. And the winter that came upon father and son in a disused barn was the coldest anyone could remember. Firewood was almost impossible to procure. Neither Ally nor he had warm clothes. Only a laboring job chopping up a hulk earned him wood and saved them both from death by hypothermia. Crèvecoeur was revisited by the hysterics and poisoned dreams to which he had succumbed in prison. He shuddered with a violent palsy, became spasmic with seizures as his little boy looked on.

Spring came. It was 1780. Friends rallied, and so did the traumatized American Farmer. He stopped shaking, did a little genteel surveying of the New York churches. Should he collapse and die, his friend Seton promised he would bring Ally up as his own child. But Crèvecoeur did not die. Instead, he and the boy recovered enough to travel. In September 1780, father and son boarded a ship bound for Dublin, doing something expressly against the message of his book, voyaging away from America, back to the Old World.

One of the victims of Crèvecoeur's ordeal had been the bulk of his manuscript. It may have been on the ship that he managed to reconstruct it, although the voyage was stormy. In London, he found a publisher who paid him thirty guineas for it; neither mean nor princely, but undoubtedly, given the circumstances he had endured, a boon. His publishers, Messrs. Davies & Davis, who also obliged Dr. Johnson, held out the further incentive of a "gift" should the work find favor with the public. That it did so, right in the midst of the war that would lose Britain its most prized colonies, might seem puzzling, except that the most memorably beautiful passages offer precisely a prelapsarian vision of an American idyll, meanly attacked by the stinging hornets of politics. Across the Channel, of course, where the book and its author were also welcomed, it could be read as a message from France's latest ally; a picture of the American promise, once liberation from the accursed British was complete.

Much to his surprise, then, Crèvecoeur turned out to have written a useful, even a politic book; a book that could make readers see the America of hummingbirds, dewdrops, and honey mead; the farmer's wife sewing or churning her butter; the husbandman bringing the lowing herd to the milking stall. Europeans, who at exactly this moment were going through much economic distress and turmoil in their own countryside, already craved this America, one in which the slave and the black freedman were noble, and the Indian sachems wise and hospitable. The book rapidly went through new editions in Dublin and Belfast, Leipzig, Berlin, and Leiden, so that when Irish, German, and Dutch peasants and townsmen wondered what it might be like to emigrate, to make that passage from the imprisonment of natal circumstance to a free life, it was Crèvecoeur's vision that sustained them. It was so successful that complaints were made in print in England that the *Letters* were a propaganda stunt designed to lure British artisans and farmers away from their home country to America. The "excellencies"of the United States, John Bristed wrote, had been exaggerated "as the abode of *more* than all the perfection of innocence, happiness, plenty, learning and wisdom than *can* be allotted to human beings to enjoy." Crèvecoeur had not written a guide to emigration, they said, he had fabricated a romance.

He certainly lived one: the prodigal wanderer, returning home to Normandy after twenty-seven years, falling into the embrace of aged parents; the shy Ally ushered forth to meet *Grandpère, Grandmaman*, St. John realizing he must call himself Crèvecoeur again; struggling with his mother tongue. Even before he became a celebrity with the publication of the French edition of his book, Crèvecoeur rode a wave of Americophilia (carefully omitting any hint of the loyalism that had helped him find publisher and public in London). Still better, he could discourse freely on the crops that were the rage of innovation-minded Norman landowners. A small book needed on the history and prospects of the potato? Rattled off in a few weeks. Lectures to the local agricultural societies in Caen and further afield? A pleasure.

Five sailors washed up on a Normandy beach without a word of French between them. Suspicions were high. St. John the anglophone was sent for to translate. They were not English, he discovered, but American seamen who had been captured by the redcoats and taken back across the ocean to a prison somewhere in the south of England. They had broken out, found a small boat, and survived the hostile Channel crossing. Crèvecoeur told them his own story, dwelling on his fears for wife and children. In gratitude Lieutenant Little promised that on returning to the United States he would have a relative living in Boston go to Orange County and make inquiries of Mehitable and the two children. In the meantime, the lieutenant took a bundle of letters from Crèvecoeur to be sent on to the family as soon as their safety and whereabouts were confirmed.

This was kindness repaid with interest. Unable to travel himself until hostilities were ended, Crèvecoeur did his best to enjoy his

French renown. He hobnobbed with *les Grands* of the Paris literary world. The ex-minister to the king and liberal reformer, Turgot, took him to Paris, where, twice a week, he frequented the salon of Mme d'Houdetot, Rousseau's old (unrequited) inamorata, and let himself be admired as the cultivated cultivator; their kind of American: francophone. He spoke of locust trees with the greatest of the botanists, Buffon, and proposed the introduction of sweet potatoes into France, where he thought they might become a staple food of the common people. He knew he was an exhibit—a specimen of a New World that French reformers hoped might be made the norm on their own side of the ocean. Agriculture and the cause of humanity were mutually nourishing as if the intelligent application of manure could guarantee a harvest of liberty along with the wheat.

All that was very fine. He was glad to be of use. But his mind drifted back west across the sea. It was four years since he had set eyes on his wife and on Fanny and Louis, and nothing had been heard from them. He hoped and feared. With the provisional articles of peace signed in Paris early in 1783, it was finally safe to travel again. He obliged the minister of the navy, the duc de Castries, by penning, in seven weeks, a rich report on the condition of the United States and the prospects of oceanic commerce between the two countries. In appreciation, word came from Versailles that he would be one of the first French consuls to the United States. Who could better embody the natural fraternity between the two nations? He was even given free choice of the town to which he might be posted. He chose New York. In early September 1783, armed with his papers and his nervous expectations, Crèvecoeur boarded ship at the port of L'Orient on the Breton coast, the departure point for the regular France-America packet service he himself had proposed.

When Crèvecoeur imagined a bustling traffic between France and the eastern seaboard of America, he thought of a swift passage west of three to four weeks. His own in the autumn of 1783 took fifty-four days of the purest Atlantic hell. The *Courrier de l'Europe* was buffeted by vicious gales, swamped in the billows, and blown far off course more than once. Crèvecoeur was sick most of the time and occasionally collapsed in epileptic seizures of the kind that had convulsed him in the British prison in New York. When the ship finally passed the

Narrows and weighed anchor at Sandy Hook on 19 November, exhaustion competed with his trepidation. New York Harbor was a picture of chaos. Washington's troops were camped on Harlem Heights, poised to enter the city. British ships were being hastily loaded up with the human remnant of their lost American empire: loyalists; slaves freed by their service with the British. On one of those ships his in-laws, the Tippets, had departed for good to recreate a new dynasty in Canada. But Seton, his fortunes badly damaged and in some fear of being identified as a Tory collaborator-magistrate, had decided America was and would be his home. Fires were breaking out on Manhattan, each side, exultant Patriots and bitter loyalists, blaming the other for the damage. Through all this smoke, terror and hatred staggered Crèvecoeur, frantic for news. He had written ahead to Seton, informing him of his arrival. But the first to make himself known to Crèvecoeur was someone Seton had sent to the dock to look for him and who, without any care or ceremony, blurted out the news that his house was burned, his wife dead, his children missing.

When Seton caught up with him, Crèvecoeur was prostrate with anguish. All his friend could offer by way of comfort was to say that he had been looking for the children and to promise that he would never abandon the search. And then, in mid December, at the Post Office amid a heap of correspondence jumbled up in the hurly-burly of the evacuation, the desperate Crèvecoeur finally discovered the truth. Waiting for him were letters from the man Lieutenant Little had sent to New York to try to contact Crèvecoeur's family, Captain Gustavus Fellowes. Somehow, the package had been sent to London rather than France and then, failing to find the addressee, had been returned to New York to await possible receipt. Sorrow was confirmed. Mehitable was indeed dead and the farm razed in an Indian raid, though whether they were on the warpath for Patriots or loyalists was unknown. But the children were alive, taken in by a neighboring family near Chester, where Captain Fellowes had found them in a pitiful state; without shoes or stockings in the bitter cold, barely fed and quite sick. Fellowes's response to what he beheld was instantaneous. He and his wife had seven children already; now they would have nine. The farmer's wife, tearful, remorseful, wept at giving them up, wrung her hands with the distress of her own situation. Fellowes consoled her as best he could, absolved her of any guilt but with no more ado, bundled the eleven-year-old Fanny and the seven-year-old Louis in bearskin blankets and set them in a sleigh that sped north over the icy roads to Boston. So there would be a reunion.

It came about in the early spring of 1784, once the roads were passable and Crèvecoeur had shaken off his epileptic seizures. At Fellowes's door in Harvard Street, he announced himself nervously, but he need not have worried. The children who came to his embrace were much grown: Fanny thirteen now and Louis nine, but they were still a family. Brought together, they stayed in Boston for some months before Crèvecoeur took them back to New York to take up his post as consul, instituting the ocean service, seeing to imports and exports (alfalfa, sweet potatoes), finding a way to pick up a semblance of domestic life. Gentle honors came his way. Ethan Allen, the governor of Vermont, named the town of St. Johnsbury for him, but then Allen was himself an ambiguous patriot, wanting first and foremost sovereignty for his state and prepared to ask Canada for it, should Congress not agree.

For a few years, the Crèvecoeurs/St. Johnses were truly if not quite easily of the two worlds: in France again between 1785 and 1787, where the children learned the language at school; then two years in New York, where he helped establish the first openly Catholic church, St. Peter's on Barclay Street. Crèvecoeur had been in Philadelphia to witness the making of the American Constitution and, stirred by what Jefferson was reporting from Paris, believed he might see something of the same dawn in France. In the summer of 1790, the closest moment to a revolutionary honeymoon that France experienced, Crèvecoeur returned and for a moment thought that there might indeed be a dawning of some universal republic of liberated humanity. Was not the declaration of the Rights of Man and Citizen addressed to the whole world (much like his own work)? But it took little time for the violence of the Revolution to shake his confidence, and whenever he could he took refuge at Pierrepont from the brutal politics of Paris that engulfed so many of his friends, both French and American. In 1792, the foreign minister Bertrand de Moleville ordered him back to the United States, but Crèvecoeur replied that, being nearly sixty, in poor health and with no funds at his disposal, he needed to secure whatever little he could and resigned his post. As the monarchy fell to a further revolutionary upheaval and the Jacobin Terror became the order of the day, Crèvecoeur had reason to regret the decision, but it was now too late to effect an exit without being suspected of espionage or outright treason. Invoking the doubtful honorary citizenship bestowed on him in Vermont would only have made matters worse. Already his son-in-law, Otto, suspected of being too friendly with foreigners, was in prison.

Crèvecoeur would survive the worst, but he would never recover, except in print, the bucolic idyll he had made famous in *Letters from an American Farmer*. He journeyed up American streams now only in his mind and in his papers, writing his *Voyages dans la Haute Pennsylvanie et dans l'Etat de New York* from Normandy and the Seine Valley. There was no interest at all in England and little more in France. Trails were being blazed elsewhere, lit by the campfires of Bonaparte's armies. But although Crèvecoeur's own dream of a republic that gathered all the *misérables* of the world to its bosom was belied by his own difficult life, others wanted to make a go of it; first his younger son, Louis, who bought 200 acres in New Jersey, built himself a log cabin, and attempted a rural subsistence. But the father's paradise eluded the son. Louis survived brutal winters in his log cabin only by living on the flesh of frozen hogs. After two years, his father pleaded with him to come back to France.

But even though its author became ever more remote from his American destiny, what he had written in the 1770s and published in the 1780s became real in ways he could never have anticipated. Thousands then tens and eventually hundreds of thousands who had never heard of Crèvecoeur, much less read him, were nonetheless moved by the vision that came to him of an American "resurrection": "no sooner than they arrive than they feel the good effects of that plenty of provisions we possess; they fare on our best food and are kindly entertained; their talents, character and peculiar industry are immediately inquired into; they find countrymen everywhere disseminated, let them come from whatever part of Europe."

How would such a newcomer fare? "He is hired, he goes to work . . . instead of being employed by a haughty person he finds himself with his equal . . . his wages are high, his bed is not like that bed of sorrow on which he used to lie . . . he begins to feel the effects of a sort of resurrection; hitherto he had not lived but simply vegetated; now he feels himself a man because he is treated as such; the laws of his own country had overlooked him in his insignificancy; the laws of this cover him with their mantle. Judge what an alteration there must arise in the mind and thoughts of this man. He begins to forget his former servitude and dependence; his heart involuntarily swells and glows; this first swell inspires him with those new thoughts which constitute an American . . . If he is a good man he forms schemes of future prosperity, he proposes to educate his children better than he has been educated himself; he thinks of future modes of conduct, feels an ardour to labour he never felt before . . . Happy those to whom this transition has served as a powerful spur to labour, to prosperity, and to the good establishment of children, born in the days of their poverty . . . who had no other portion to expect than the rags of their parents had it not been for their happy emigration."

That, at any rate was the idea.

26. The German threat

Crèvecoeur was fond of seeing America as the regenerative asylum "of low indigent people who flock every year here from all parts of Europe." The "promiscuousness" (his term) of its mix of nations he thought, in principle, entirely benign. Observation, however, modified this indiscriminate enthusiasm. The people who made up this new country were all Nordically pink: "a mixture of English, Scotch, Irish, French, Dutch, Germans and Swedes." (Though Crèvecoeur was passionate on the subject of the treatment of both Indians and slaves, it never occurred to him, of course, to count them as members of the American nation.) But even within the relatively narrow range of peoples who had become American, he admitted, even as he insisted on the transformative effect of the country on all of Europe's poor, that some actually did better than others. Scots like his "Andrew the Hebridean" did well but the Irish were an altogether different story. When he wrote, there were probably around 400,000 Americans of Irish stock, but overwhelmingly they were Ulster Presbyterians. Yet Crèvecoeur's stereotype anticipates all the stock traits ascribed to the Irish who would sweep into America in the nineteenth century. "They love to drink and to quarrel; they are litigious and soon take to the gun which is the ruin of everything; they seem to labour under a greater degree of ignorance in husbandry than the others; perhaps it is that their industry had less scope and was less exercised at home . . . the poor are worse lodged . . . than anywhere else in Europe; their potatoes which are easily raised are perhaps an inducement to laziness; their wages are too low and their whisky too cheap." Thus in the mind of the greatest enthusiast of an immigrant America, the caricature of the shiftless, violent Irishman was already present.

The opposite of the Irish were not the English or the frugal Scots, but the "honest Germans" who, for Crèvecoeur, were the epitome of what it took to become true Americans. Escaping from the petty tyranny of some tinpot martinet, they arrive in America, "observe their countrymen flourishing in every place; they travel through whole counties where not a word of English is spoken and in the names and the language of the people they retrace Germany." But by "dint of sobriety, rigid parsimony and the most persevering industry, they succeed." From their labors have come "the finest mills in all America, the best teams of horses." They were, for Crèvecoeur, an American success story.

But others had felt differently about the Germans and continued to do so, none more sharply than Benjamin Franklin. Of all the Founding Fathers, none had a more richly developed sociological sense of America's destiny, well before its war for independence. For Franklin, America was not just the unfolding of an idea of liberty. Its physical character—continental space, and a population that would double every twenty-five years—constituted sovereign reality. Against these two mutually sustaining facts of social economy, British pretensions to exercise authority in the same manner that it exercised it over, say, Wales, were absurd and doomed to fail. Franklin attempted to explain this unstoppable American future to members of Parliament in the 1760s but was met mostly with bored incredulity. Only in Scotland, where social philosophers like Lord Kames and Adam Ferguson were rewriting the laws of history in much the same way, did Franklin's projections receive an attentive hearing.

The frustration that Franklin felt at his message falling on deaf ears was all the more acute because he saw America essentially as a transoceanic Britain; even more narrowly perhaps, as a new England. (He had been born, after all, in the most racially homogeneous place in America: Massachusetts.) The Scots and Irish might find a place, but America's imperial destiny would prosper insofar as it retained its Anglo-Saxon racial core. Part of his objection to slavery was that it diluted that essence. Quite apart from the multitudes of captive Africans themselves, the temptations of slave-owning sapped the reserves of can-do northern energy and replaced that vigor with degenerate indolence. That might suit the Iberian races who had first brought slavery to the western hemisphere, but it was a poor lookout for those whom Providence had entrusted with northern America. But then all kinds of immigration made Franklin uneasy. He rather hoped that the great population boom he projected for America would be served by lively natural increase. But Franklin's attitude to open immigration was, to say the least, ambiguous.

Unlike Crèvecoeur, Franklin was not an admirer of the Germans of Pennsylvania. They were coarse, and there were far too many of them. "Those who come hither," he wrote, "are generally of the most ignorant and stupid sort of their Nation . . . why should Pennsylvania, founded by the English, become a Colony of Aliens who will shortly be so numerous as to Germanise us, instead of our Anglifying them and will never adopt our Language and our Customs?" It was Benjamin Franklin, then, whom we think of as epitomizing the cosmopolitan spirit of the Enlightenment, who as early as the 1750s sounded the alarm about the imperiled state of Anglo-American culture. It was expansive, inventive, gregarious Franklin who was the founding father of American paranoia. The most stridently anti-immigrant writers today like Pat Buchanan make sure to invoke Franklin to give their ethnic phobias a venerable pedigree, although they get him badly wrong in one important respect. It was the breadth of his intellectual imagination, not its narrowness, that could make him quirkily inhospitable. A sociological realist, he thought the ideals of universal brotherhood, admirable in themselves, were just that, and that human reason was always having to contend with tribal instinct and inherited affinities. Well-meaning attempts by Franklin biographers to pass off a notorious passage at the conclusion of his *Observations Concerning the Natural Increase of Mankind* (1751) as tongue-in-cheek drollery are unpersuasive. Franklin may have been layering the confession with a patina of roguish brazenness, but what he wrote was what he meant. "Why increase the Sons of Africa by planting them in America," he wrote, "when we have so fair an opportunity, by excluding all Blacks and Tawneys, of increasing the lovely White and Red? But perhaps I am partial to the Complexion of my Country, for such Kind of Partiality is natural to Mankind."

Surprisingly, most of the German "boors" whose presence Franklin found threatening and among whom he numbered the "swarthy" were that notoriously dusky lot, the Swedes. Only the true Saxons counted, along with the English, as true lily-whites whose numbers "I wish were . . . Increased." The rest of the Germans, however, were a distinct threat to the American future. "Why should the Palatine boors be suffer'd to swarm into our Settlements, and by herding together, establish their Language and Manners to the Exclusion of ours?"

Whether Franklin became so vehement on the subject because, or in spite of, his short-lived experience as the first editor of the German-language *Philadelphische Zeitung* is hard to say. But there is no doubt that he was reacting against the advice given to German immigrants by the journal's proprietor, Christopher Sauer, that, as much as possible, they should keep themselves to themselves. Sauer advised the Germans that it would be in their interest to vote alongside the Quakers (and thus against deists like Franklin and Episcopalian English-stock Pennsylvanians); and to steer clear of "involvement with English-speakers that might endanger our language, our families, our customs and our faith." In response, the Society for Promoting Religious Knowledge and English Language among the German Immigrants in Pennsylvania was created to sponsor and monitor free schools. Eleven were operating by the mid-1750s, but the German community saw them as intrusive, mounted a campaign of resistance and the experiment at educational integration proved short-lived. More draconian proposals from the colony's Committee for German Affairs were made to ensure the Germans integrated themselves properly into Pennsylvanian society: forced intermarriage, the suppression of presses that published only in German, a prohibition of German-only schools, and the requirement that all legal documents be printed solely in English. For Franklin this was a step too far, and it belat-edly brought out the ecumenical side of him. Only "methods of great tenderness should be used," he wrote to his London Quaker scientist friend Peter Collinson, "nothing that looks like a hardship [should] be imposed. Their fondness for their own Language and Manners is natural; it is not a Crime."

From the start, then, the assumption that America's unique alchemy was to make *E pluribus unum*, one nation from many, begged the question for many Americans about the cultural compatibility of some elements of the *pluribus*. In the year of the opening of the constitutional convention, 1787, John Jay (who was to become chief justice of the United States) clarified in a Federalist Paper his vision of who was and who was not an American. "Providence," he wrote "has been pleased to give this one country to one united people, a people descended from the same ancestors, speaking the same language, professing the same religion, attached to the same principles of government, very similar in manners and customs." In his *Notes on the State of Virginia*, Jefferson made much the same point in the chapter on "Population" where he questioned the wisdom of "rapid . . . importation of foreigners." Since, Jefferson argued, American government was "peculiar" in its adoption of the "freest principles of the English constitution," it would be threatened by a mass of emigrants from the absolute monarchies, who even in escape "will bring with them the principles of the government they leave, imbibed in their early youth." And even "if able to throw them off, it will be in exchange for an unbounded licentiousness, pass[ing], as is usual, from one extreme to the other. It would be a miracle were they to stop precisely at the point of temperate liberty. These principles they will transmit to their children. In proportion to their numbers they will share with us the legislation. They will infuse into it their spirit, warp, and bias, its direction and render it a heterogeneous, incoherent, distracted mass."

So while people in Belfast and Leipzig were reading Tom Paine and Crèvecoeur and imagining the social miracle by which the most oppressed peasant or laborer could be transformed in America into a free citizen, those who would greet them from the seats of power on the other side of the ocean were having serious second thoughts about the immigrant romance and looking for ways to weed out the more, from the less, acceptable newcomers. This cautionary approach to immigration would remain deeply lodged in the American mind, even as its public image was one of the indiscriminate embrace of the unfortunate. If involuntary political and cultural inheritance was a concern for Jefferson, the social background of immigrants was an issue for Madison. In the congressional debate over the first federal Naturalization Act of 1790, which restricted accessibility to citizenship to "free white persons" over twenty-one who had resided in America for at least two years, Madison expressed his anxieties about merely swelling "the catalogue of people" regardless of what skills or fortune they brought to the country. "Those who acquire the rights of citizenship without adding to the strength and wealth of the country," Madison bluntly announced, "are not the people we are in want of." Put the objections of the two Virginians together, and it's evident that no riffraff, and no one trying to escape the tyranny into which they were born, need apply. Put another way, only those who didn't need to become Americans in the first place would actually be welcome.

And yet, it would be Crèvecoeur's sweet optimism and not Jefferson's and Madison's pessimism about immigration that would ultimately triumph as one of the great motifs of American history; one of the reasons why the United States should exist at all. Mass immigration would turn out to be compatible with both liberty and prosperity (indeed a condition of the creation of American wealth). But the rumble of anxiety first expressed by the Founding Fathers, that the unwashed might overrun the purity of the political nation they had made, never really goes away. Every time the American economy hits a reef, the last on the boat are usually those whom nationalist politicians want to throw from the decks.

In the first decades of the republic's existence, naturalization acts were used to sift the desirable from the undesirable. In 1795, because of anxiety about the influx of politically dubious refugees from the European revolutions, the residence requirement for citizenship was raised to five years, and in 1798 it was raised again to fourteen. The Alien Act also allowed for the scrutiny of anyone deemed a possible enemy alien and for summary deportation. It was evidently John Adams's intention to deny the Federalists' opponents the support of immigrants whose more natural allegiance would have been to Jefferson's Democratic Republicans (and who presumably hadn't read *Notes on the State of Virginia*). Which is partly why in 1802 Jefferson, even as he prohibited the admission of foreign convicts, cut the residence requirement back to five years, with an obligation to declare the intention of seeking citizenship at least three years before obtaining it. And that is where the law has stood ever since.

The date of Jefferson's act is significant, though, for quite another reason: 1802 was also the year in which negotiations began in earnest with Napoleon for the 828,000 square miles of territory that in the following year would pass to the United States. With the completion in April 1803 of the Louisiana Purchase and the acquisition of so vast a territory—in effect its continental future—the United States abandoned precisely the tight little box of monocultural English politics, farming, laws, and religion in which Jay and Madison had wanted to confine it. Jefferson, on the other hand, stood (as he so often did) at the cliff-edge of a breathtaking gamble, his instinct for the spatial transformation of the country overcoming his cultural conservatism. Perhaps there might yet be a way for America to thrust itself into the alien body of Hispanic America and still produce Anglo-American issue.

27. The Chicken Club, south Texas, July 2008

Between Brownsville and Port Isabel, where the Rio Grande flows into the Gulf of Mexico, the arid Texas landscape sinks into salty wetness, and the horizon is wide enough to see the curvature of the earth. Streams and small ponds, catching the silvery light, thread through what was until recently duneland, turning it into a dark mudscape over which herons and egrets sail looking for shrimp and usually finding them. This is the Bahia Grande: not so long ago an arid bowl of salt and dust that blew into air-conditioning fans and car radiators and let nothing much grow except a thuggish variety of yucca, not unlike a Joshua tree; spindly fibrous trunks crested by spiky crowns of leaves. The yucca are still there in their thousands, poking comically from the surface like so many prairie dogs on their hind legs. But the roots of the yucca now often sit in that saline water better suited to the black mangroves that diligent high-schoolers are planting as fast as they can in the mud. The weird coexistence of these species, arid and marine, has been the result of a Gulf of Mexico reclamation project that in 2005 flooded 10,000 acres of the Bahia, restoring to wetland what the Brownsville Ship Canal took away when it separated the Rio from its natural outlet to the sea. The pelican and the cormorant, the shrimp, the shrimp fishermen, and the tourists on South Padre Island out on the causeway are all happy about this. It's hard to find anyone around Brownsville, a town which along with its sister on the other side of the river, Matamoros, lies at the heart of MexTex border history, who isn't. The new-old landscape of the estuary seems to hum gently of an older, slower world.

A few miles inland, at the edge of the zone where razor grass, sage, and prickly pear take over again, there's more singing and yearning for a lost place. Each week, Jesus, Arturo, Alberto, Juan, and a few friends meet in a backyard to barbecue chicken and play Norteamericano music; the music of *la frontera*. This week they're kind enough to play for us in front of a barn, festooned with abandoned car parts, old guitars, and a neon Budweiser sign that picks up in magenta intensity as the evening draws in. The men are all in their late sixties and early seventies, but look enviably younger, especially Juan, one of the singers who is the most handsome vaquero you've never met, sporting a dashing mustache and a certain look in his eye beneath the cowboy hat. Their music is famous for its aching sweetness, sung up and down the scale, the throaty pleasure and the drawn-out pain of memories of land lost, mothers, cooking, lovers. They give "Canción Mixteca" their all in three-four time; "How far from the beautiful land of my birth," they sing longingly as half chickens are hoisted by their legs from their marinade high into the light and then dumped on the grill to sizzle and char, smoke curling in the air as the band hits a high note. Hiss, *ayy, sospiro, I sigh* . . . and rangy Juan leans into the mike like it was a senorita smooch. Our director had wanted to record the Chicken Club playing acoustic guitars. Not a chance; the men are too attached to their amplifier—driveway authenticity—and as if in vindication, turn up the juice on a rowdy cicada that has just machine-gunned its way into their act.

The sounds of longing fade; cans of beer are flipped open; we all load our plates with chicken and guacamole—and none of your bright green supermarket pap, either. Ricardo, our director who grew up in Chile, asks the musicians straight out whether they think of themselves as American or Mexican. He speaks to Jesus, Alberto, Juan, and Arturo in Spanish, the only option, even though some of them have been in Texas for thirty or forty years. One by one they all answer, forthrightly and without any defensiveness, "Mexican." And they smile as they say this, seeing no contradiction between this profession and their American citizenship. For they are fiercely loyal to the United States as well; and to the next question, what do they think of the country,

they answer, in effect, "the world": a good place, the best; the country
where a man may make a decent living for his family; the country that
is still respected abroad. For the Chicken Club these two loyalties are
not in conflict; they are mutually reinforcing. And they have little or
no sense of their people being at the eye of a political storm; of being
accused of staging a "reconquista," taking back the country Mexico
lost to the United States in the war of 1846–48. How can you take
back what was never really lost? Lost maybe to the Mexican state;
but the north country has always been a piece of Hispanic America,
and a piece of Anglo-America too. What's the difference: They speak
Spanish; their children are bilingual; their grandchildren speak mostly
English. Things shift and move in the borderland; back and forth like
the battered old pickups they drive over the Brownsville bridge without
any fuss, without any ceremony.

28. The immigrant problem in Texas

The trouble with the immigrants, of course, was that they were clannish, "lazy people of vicious character," overfond of hard liquor and
prone to let black people slave for them while they sat around drunk
in the heat. Many of them were fugitives from justice in their own
country; debtors on the run or worse. "GTT"—Gone to Texas—was a
euphemism for being on the lam. They were ignorant of the language,
picked fights, ogled the women (whom they then reviled as whores),
and huddled together in rickety little towns, the houses not laid out
in proper streets in the Mexican manner but scattered about "in
an irregular, desultory manner." Their "wretched little stores" sold
whiskey, rum, coffee, and sugar to their own kind with some rice
and flour—and of course ball and lead, for they were all trigger-
happy. There was no doubt, thought the young artillery officer José
Maria Sanchez on a tour of inspection of Texas in 1828, that Anglo-
Americans were unpromising material for integration into the free
Mexican republic. If they weren't stopped soon, they would swamp
the native population and culture. But it seemed impossible, not just
to arrest the immigrant tide, but to prevent them from taking land.
"They immigrate constantly, take possession of the places that suit
them best without asking leave or going through any formality other

than that of building their homes." When they were discovered settled on someone else's land, the original titles being impossible to find, it became impossible to dislodge them.

Sánchez's senior officer, a veteran of the Mexican War of Independence against Spain, an engineer and scientist, General Manuel de Mier y Terán, had been sent by the president to survey the border with the United States. He was less dismissive of the social caliber of the newcomers and therefore more anxious about the future. From Nacogdoches he wrote President Victoria that only in San Antonio de Béxar was there a substantial Mexican population capable of holding its own against the incoming tide of American immigrants. In most of the other towns they were outnumbered almost ten to one. And since the Americans either educated their children in their own schools or sent them home for schooling, the cultural divide was only going to grow wider. An alien culture had been embedded in Mexican soil, and pruning its vigorous shoot would just encourage a growth spurt. Worse, the Americans were clever enough to exploit the grievances that Tejanos (the local Hispanic population) had with their state government in distant Coahuila and the national authorities in Mexico City. Together, the unlikely alliance might agitate for Texan autonomy or even independence. The Anglos had a genius for converting personal grievances into political umbrage. They all seemed to think they were Thomas Jefferson. "Among these foreigners are . . . honest labourers, vagabonds and criminals but honourable and dishonourable alike travel with their political constitution in their pockets demanding the privileges, authority and offices which such a constitution guarantees." But Jefferson was, of course, an eater of territory, none hungrier; a true American. The real problem, Mier y Terán tried to tell Mexico City a year later, was that Texas was "contiguous to the most avid nation in the world," one which was very unlikely to stay put within its frontiers.

This sudden disenchantment hurt, since it had been the Mexicans themselves who, after securing independence in 1821, had thought Americans might be usefully tame colonists who would populate the arid regions of the north. And how persuasive they had been, those charming wolves! No sooner had Mexico won its liberty than the first of the pack, Moses Austin, erstwhile lead magnate and founder of Herculaneum, Missouri, had thought to recover from a

banking debacle by taking his chances in Texas. In the backwash of
the economic meltdown of 1819, the Austins were just one of countless families on the run from the consequences of their improvidence
who thought all would be well if only they could land the perfect
sweetheart deal. To the Mexicans it looked like a good match. They
had too few entrepreneurs, and land to spare. Hitch the one to the
other; watch the seedlings grow; in would come eager new migrants,
from both south and north; the government's coffers would smile
with revenue, and everyone would be happy. What could possibly
go wrong? So after a few go-rounds Moses Austin was, in principle,
granted the status of an *empresario*. This meant he was entitled to a
sizable land title on condition of his ability to attract 300 other settlers.
But an attack by bandits on the way home to Missouri left Moses badly
wounded and led to his death shortly after. However, his son Stephen
would—after much bother—see the promised land. Two years later
Stephen Austin became one of three founding *empresarios* and was so
successful that before long he had amassed 100,000 acres. The Texan
spread had been born.

What had taken the gringos south anyway? American trappers and
hunters had long been roaming around the mountainous regions of the
northern states of Mexico for hides and beaver pelts. (Every Victorian
top hat started as a beaver.) But the real enticement was the possibility of linking the old mule-train routes from Santa Fe, with its
loads of silver, wool, and hides, to the river basin of the Rio Grande.
Steamboats would change everything. Aboard paddle steamers, those
cargoes could reach the Gulf of Mexico and once ports had been built,
could be shipped anywhere in the world. Along the route, towns would
spring up. They would need feeding, so cattle ranches would supply
them with meat, and the hides would be turned to boots and saddles.
Perhaps even cotton and rice could be cultivated in the lowlands. Who
knew where it would stop?

The plan was unoriginal, which did not mean, however, that it
was unexciting. In Edinburgh, Manchester, and Westminster, British
political economists were saying the same thing about Bengal or
Argentina. In Paris, in the Grandes Ecoles, the French were developing
a developer's appetite for Algeria and Egypt. Like their European
counterparts, the Americans who cast hungry eyes on northern Mexico
and made sententious noises about the March of Progress required

certain conditions to be met—legal and political—before they would
be able to realize the ambitions that would benefit all, rich and poor,
natives and newcomers, hosts and guests, who together would rise on
the swelling tide of Improvement. What was more (another standard
lever of territorial insertion), the newcomers would supply security
for the native people against the depredations of marauding Indians;
in this case primarily the Comanche, who were certainly a tough nut
to crack. So, they liked to argue, everyone would gain from their
immigration. The parched land would receive the boon of capital and
the occupation of grazing herds; the "peons" (whom the Americans
regarded as just barely human at all, so steeped were they in sloth
and filth) would get a living wage; and the local elite would suddenly
find themselves on the highway of the world's traffic. But first there
had to be secure law and hospitable politics. The problem in Texas,
New Mexico, and California was that there was nothing between
the informal law dispensed by local alcalde magistrates with no legal
training, and the capricious interference of the central government.
Worse, juries were unknown. When the immigrants tried to assert
their own understanding of law and were rebuffed, it naturally caused
trouble. Thus came into being, for a few weeks, the sublimely named
Republic of Fredonia. Its president was not Groucho Marx but an
empresario called Haden Edwards who had been defeated in court over
land claims. What Mexico would not grant, the sovereign state of
Fredonia would bestow, so Edwards ran up a flag of his own design,
red, white, and blue of course, inscribed with the motto "Freedom,
Justice and Independence," before Fredonia was extinguished by the
arrival of Mexican troopers.

As Mier y Terán had observed, the immigrants were quick to
invoke the rhetoric of the American Declaration of Independence
whenever it suited them; another irony since the Mexican freedom
war had been consciously inspired by Washington, the Declaration of
Independence, and the Constitution. It was a commonplace among
the American Texans to complain about having to knuckle under to
a Mexican regime that, for all its lip service to liberalism and reform,
was still, they said, despotic, arbitrary, and priest-ridden. But when,
in 1829, that same Mexican government did something that took the
American Declaration of Independence's proclamation of liberty and
equality more seriously than the United States and abolished slavery,

it was taken as a hostile act, aimed at American immigrants who were overwhelmingly from Southern states and so had brought slaves with them into Texas. This act also made the Americans within the United States sit up, as Mexico—and Texas in particular—looked as though it might turn into a southern refuge for runaways. Compounding the effrontery, a year later, in April 1830, the Mexican government further decided that the only way to preempt the conversion of Texas into a de facto annex of the United States was to shut down American immigration altogether. Americans were welcome to immigrate elsewhere in Mexico if they chose. General Mier y Terán was handed the unenviable job of enforcing the law.

He knew that both the antislavery laws and the immigration ban were a dead letter without a substantial Mexican border force to keep out the Yankees. Mier y Terán built a string of forts from which vigilant anti-immigration patrols were sent out to scout *la frontera,* but it was as porous as it is today and gringo smugglers—white *coyotes*—knew the territory well. The beleaguered commandant then attempted, without much success, to bring Mexican settlers into Texas as a counterweight to the Anglos. But during the four years between the imposition of the ban and its revocation, the rate of illegal Anglo immigration soared, so that by the time of the Texan War of Independence (1835–36) there were around 40,000 of them along with their slaves and the fate of the region was sealed. It was just a matter of time before some grand 1776-like public utterance was made. Mier y Terán knew it; and between being unable to concentrate the attention of the Mexican government on what was inexorably happening and the tide of immigrants from the north, he was eaten up by despair. Seeing the loss of Texas, the loss of all north Mexico, he leaned the hilt end of his sword against a wall in 1832 and fell hard on it. The day before he had written to a friend: "A great and respectable nation of which we have dreamed . . . can never emerge from the disasters which have overtaken it . . . we are about to lose the northern provinces. How could we expect to hold Texas when we do not even agree among ourselves . . . *En que parara Texas? En lo que Dios quierera.* What is to become of Texas? Whatever God wills."

Not all Mexicans were so pessimistic or so prescient. There were a significant number who despised the aggressively centralizing regime of General Santa Anna in Mexico City as much as the Americans and were prepared to suspend disbelief and join them in the common cause of an independent Texan republic. Such a liberal state, they believed, could interpose itself between Mexico and the United States. But the declarations of independence issued from local Texan "committees of safety" were imprinted with the American view of what had happened and what was at stake. The San Augustine declaration, for example, stated that they had settled "an uninhabited wilderness . . . the haunt of wild beasts and hostile savages" but had been rewarded with the outrage of liberated Negroes and, through the 1830 immigration ban, the separation of families. It was time to recognize that "Anglo-Americans and . . . Mexicans, if not primitively a different people, habit, education and religion have made them essentially so. The two people cannot mingle together . . . And as long as the people of Texas belong to the Mexican nation, their interests will be jeopardised and their prosperity cramped."

For Tejanos like José Seguin and José Antonio Navarro who fought on the Texan side of the war, this was ominous. Their ardor was for a bicultural free liberal republic where Catholics and Protestants could each worship in their own way; two languages would be spoken and two peoples engage in a fraternal experiment on American soil; a sweet dream. What they got instead was reality: a dependency of the white United States and more particularly of the slaveholding South. The punitive massacres at Goliad and the Alamo that Santa Anna had inflicted on the Texan rebels before being routed by Sam Houston's army at San Jacinto had made a vindictive aftermath inevitable. Though the second president of the Texan republic was a Tejano, most of its Anglo population thought of its existence as merely a prelude to entry into the Union. Since that future was all but certain, they had no compunction in dispossessing as many Mexicans as they could; clearing them from whole regions just as Indians like the Cherokee, Creek, and Choctaw had been "removed" from their own ancestral homelands by President Andrew Jackson and his successors. Bitterly disillusioned, men like Seguin saw his hometown of San Antonio de Béxar, where he was mayor, become "the receptacle of the scum of [American] society . . . At every hour of the day and night," he wrote, "my countrymen ran to me for protection against the assaults or exactions of these adventurers. Sometimes by persuasion I prevailed on them to desist; sometimes also force had to be resorted to. How could I have done otherwise? Were not the victims my own countrymen, friends and associates? Could I leave them defenseless, exposed to the assaults of foreigners who, on the pretext they were Mexican, treated them worse than brutes?" In 1839, a hundred families were expelled from Nacogdoches. In Matagorda County a meeting, typical of the time, simply ordered a mass expulsion of Mexicans. A newspaper conceded that "to strangers this may seem wrong but we hold it to be perfectly right and highly necessary . . . in the first place there are none but the lower class of 'peon' Mexicans in the county, secondly they have no fixed domicile but hang around the plantations taking the likeliest Negro girls for wives and thirdly they often steal horses and those girls too . . . We should rather have anticipated an appeal to Lynch law than the mild course which has been adopted." The same would have happened later in San Antonio had not the local German population refused to participate.

So what was this Texas? Neither fish nor fowl, Americans both within and without its borders complained. And more important, were its people, the Texans, to be a hybrid nation? In the arguments over the fate of Texas and the border country the infection of race politics was everywhere. The American majority had no intention of being trapped in a place in which slavery had been made illegal, and which, since for some reason the Tejanos and Mexicans seemed to consort with blacks, free or slave, with no trouble at all, would become populated by a bastard race of mestizos! In such a place, the very virtues that made America America—vigor, enterprise, strength—had no chance of prospering, and Texas would be dragged down to just another "swarthy" backwater ruled by the siesta and the Mass. The only solution was annexation by the United States so that a proper social order—white sovereignty and black servitude—could be instituted.

That was what Andrew Jackson, the old Democrat frontiersman from Tennessee (as he advertised himself), wanted, pushing the United States further into the transcontinental destiny that Jefferson had charted. But Jackson left office in 1837, having declined to push the annexation issue. Perhaps this was because, for all the heroics on which his reputation had been founded in the 1812 war against the British, Jackson flinched before another war on two fronts: against Britain over the northern boundary of Oregon, and against Mexico over Texas. There was, after all, no regular American army of any size, a volunteer force was an unknown quantity, the militia were a crapshoot, and the West Point officer corps were mostly engineers. The Mexican government had never accepted the separation of Texas as legitimate and had made it clear that any annexation would be taken as an act of war, so the old man held off. Jackson's successor, Martin Van Buren, was still more cautious, not least because the story of Texas had now become inextricably linked with the fate of slavery in the Union. Annexation of Texas or resistance to it became a proxy rehearsal for the Civil War. Another president, John Tyler, a slaveholder, wanted to take the state to protect and enlarge slavery within the Union. For the same reason, abolitionists like the ex-president John Quincy Adams (John Adams's son) were against it for what it would do to the United States overall.

Enter, at this critical juncture around 1840, the real joker in the pack: imperial Britain. In London, an international antislavery convention had been addressed by no less an eminence than Queen Victoria's husband, Prince Albert, and featured a heavy presence of America's most fervent abolitionists like William Lloyd Garrison and Theodore Weld. Seven years before, in 1833, Parliament had abolished slavery throughout the empire, and the Royal Navy was now enforcing the law on the slaving coast of Africa, apprehending slavers and liberating their human cargoes. All this was taken by the politicians of the American South like John C. Calhoun as a profoundly hostile act against the United States. Beneath the sanctimonious self-righteousness, the real goal of British strategy, they judged, was the sabotage of the American economy. The growth of an Atlantic republic, destined to become a competitor for the global riches and power enjoyed by Britain, had to be nipped in the bud. Statesmen like Jackson looked at the map, knew what they knew about the selectively righteous British, and felt the nip from a pincer movement, an expansive Canada to the north and a British satellite to the south.

With the arrival of the Man in the White Hat (as he was known) in Galveston on the Gulf of Mexico in 1842, this scenario could no longer be dismissed as neurotic fantasy. Charles Elliott embodied everything that American nationalism feared and despised: Machiavellian opportunism dressed up as Victorian liberalism and polished with a patina of understated charm. After a promising career in the Royal Navy Elliott had become official Protector of Slaves on the Guinea coast. In China he had attempted to end both the Opium War and the opium trade by negotiations with imperial commissioners, a reasonable course that won him the enmity of the traders but also the possession of Hong Kong. Now he was chargé d'affaires to the Texan republic and rapidly became close with both its former and present presidents, Sam Houston and Anson Jones. Elliott's proposal was that Britain should broker a peace directly between Mexico and the Texan republic in which the former would accept the latter's independence, on condition of annexation to the United States being ruled out. Thus Britain could pose as peacemaker while sticking its finger in the eye of American expansionism and, doubtless, getting the usual considerations of trade and port facilities that it had exacted in South America and the China Sea. Galveston would be Hong Kong on the Gulf, and before you knew it, there would be clippers and docks, schools, limestone churches, and a decent little opera house, and vaqueros from Yorkshire would pause from branding the steers to sip a refreshing cup of tea. From the Texan side, such a deal was not altogether to be precluded, since a Mexican army of 30,000 was ready to march, while an American army was not, at least not yet, mobilized. Anson Jones heard Elliott out and extended the dalliance even in the face of most Texan Americans, who were hot for union.

Congress was faced with a double disaster. A permanently independent Texas, guaranteed by Britain, Mexico, and possibly France as well, would turn into an American nightmare: a home for runaway slaves in which the likes of Charles Elliott could wave the flag of Christian humanitarianism, creating a refuge even more subversive than Canada because closer to the southern heartland. Worse still, it was rumored that German farmers were growing cotton on the Rio Grande without benefit of slaves so that an alternative source of the raw material might be available to British manufacturers. The economic empire of the British would have suddenly acquired a stupendous strategic extension, running all the way from southern Oregon down through the Rockies to the Santa Fe Trail and on via the Rio Grande to the Gulf of Mexico. Everything that Britain had lost in 1783 might be recovered sixty years later. On the British side there was a good deal of jolly hand-rubbing. Who cared about Delaware when British California was in the offing?

From his hermitage near Nashville, the ailing Jackson saw the

British threat with chilling clarity. Suddenly Elliott's personal history—prosecutor of slave traders and China hand—all became part of a devilish British plot to seize, in all but formal title, not just the Gulf of Mexico but the American Pacific! Already, Chinese and Japanese trade and markets were seen as a long-term prize of immeasurable value. A race was on, as Stephen Douglas, Lincoln's victorious opponent for the Senate seat of Illinois, put it, "for the maritime ascendancy of these waters." Had the frontier, the engine of American history, consumed so much time, so many lives in the move west along the Oregon Trail only to be headed off at the pass to California by the damnable old foe, Britain? A year before he died Jackson warned that to sit still in the face of British geo-economic political designs on the Texas-California territories would be to submit to an "iron hoop" locked about the neck of the American future that would "cost oceans of blood to burst asunder."

But 1844 was an election year, and James K. Polk, a small-town lawyer from Jackson's Tennessee, was running as Old Hickory's protégé and heir; the flag-waver of nationalist Democrats, committed to the annexation of Texas. The British gambit, which had been meant to give the American war hawks pause, had the opposite effect, injecting into the campaign and public debate a note of unprecedentedly fierce nationalist zeal, voiced even by those, like Ralph Waldo Emerson and Walt Whitman, who had no sympathy for the slave-owning South. Emerson could not have been happy that in February 1845 Congress passed a resolution admitting Texas to the Union (signed by President Tyler in the last week of his term), leaving the matter of slavery to the Texans themselves. In August 1845, local Texan conventions designed to set the seal on annexation adopted an article in their own constitution forbidding any act of emancipation without the consent of slave owners and mandating compensation to them for any losses. But, Emerson may have reasoned, that decision might still be reversed. The paramount fact was the steam-driven locomotion of American history pointing south and west right across the continent. In an overexcited speech at the Boston Mercantile Library in 1844, Emerson called on the "Young American" to embrace the "sublime and Friendly Destiny" of the nation. "The bountiful continent is ours," Emerson proclaimed, "state on state, territory on territory, to the waves of the Pacific sea." If biting off half of Mexico, digesting

Hispanic America into the system of Anglo-America were the way for that destiny to be accomplished, so be it. For then the slumbering sombrero despotisms of that world would be awakened by the rough kiss of industrial democracy. Roll the Conestoga wagons (3,000 were on their way west to Oregon as Emerson spoke), sound the bugles, hitch up the caboose!

The campaign season that ended up electing President Polk and the first year of his term stirred the press into a lather of nationalist elation, in which the weighty phrase "manifest destiny" was heard for the first time. John Louis O'Sullivan, who published Whitman and Emerson, among others, in his *Democratic Review*, took umbrage against British interference, intended as it was to check "the fulfilment of our manifest destiny to overspread the continent allotted by Providence for the free development of our yearly multiplying millions." He and his cowriter, the remarkable woman journalist (and Texas land speculator) Jane McManus Storm, would repeat the phrase like a mantra until it became an almost obligatory item in the democrat-nationalist repertoire. Writing in both the penny paper the *New York Sun* (where, at the age of thirty-nine, she became national news editor) and the *Democratic Review*, Storm and O'Sullivan may well have reached a readership of quarter of a million readers and in July 1845 simply ordered the opposition to Texan annexation to desist. "It is at last time for common sense to acquiesce with decent grace to the inevitable and the irrecoverable. Texas is now ours." For O'Sullivan, the moment was all about the demographic future of America. Texas, California, and New Mexico had to be part of the United States, because the population of the nation would swell to no less than 250 million, dwarfing the decrepit empires of the Old World. And when O'Sullivan and Storm thought of that immense population they certainly did not have a "swarthy" hue in mind. The West would be white.

But neither Storm nor O'Sullivan wished to accomplish this act of territorial incorporation through war. Although the Mexican government had broken off diplomatic relations after the American declaration of annexation, Polk had sent an emissary to Mexico City in December 1845 to propose assuming Mexican debts in return for their consenting to the annexation of Texas, and further proposed buying California and New Mexico. The advance groundwork had not been well prepared. A fiercely conservative government had just come to power in Mexico, took the purchase offer as insulting, and buried any earlier thoughts of letting Texas go. Storm suspected Polk of bad faith in that he had sent his agent knowing full well he would be spurned, thus giving the president a pretext for hostilities. Writing in the *New York Sun* under the byline of "Cora Montgomery" that she would use as war correspondent, she denounced the "class of politicians that are anxious to bathe the country in blood to win notoriety and office for themselves." She and O'Sullivan wanted it both ways: maximum territorial expansion with minimum casualties. Having urged everyone forward, O'Sullivan now skidded to a halt. "We are," he wrote in May 1846, "on the threshold of a long, troublesome, destructive and expensive war."

Polk and his gung-ho secretary of the treasury, the Mississippian champion of slavery Robert J. Walker, were undeterred. Affronted by the additional threat to California, the Mexicans had declared that the issue would be settled by arms, and Polk was only too happy to oblige. A call went out to "Young America" for 75,000 volunteers. Enormous numbers enlisted, especially from the southern and Appalachian states. The veteran General Zachary Taylor was ordered into territory between the Nueces River (hitherto accepted as the Texan boundary) and the Rio Grande, which the United States now claimed to be its southern frontier. There he was attacked by a Mexican force, providing Polk in his declaration of war with the justification that "American blood had been shed on American soil." It took a young Abraham Lincoln, brave to the point of foolhardiness, to point out that as far as the Mexicans were concerned land up to the Nueces was indisputably theirs, and thus it was they, and not the United States, who could consider themselves the victims of invasion.

No one ever accused Abraham Lincoln of running after popularity. For many of the young soldiers, from the new immigrants of Massachusetts to Kentucky farmers, the battles were their own proving ground as well as a triumphant demonstration of the racial superiority of Anglo-America over the "mongrel" breed of the Mexicans. Their technology-driven society manufactured the guns which, at the battle of Palo Alto on the Texan chaparral, could outload and outshoot the Mexicans by three to one. The outcome, the thunderers back at home insisted, was inevitable, a contest of unequal races. Whiteness, Protestantism, and superior technology were all interconnected even though no one could give a coherent explanation as to how.

The converse, though, Americans thought, was self-evident. The Mexicans were bound to lose, wrote James Gordon Bennett in the *New York Herald*, because of the "imbecility and degradation" of their people due to "the amalgamation of races." But what was good for the war might yet prove bad for the peace. When Mexico City surrendered to General Winfield Scott, and his officers and men took a good look around at what and whom they had conquered, the popularization of the idea that the "mongrel" mestizo race had succumbed to the racially superior army from the north generated an especially ugly debate between territorial maximalists who thought History beckoned the United States to swallow the entirety of Mexico, and the racial purists who saw this as inviting the infection of miscegenation into the pure body of Anglo-America. It was all very well for troops to amuse themselves in cantinas with black-eyed senoritas in their beguiling décolletage (about which soldiers wrote with the compulsive mock-prurience of the guiltily aroused). It was quite another to expose white America to the Mexican lasciviousness that would be its undoing. Sex and priesthood played darkly (often in the same sentence) on American minds. Lieutenant Ralph Kirkham, who was with Scott's army in Mexico City, wrote to his wife in New England: "I suppose there is no nation on earth where there is so much wickedness and vice of all kinds . . . instances are common of men selling their wives and daughters. The clergy, generally, are very immoral and ready to stoop to the very lowest acts of villainy and wickedness." So although there was an influential body of opinion, both in Congress and the press, in favor of annexing the entirety of Mexico, the guardians of race, for whom America was Anglo-Saxon or nothing, warned of dire cultural consequences. John C. Calhoun, the most militant defender of the rights of the slave states, was adamant in his opposition to all-Mexico annexation. Latin America's sorry condition, he thought (not entirely accurately), was precisely its peculiar habit of "placing these coloured men on an equality with the white race." Why would the United States import such an error into its own social constitution? "Ours, sir, is the government of the white race."

Then there were the political consequences to be considered, given that the peace treaty conferred the same rights, including the right to vote, on any Tejanos opting to remain in the newest territories of the Union. An old ranger encountered by the landscape architect Frederick Law Olmsted (the man who with the Englishman Calvert Vaux would create New York's Central Park) during a "saddlebag" reporting trip to Texas in 1854 for the *New York Daily Times* spoke for many: "Mexico! What the hell do we want of it? It isn't worth a cuss. The people are as bigoted and ignorant as the devil's children. They haven't even the capacities of my black boy . . . You go any further into Mexico with surveyors' chains, you'll get Mexicans along with your territory and a damned lot of 'em too. What are you going to do with 'em? You can't drive 'em out because there ain't nowhere to drive 'em. No sir, there they've got to stay and it'll be fifty years before you can outvote them." Polk might almost have been listening to the ranger. When peace terms were imposed on Mexico, more or less at gunpoint, lopping away half its territory, he made sure it was the half with the least Mexicans in it.

Voices were raised in dissent at this spectacular increase in American land, a stretch of territory that took in not only California, New Mexico and what would be Arizona, but also large areas of Colorado, Utah, Nevada, and Wyoming. They were precious few, but the penetration of their fury went beyond the poverty of their numbers. Lincoln's sarcasm at the transparent hypocrisy by which the United States had made its casus belli was matched by the smoke going up from beside the placid banks of Walden Pond. "How does it become a man to behave toward this American government today?" asked Henry David Thoreau. "I answer that he cannot without disgrace be associated with it . . . when . . . a sixth of the population of a nation which has undertaken to be the refuge of liberty are slaves and a whole country is unjustly overrun and conquered by a foreign army, and subjected to military law, I think it is not too soon for honest men to rebel and revolutionise. What makes this duty the more urgent is the fact that the country overrun is not our own but ours is the invading army."

As has often been said, at the time when Hispanic America became Anglo-America, Mexicans did not cross borders; the borders crossed them. The question, after the war, when the Treaty of Guadalupe Hidalgo was signed, setting the southern boundary of the United States at the Rio Grande, was whether or not those who had been Mexican and were now American would be treated with all the rights

of citizenship formally promised them. But even before the treaty was
ratified there were ominous indicators. Article X, which protected older
Mexican land titles, was stricken from the treaty by the United States
Senate lest it put in question the later claims of Anglo-Americans made
during the period of the Texan republic. To pacify Mexican anxieties,
Secretary of State James Buchanan told his counterparts that if there
were valid titles they would always be upheld in American courts. To
revive ancient and specious claims against settlers who bought property, said Buchanan, would be "an act of wanton cruelty." Nonetheless
those former Mexicans who remained in the ceded territories—and
there were tens of thousands of them—were promised "all the rights
of citizens of the United States."

What ensued of course was grimly predictable: the force of conquest
imposed on a helpless people; the same that had occurred during
the Texas republic only more so: evictions, dispossessions, physical
intimidation, lynchings. Mexican cartmen were attacked by gangs of
masked armed men who meant to ensure that Anglos would have a
monopoly of the local carrying trade. When the numbers of those killed
in the "cartman wars" rose to seventy-five, the Mexican Embassy in
Washington made a formal protest and the secretary of state wrote a
stiff letter to Governor Pease of Texas about the "violations of rights
guaranteed under the law" and urging "energetic measures to punish
the aggressors."

When Olmsted arrived in 1854, he found a society of conquerors
and subjects. The protections of Guadalupe Hidalgo were already a
joke. "Ignorant of their rights and of the new language," Olmsted
wrote, the Tejanos had "allowed themselves to be imposed on by the
newcomers who seized their lands and property without a shadow of
a claim and drove hundreds across the Rio Grande." He had hardly
been there a few days when he ran into a white woman who let it be
known that she regarded the Mexicans "not as heretics or heathens
to be converted with flannel and tracts but rather as vermin to be
exterminated. The lady was particularly strong in her prejudices [saying
that] white folks and Mexicans were never made to live together
anyhow and the Mexicans had no business here. They were getting so
impertinent and were so well protected by the laws that the Americans
would just have to git together and drive them all out of the country."
That process, Olmsted believed, was already under way. "Last year a

large band of Texas Free Companies plundered and burned in mere wantonness a peaceful Mexican town on the Rio Grande; 400 United States troops listening to the shrieks of fleeing women and looking on in indolence. This has passed without a rebuke and with entire public and official indifference."

San Antonio was the town where Olmsted, with his shrewd eye and ear, tested the Texan future. With a population of some 10,000 in 1850, it was the one place where the Tejano population had mostly stayed put after the war, believing, after all, it was still their city. But what the increasingly mordant Olmsted saw was "the first of a new class of conquered cities into whose decaying streets our rattling life is to be infused." The hard-bitten Yankee reporter became romantic, drawn to a place that, from the way it went about its days, resisted wholesale Anglo makeover. The "easy lollopy sort of life" others brushed aside as a barrier to hustling progress, Olmsted, who was not especially lollopy himself, saw had "been adopted as possessing on the whole the greatest advantages for a reasonable being." In San Antonio he let go his puritan striving. Instead Olmsted's enthusiasm rose to the sound of drums and trombones heralding the "Mexican mountebank" street entertainers who played three times a week, wearing glittering short coats and spangled tights. He walked among the crowds of children and adults, his hands happily greasy with tamales, letting the sounds and smells take him. The mountebanks were a brilliant contrast to the "thin local company" of tragedians who oiled their hair and flourished rapiers for the Anglos, to the usual accompaniment of "peanuts and yells."

Though Mexican laborers earned a paltry eight dollars a *month*, Olmsted admired the way their lives in the town had stayed unchanged. The ruined Alamo already stood as a holy place for the Anglo-Texan version of their history, but surrounding it were streets in which house doors and windows stayed open; where cats, dogs and gamecocks strutted to and fro; and strangers like himself were received "with gracious and beaming politeness and dignity." In the warm evenings Olmsted took pleasure watching the affection lavished on children as they walked or ran with the family promenades, as serenades would suddenly burst forth on a street corner.

There was a good deal of picturesque sentimentality and condescension in Olmsted's portrait of Tejano San Antonio, and, looking for good news, he may have exaggerated the degree to which Mexicans made "no distinction from pride of race." But in San Antonio he did at least find a "jumbled" America that he thought ought to be celebrated rather than abhorred as degenerate in keeping with the race theory of the age. And the big surprise for Olmsted was the part that Germans played in this cultural mosaic. Over a third of San Antonio was German immigrant, and there were another 5,000 in the settlement town of New Braunfels some fifteen miles away. Their Texan immigration had begun in the mid-1840s, when much of the south German countryside suffered from the same potato blight and tenant insecurity that drove the Irish to leave for America. Another kind of natural disaster had taken its toll on the German countryside: catastrophic flooding in the Rhine, Mosel, and Elbe valleys that had made land unworkable and wiped out harvests years in a row. Governments in both Britain and the German states had actually subsidized emigration in calculated attempts to lighten the burden of the poor on their middle-class taxpayers; a practice that hostile American nativists would characterize as "dumping." Those same governments in Germany were equally glad to be rid of unwanted troublemakers from the failed revolutions of 1848–49: journalists, professors, physicians, poets—the usual crowd.

To Olmsted it was perfectly clear that it was immigrants brought to Texas by remote relatives of Queen Victoria (the princes of Leiningen and Solms-Braunfels) who showed the way to cultural neighborliness. While less racist toward the Tejanos than, the Anglos, the Germans had created their own world in town and country: reading the *San Antonische* or the *Neu-Braunfels Zeitung*; living first in log cabins and then in sturdier brick houses; opening hotels where the rooms were embellished with dark oak chests and wardrobes, and lithographic prints of American and German scenery on the walls. Olmsted noticed that some of the farmers practiced the kind of intensive agriculture that had worked in Germany, irrespective of the acreage spreading away to the Texan horizon, and rather admired them for it, notwithstanding the ridicule they sometimes got from Anglo ranchers. It all seemed a fine little version of America. The farmers' wives churned butter and cured famously savory ham; the children were sent to Lutheran or Catholic free schools; agricultural societies, mechanics' institutes, horticultural clubs, and of course Harmonie societies all flourished. All of which might have vindicated Benjamin Franklin's fears that the Germans would create their own closed-off world within Anglo-America; impervious if not hostile to the political values embodied in the Constitution. But that was not the little Germany that Olmsted reported on in south Texas. Instead it was the Germans who best managed to square true American republicanism with liberal coexistence alongside the older Tejano society. When an ugly race riot flared up in San Antonio in 1854 and the sheriff called on a posse of 500 volunteers to clear the town of Mexicans, it had been the young Germans who had balked and shamed the Anglos into desisting, saying, "it was not the right republican way." Even more provocative to the Anglo majority that had, after all, become American to protect slavery, were German farms producing cotton with free labor, even daring to send the crop to market clearly labeled as such.

For the Germans, being Texan was about two things: the possibility of a better economic life, but most of all about freedom. Olmsted spoke to a shoemaker who admitted that he had less "comfort" than in the old country but when asked why he liked it in America, replied "because here I am free. In Germany I cannot say how I shall be governed. They govern the people with soldiers. They tried to make me a soldier too but I ran away." He planned to return to Germany to find his sweetheart and bring her back to Texas. Won't you be arrested? Olmsted asked. Oh no, replied the shoemaker, full of simple American faith, "for then I shall be a citizen."

29. The German threat—again

The shoemaker of New Braunfels was lucky not to be in Louisville, Kentucky, the following August, or he might have had his belief in American justice and freedom badly shaken. In fact, he might not have gotten out of town alive. On 6 August 1855, in Louisville, a riot destroyed houses and stores in the German and Irish sections of town and killed at least twenty-two immigrants. Some were burned alive in their houses; some were stoned, others knifed or lynched. Only the mayor of the city, commandeering churches as refuges, stood between the rioters and a much more deadly toll of victims. Like a similar three-day riot in Philadelphia in 1844, the violence was above all anti-Catholic, but anyone with a German name, be they Lutheran or even Jewish, was a likely target of attack. The Germans of Texas were fortunate, in the climate of hysterical xenophobia that swept through the United States in the mid-1850s, that angry Anglo-Americans had someone else to ride: the Tejano Mexicans.

The paradox was this: at the same time that immigration to the United States reached a peak, with 655,000 arriving in the single year of 1855, in hope of a new life beyond the reach of despotism and destitution, American cities were in the grip of nativist hatred. The percentage of the total American population constituted by the foreign-born rose in that year to 14 percent, a proportion that would remain more or less constant right through the nineteenth century until the restrictive legislation of the 1920s reduced immigration to a quota-monitored trickle. A full 50 percent of the newcomers in 1855 were German; a majority of them Catholic, but outnumbering the more highly publicized Irish immigration two to one. In cities like Louisville, Cincinnati, Milwaukee, St. Louis, and Chicago, they built bigger versions of New Braunfels, with German-language newspaper presses, German-language schools, and their own churches, which did not, however, prevent them from creating beer gardens where bands oompahed and steins flowed freely on Sundays, much to the horror of temperance-minded Protestants. Worse still, German Catholic schools replaced the King James Bible with the Counter-Reformation, Jesuit-approved Douai Bible.

For the nativist journalists and politicians who claimed to embody authentic American values, the addition of south German Catholics to an already swollen Irish Catholic population in cities like Boston and New York spelled the doom of democracy in the United States. "A Romanist," one of those journalists wrote, "is, by necessity, a foe to the very principles we embody in our laws, and a foe to all that we hold dear."

These views were not gutter politics. As early as 1834, Samuel Morse (painter and inventor of the telegraph, from Charlestown, Massachusetts) had warned, in a series of letters to the New York Observer, of a popish conspiracy to undermine the American Constitution. Morse was said to have had a Protestant epiphany in Rome when, refusing to remove his hat in the presence of the pope in St. Peter's Square, a

Swiss guard had knocked it off his head. The affronted head became hot with indignation. "Surely," he wrote, "American Protestant freemen have enough discernment to see beneath them the cloven foot of this subtle foreign heresy." Hordes of illiterate credulous Catholics were being mobilized—from Austria, apparently—to invade America and enthrall it to "a system of darkest political intrigue and despotism." Collected as *Foreign Conspiracy Against the Liberties of the United States*, Morse's published letters did little to help his own run for mayor of New York in 1836, but they lit a fire under influential preachers like Lyman Beecher, the Presbyterian clergyman whose Boston sermon against the Catholic invasion of the West was duly followed by the burning of an Ursuline convent in that city. Beecher was the father of Harriet Beecher Stowe and himself an ardent abolitionist and temperance reformer; all of which interestingly complicates anti-immigrant history, for the likes of Morse and Beecher and many that followed believed themselves to be acting in defense of liberal democracy, and against Catholic reaction, when they demonized Irish and German immigrants. The revolution they looked to was as much the "Glorious" English Revolution of 1688 as the American revolution. The latter, they reasoned historically, was the fruit of the former. The Glorious Revolution of 1688 had, after all, removed the Catholic James Stuart from the throne and replaced him with Protestant Dutch William, and had been necessary for the survival of English parliamentary liberty. Had not Lord Macaulay (immensely popular in the United States) said as much? Sometime in the eighteenth century, Hanoverian oligarchy and "ministerial despotism" had irreversibly perverted that precious constitution. It had been left to the Founding Fathers to rescue English Liberty from the British and give it a second life on the far side of the Atlantic.

Now that liberty was being threatened by something much worse than a Stuart monarch: millions of unstoppable foot soldiers of the pope. Following Pius IX's declaration that freedom of conscience was anathema, this army of immigrants was poised to install, through sheer weight of numbers, a Catholic absolutism in America. By refusing the reading of the King James Bible, they had begun the process of indoctrinating their children and inoculating them against the Constitution. There was worse. Ignorant, they bred like rabbits, lived in filth, and generated disease and crime. Already poor or semi-criminal when they came, they were turning American cities into verminous tenements where the rum-hole and the criminal gang ruled. They ran, ratlike, in packs to the polls, and because they were useful to the unscrupulous ward bosses of the Democratic Party, were rewarded with disproportionate positions in the police, so that crime and "law enforcement" were indivisibly part of the same racket. Worst of all, for Morse and Beecher and the much more powerful politicians who built an entire popular movement on this anti-Catholic creed, the Irish in particular were notoriously hostile (not just indifferent) to the sufferings of the enslaved Negro. Put all this together and what did you get? A whiskey-soaked, priest-governed, black-hating, socially delinquent city swarm, numerous enough to impose their will at the polls. Good-bye liberty; farewell America.

Catholics couldn't win. Hispanics in the South were attacked by the defenders of Anglo-Saxon America as being altogether too friendly with blacks; while in the North the Germans and Irish were attacked for not being friendly enough. The tide of hostility that rose alongside the immigration figures in the late 1840s and early 1850s initially took the form of semi-secret "lodges" like the Organization of United Americans and the Order of the Star-Spangled Banner. Though "mechanics" (artisans) and workmen supplied many of the rank and file, members of the professional and commercial classes joined in droves in the big eastern and midwestern cities. To qualify, members had to take an oath that they were Protestant and the children of Protestants and when asked by outsiders about their organization were to reply that they "knew nothing," hence the name of what became the briefly formidable political party of the Know-Nothings. At their height they had a million members and elected the mayor of Chicago, who immediately banned immigrants from city posts. The Know-Nothings were pressing in the first instance to make the waiting period before naturalization a full twenty-one years, corresponding to the age of maturity. In effect this was to confine citizenship to those who had either been born in the United States or managed to demonstrate their unswerving allegiance over a long period. Paupers and criminals were to be denied admission, and some of the more extreme of the Know-Nothings wanted to deny Catholics both office and the vote.

They did not have the floor to themselves. Other high-minded members of the New England intellectual elite were equally resolute in defending the Crèvecoeur and Tom Paine tradition of asylum. In 1851 Edward Everett Hale wrote a series of articles for his Boston newspaper the *Daily Advertiser* in which he depicted the helpless Irish peasants, victimized by cruel landlords and British heedlessness, as standing with their backs to the sea at Galway, finally driven to the ships "by a charge of bayonets." If the survivors were poor and wretched, Hale said, that was all the more reason for America to do everything it could to give them a new life. "The state should stop at once its efforts to sweep them back; it cannot do it; it ought not to do it. It should welcome them, register them, send them at once to the labor-needing regions; care for them if they are sick."

In the early 1850s, though, this was a minority view. And in a peculiar moment of party political giddiness, Know-Nothing prejudices coalesced into an actual political party and program. The Democrats were tagged as the party of immigration, and as the party that would leave the slave South alone. Against them were the heirs to Hamilton's Federalists, the (slightly) higher-minded Whigs. But Whig unity collapsed over the tactics to be adopted to preserve a nation bitterly divided over slavery. The Whig president Millard Fillmore, who had come to the White House after Zachary Taylor had died from his iced milk on 4 July, believed he could manage the pressures pulling north and south. His compromise was to admit California to the Union as a free state but allow federal agents to enforce the Fugitive Slave Act, hunting down runaways and returning them South. Instead, Fillmore succeeded in destroying his own party. Horrified northern Whig abolitionists searched for somewhere to go. The Protestant intensity of the Know-Nothings—their devotion to abolitionism and temperance— gave them a place to pitch their tent, and into that camp they poured in their hundreds of thousands. So the most peculiar political party in American history was at once violently xenophobic and the friend of African Americans. Either way, they were no fans of Catholics.

The head-spinning contradiction only makes sense through the Know-Nothing insistence that both attitudes were the Protestant way. The reformed religion presupposed an educated Bible-reading Christian in personal communion with his God—blacks most certainly included—just as true American democracy presupposed an informed, educated man in communion with his vote and the spirit of the Founding Fathers. Both were inimical to receiving orders from priests and popes.

Once a new Republican Party became organized on the ruins of the Whigs, committed to halt, even at the cost of conflict, the spread of slavery in the Union, the Know-Nothings had served their turn and fell apart as quickly as they had arisen. But they left a bad smell behind. The view expressed by one militant Know-Nothing, Daniel Ullmann in *The American* in April 1855, would turn out to endure well beyond the life of his quixotic party: "Where races dwell together on the same soil and do not assimilate they can never form one great people, one great nationality . . . [America] must mold, absorb the castes, races and nationalities into one homogeneous American race."

30. The importance of Fred Bee

There's a photograph of the colonel, in front of a tent, standing next to Wong Sic Chien, the Chinese consul from New York, and their colleagues in the Investigation Commission, taken in Rock Springs, Wyoming: a redbrick coal town on the upmost reaches of the Colorado. It's still a Union Pacific freight yard, linked to its bigger neighbor Green River. With gas at four bucks a gallon, business is looking up, and the chains of freight wagons, many of them with Japanese markings, roll through the yards for a good fifteen, twenty minutes a time. Looking down on Rock Springs from surrounding hills are the usual strip malls, Rite Aid, Starbucks, KFC; tree-lined streets with their natty two-car suburban driveways. From this eminence modern Rock Springs surveys the relics of a lost industrial Wyoming: the Beaux Arts or Flemish-gabled banks half boarded up; more churches than a decayed coke town rightly needs, and between them Asiatic manicure parlors and funky tattoo stores; a gesture of gentrification here and there: the craft-ale bar, scrubbed pine and stainless steel brewing vats. No Chinese restaurants, not that I saw: a bad memory that won't quite go away; twenty-eight Chinese massacred in a hate riot in 1885. The crime for which they paid with their lives had been to decline to join a strike organized by Welsh and Swedish miners with whom they shared the pit and the town. No love lost. When it came to it, the Welsh colliers set Chinatown on fire, shot up anyone trying to leave; 400 were driven into the hills, where more died of exposure attempting to get to Green River.

That was the sort of thing that got old Fred stirred up. He reckoned someone had to be, and he was, after all, Chinese consul for San Francisco; though he didn't much look the part with his white whiskers, old-fashioned high-collar coat, and plug hat. In his time, Fred Bee had been many things, the sort of things you would expect from a Placer County man: soldier, founder-investor in the Pony Express, the Sausalito Land and Ferry Company north of San Francisco, and, in 1858, the Placerville-St. Joseph (Overland) Telegraph Company, the parent of Western Union from which he had made a nice little pile. But Fred Bee wasn't made for a quiet, pipe-sucking kind of life, and since he was an attorney, he spent most of his time going to the law for the Chinese Consolidated Benevolent Association. He figured it was a matter of upholding the dignity of the country; the integrity of its justice. The Burlingame Treaty of 1868, which had confirmed the right of the Chinese to "expatriate" themselves to the United States (there being an awful lot of railroad track to lay at that particular time), had reciprocally allowed Yankees to go and trade whatever they liked—tobacco or souls—in Ch'ing China. The treaty had also guaranteed that these migrations would be voluntary. Chinese migrants would be free to come and go within the United States as peaceably as they wished. Though Americans in China insisted on their own extraterritorial jurisdiction, not being keen on imperial law, the Chinese migrants were to enjoy the full protection of the law afforded to citizens, even though they had been permanently disqualified from naturalization.

If America was now a continental nation, courtesy of the meeting of the Union Pacific and the Central Pacific in May 1869, it was the Chinese who had made it so; not just the railroad graders and exploders and masons, but the multitudes of men who made a rough life bearable: the launderers and the dry-goods merchants and woodcutters and sometimes camp girls too. And instead of thanks, what they had got, Fred Bee noticed, was the smell of their Chinatowns burning to the ground, lynch mobs, and summary orders to leave town sharpish. So they went at gunpoint, the mobs wagging lengths of rope at them in Rock Springs, women laughing and clapping as the Chinese miners shuffled into the darkness pushing their sad, overloaded handcarts. This had happened in Rocklin and the rest of the Sacramento Valley, in Eureka and Truckee and Tacoma, Seattle and San Jose and countless other places up and down the West Coast; and governor after governor had turned a blind eye, knowing mayors and police chiefs were in cahoots. The least that decent America could do for these innocents, Colonel Bee thought, was to make reparation for their losses and their suffering. It had been Rocklin in 1877 that had first got him aggravated. But the story had begun long before that.

It had begun with the Great Greaser Extermination Meeting, as the California gold miners called it. Their problem then had not been with the Chinese but with the Latinos who were sitting right on the American Eldorado and had the cheek to imagine they might get a share of it. It hadn't been one of the Mexicans or Chileans or Californios who had made the first strike, after all. It had been James Marshall working down by the tail race of the sawmill he was building for Mr. John Sutter on an icy morning in January 1848. But the "greasers" had got their hands on gold around Sonora. It's true they had actually built the town in the first place, naming it after their homeland in northwest Mexico, meaning to search for the gold deposits they knew were there. It's true that they had been mining for generations before California had seen hide or hair of the Anglos. "Placer" mining, the sifting of gold specks and nuggets from the pack of debris and dirt that had been eroded away from rock veins, was an old Mexican technique. *Placera* was Spanish for alluvial deposit.

For the Americans who rushed to the southern Sierra Nevada in 1848 and 1849, the prior presence of so many Chileans, Argentinians, and, more inconveniently, Californios whom the new-minted Treaty of Guadalupe Hidalgo had now decreed to be Americans, was an irritation. But they could be dislodged without too much trouble by tried-and-true methods of threats, assaults on the camps, the occasional lynching and race riot. Hence the Great Greaser Extermination Meeting summoned to coordinate all these efforts. But when the Chileans had the gall to organize and arm themselves in defense, institutional means could do the trick. A tax was imposed on "Foreign Miners" of three dollars a month. Although Californios and Mexicans were now part of the postwar United States, merely speaking Spanish was often enough to be obliged to pay the punitive duty. It worked. At the end of 1849 there were 15,000 Hispanic miners working the placers around Sonora and elsewhere in the southern Sierra. A year later just 5,000 remained. It would not be their gold rush.

In 1852, the Foreign Miners Tax was reissued with a specific proviso exempting anyone who might in due course become a naturalized citizen. This meant the duty was aimed at those who, it had been decided, could never achieve citizenship; who had been declared constitutionally unassimilable. This meant the Chinese. There were only a few thousand of them in California at this time, but already the defensive locals were beginning to dread the hordes to come. Unless of course they were the ones making money from their trans-shipment, like the labor-broker Cornelius Koopmanschap of San Francisco and Canton. He and shippers like him who knew south China well had struck up a profitable relationship with their counterparts in the Pearl River Delta, the deepest reservoir of Chinese emigrants to the United States. Their province of Guangdong was the one most deeply penetrated by Western guns and money. British victory in the Opium War had opened Hong Kong and Canton, and the endless cycle of misery, famine, epidemic, and civil war had created a well of desperation on which the labor merchants could draw for recruits. The men who went to the Gold Mountain, as the United States was now known, would send the money they earned back to their homes and villages, followed in due course by their triumphant selves. And though the anti-Chinese campaign always called them "coolies," they were not going as oriental slaves. But the terms on which they emigrated, agreeing to pay back the merchants who put up their transport and medical costs (with a healthy markup) from future earnings, meant they were not exactly free either. The emigrants were indentured laborers, subjected to the debt sovereignty of the merchants and, once they got to California, the Chinese Six Companies society that officially managed the interests of the community.

But still, it would be Gold Mountain. "They want the Chinaman to come and will make him welcome" promised the flyers in China. "There will be big pay, large houses and food and clothing of the finest description . . . It is a nice country without mandarins and soldiers. All alike; big man no larger than little man."

According to the Six Companies, by 1855 there were more than

40,000 Chinese immigrants in California. Apart from the teenage girls who were often abducted into prostitution or sold by their families, they were all male; and not all of them came to work as miners. From the beginning there was an astute realization among the Chinese communities that the bachelor world of the miners would require a host of goods and services that women normally provided, other than sex: laundry, groceries, cooking, hostelries, firewood, fresh produce, fish. Those were all menial jobs that had mostly been beneath male dignity in south China but would provide a steady living if the gold, literally, did not pan out.

They came in shiploads, packed so tightly into holds they might as well have been slaves, and subjected to the ferocious discipline of the merchant shippers. When they got to the Sierra, abuse and intimidation immediately followed. In 1852 a gang of 60 white miners attacked 200 defenseless Chinese men at their camp on the American River in Tuolumne County and then went on to take the assault to another 400 downstream. Repeatedly thereafter they were threatened at gun- or knifepoint. But many still persisted, following the hope of fortune. Though a California law had prohibited the Chinese from legally filing claims, some took over mines abandoned by whites and, because they came from water country at home, understood the flumes well enough to do better. Others went farther afield, to Nevada, Oregon, and Washington, and up the Rockies to Idaho, Wyoming, and Montana. By 1870, a quarter of all miners in the West were Chinese, and paranoid hatred stalked them wherever they went.

But the business opportunities for the merchants who were supplying Chinese labor were just beginning to open up. In 1862 Congress, at Lincoln's behest, passed the funding for a transcontinental railroad, and work began in 1863. The timing was not accidental, coming as it did in the middle of the war. If the politics of slavery were tearing the Union apart, the railroads would make an end run around the strife and knit the country together again. The westward line of the Union Pacific was dominated by Irish labor, using picks and mules, and living in some of the rowdiest work camps in America. Their opposite number on the Central Pacific, moving east from Sacramento, faced the most daunting challenge with a steep rise of 7,000 feet in just a hundred or so miles, from the Sacramento Valley to the summit of the Sierra Nevada.

The Central Pacific was made possible by a consortium of four major investors: Collis P. Huntingdon, Mark Hopkins, Leland Stanford, and Charles Crocker. As governor of California in the year of the Railroad Act, Stanford had spoken out against Asia "with her number-less millions" sending "to our shores the dregs of her population." There could be no question, Stanford said, sounding a note that would become academic orthodoxy among many social scientists in academia thirty years later, "that the settlement among us of a degraded and dishonest people must exercise a deleterious influence upon the supe-rior race."

But among the Central Pacific partners Stanford wasn't the one with the urgent job of finding adequate labor to build the line. Casualties were already high, wildcat walkouts common. Crocker needed a long-term labor force of at least 5,000 if the job was to be completed on time, and as of winter 1865 he had just 800, mostly Irish. It was his older brother Edwin Bryant Crocker, the company attorney, who suggested the possibility of employing Chinese workers who had been used for construction on the admittedly much easier California Central Railroad. They could be dependably delivered by the Six Companies, worked well in teams, and were said to be "docile and industrious." The superintendent of works, James Strobridge, hated the idea of being "boss of the Chinese," believing they were altogether too fragile for such work, buying into the received wisdom that because Asians had less body hair they were somehow more effeminate than European men. But Strobridge had little choice. He was down to a mere 300 workers, most of whom, when a section of track was finished, would disappear on a drunken spree with few returning. A trial gang of fifty Chinese were recruited from the towns around the mountains and used by the skeptical Strobridge to load dump carts with rock debris and then drive them. Satisfied with their labor, he then gave some of them picks to work on easier excavation. With every job escalating in difficulty, the Chinese exceeded his expectations.

By the autumn of 1866 3,000 had been hired, and Stanford, who not long before had been so insulting about the damage done to California by Asian imports, now took every opportunity to sing the praises of "his" Chinese. By 1867, they were 75 percent of the Central Pacific workforce; at its peak between 10,000 and 12,000 men. They were a tunneling army of working prodigies, comparable to those that built the Great Wall, an analogy much invoked by the proud managers of the line.

There were so many ways to die for the Central Pacific. Blasters hanging in perilous baskets at the rock face would swing back to the explosion after drilling the stone and lighting the fuse. Sometimes the nitroglycerine they used was so volatile and violent that the explosion stoned them with flying boulders. Many were taken by avalanches. Five were killed near the Donner Summit on Christmas Day in 1866. The company had offered its workers canvas tenting, and the Irish and the Cornish ex–tin miners used them. But with good reason the Chinese thought the tents more likely to be buried beneath the forty-feet drifts that accumulated when the mountain wind got up, and preferred a molelike subterranean existence, excavating long tunnels, both for work and shelter, some wide enough to take a two-mule sled. But inevitably sometimes an avalanche blocked the entrances to the warrens, or the chimneys and the workers were buried alive inside, their bodies irrecoverable until the spring thaw.

Was this a Chinese or an American microworld? The workers spoke Cantonese and were supplied with their own kind of food: abalone, dried mushrooms, cuttlefish and oysters, salt cabbage and pork, and plenty of rice. In spring and summer, if they were at all near any of the little market-garden villages that settled along the route, there were fresh vegetables, beans, and onions. Their standard beverage was green tea, brewed with snowmelt or scrupulously boiled river water, served from an iron pot beside which stood the huge whiskey casks demanded by the Cornish and the Irish. "I never saw a Chinaman drunk," said Strobridge in a later testimony to the California Senate on the moral effects of Asian immigration. China tea saved lives, since those who gulped down water straight from the polluted streams paid heavily with violent and sometimes fatal dysentery. The Chinese were also fastidious about their hygiene. The cooks assigned to each crew of a dozen or so workers boiled water in the emptied black powder kegs that were used for daily sitz baths after the last shift. For a dawn-to-dusk day, six days a week, they were paid around thirty to thirty-five dollars a month (in gold), which went directly to the crew boss responsible for buying provisions. (The Irish and Cornish got room and board free.) Those expenses left the workers with a bare twenty dollars. By 1867, they had become American enough to decide this wasn't enough. Two thousand of them struck for a ten-hour day, a forty-dollar monthly wage, and the elimination of the degrading power to be whipped or confined to prevent them walking off the job. Crocker tried to break the strike by putting out a call for ex-slave labor, but few of the African Americans responded to the invitation. So Crocker, whose benevolence to the Chinese made the Irish call them "Crocker's pets," turned tough, blocking food supplies and starving them back to the job.

On 10 May, at Promontory, near Ogden, Utah, the last "golden spike" joining the two lines was driven home. The Central Pacific gangs took it as a point of honor to beat the record of their rivals on the Union Pacific, laying down the last ten miles of track in just twelve hours. For showtime, those who performed the finishing touch were Irish, a crew of about a thousand Chinese arriving by train from Victory to Promontory, an hour or so before the ceremonies. But they were needed all the same, for the dignitaries like Leland Stanford had so much trouble getting in the last spikes that Chinese workers in blue cotton duck coats and trousers helped them by starting the hammering, leaving the management to apply the last dainty taps before they lit the cigars. Strobridge, who had been so reluctant to hire the Chinese and was now their great champion, invited them to his railway car, laid out for a banquet, and made a great fuss of the Chinese crew bosses, while acknowledging that the transcontinental American railroad would never have been built without Chinese toil and sacrifice.

The number of those who perished along the way can never be known. Officially, Central Pacific reported 137 deaths during the four years of construction. But on 30 June 1870, a journalist for the *Sacramento Reporter* saw a train loaded with the bones of Chinese bodies that he estimated to be at least 1,200 and commented on the discrepancy between the official statistics and the wagonload of remains. The bones were being carried west to San Francisco along the track the Chinese had laid, to be shipped home to rest among their ancestors.

Coast to coast, the railroad unification of North America was greeted as a second revolution; the necessary completion of the first, almost a century later. When the news was relayed, fireworks burst in the sky above Frederick Law Olmsted's Central Park, the Liberty Bell was rung in Philadelphia, and in San Francisco people started drinking. When the city sobered up, it lost no time in passing anti-Chinese legislation, just in case the Asiatics were deluded enough to expect a vote of thanks. Life in Chinatown was made as miserable as possible. The wide basket-carrying shoulder yokes used to carry vegetables or laundry were banned from the streets as a hazard. A "cubic air" regulation was enacted requiring 500 cubic feet for every inhabitant, giving the police the right to enter any household to detect infractions. Anyone arrested for that or any other misdemeanor was now liable to have their queue cut off and head shaved, in a gesture of gratuitously aggressive humiliation. Most emblematic of all, a Chinatown fire gave the city a pretext to ban wooden laundries (in a city where almost everything was timber-framed and fires happened every day).

San Francisco—and almost every town of any size in California—wanted the "Chinee" out. The completion of the railway project meant that 25,000 from both companies were now out of work. Another economic downturn, which turned into a steep recession in 1873, only made the competition for work more brutal. The prejudices of the white working class, mostly Irish, now hardened into something like race war. The Chinese were said to be parasitically sucking wealth from the American economy to be shipped back home to Canton and Hong Kong and, by taking jobs as cigar rollers and industrial shoemakers for rates no white worker would accept, were artificially depressing the labor market. Their increased dispersion into the interior of the country through the Midwest and farther east meant, so their antagonists claimed, that before long American workingmen all over the country would have their living depressed to the level of these people who "lived like beasts." Around these grievances arose a more general caricature of John Chinaman as a monster of sinister guile, addicted to whores and opium, "treacherous, sensual, cowardly and cruel" as Henry George put it in a classic statement of the anti-Chinese case in the *New York Herald Tribune.*

The festering of all this polemical poison was too good an opportunity for writers with an eye on the main chance to let slip. In 1870, Bret Harte was working for the United States Mint in San Francisco while editing the *Overland Monthly* and taking advantage of the appetite for tall western tales which, after the completion of the railroad, was raging through the East. He had just delivered to the publisher a collection of his western stories, *The Luck of Roaring Camp*, about which his sometime friend Mark Twain would roll his eyes. In September 1870, not long before he picked up and headed back to New York, Harte published in the pages of his own magazine some rhyming verses called "Plain Language from Truthful James," which very rapidly became known as "The Heathen Chinee." The eponymous James was the narrator of a poker game in which the rip-roaring miner Bill Nye is bested by the wily, card-concealing Ah Sin. The verses were accompanied by illustrations that played to every grotesque stereotype of the slant-eyed, cackling, pigtailed Asiatic.

Back east, Harte disingenuously distanced himself from the poem, "the worst I ever wrote," and claimed it had been written to parody ignorant anti-Chinese bigotry. But he knew very well that it played perfectly to those prejudices, right down to the sing-song meter that was perfect for music hall and saloon recitation:

> Which I wish to remark
> And my language is plain
> That for ways that are dark
> And for tricks that are vain
> The heathen Chinee is peculiar
> Which the same I would rise to explain.

The Luck of Roaring Camp had already made him America's western writer, but "The Heathen Chinee" was an even greater hit all over the country, for every bad reason. Harte made sure to include toward the end a verse that spoke to all the strong feeling concerning the harm that Chinese immigration had done to the lives of honest American workingmen:

> I looked up at Nye
> And he gazed upon me
> And he rose with a sigh
> And said "Can this be?
> We are ruined by Chinese cheap labor"
> And he went for that heathen Chinee.

The last of the illustrations has Ah Sin thrown to the floor, set upon and then booted out of the door. And these were the images in not only the *Overland Monthly* but also the pocket version printed up along with railroad timetables and brochures from New York and Chicago. Bret Harte had made the crudest animosity respectable as he had made physical assault on the Chinese a matter of rib-poking glee.

So now it was just fine to lay hands on the "heathen Chinee" and to move from verbal to physical violence. After a white rancher was caught in the crossfire between two Chinese gangs in Los Angeles in 1871, the rest of the city of 5,000 went on a rampage, burning the Chinese district to the ground and killing sixteen. The next day bodies were still swinging from gateposts and crossbeams. This was just the beginning of one of the great American pogroms: an all-out ethnic cleansing of the towns of the West. It was of a piece, of course, with the assertion of racial supremacy in other sections of the nation's life: the last acts in Native American genocide, the liquidation of Reconstruction in the South. The Democratic Party, which profited from the near-hung election of 1876 to end civil rights in the old Confederacy, also seized the opportunity to win working-class votes by posing as the champions of the anti-Chinese movement. In San Francisco, Dennis Kearney, a flamboyant Irish-born demagogue with the gift of the toxic gab, used the opportunity of state senate hearings at the Palace Hotel to stir up crowds—in their thousands—assembled within hearing at a nearby sandlot. Inside the hotel, Colonel Bee as the attorney for the Six Companies was defending the moral and social reputation of the Chinese, and Charlie Crocker and James Strobridge were testifying as to their probity, industry, and heroic sacrifice on the Central Pacific. Outside, Kearney was teaching the crowd to chant his slogan 'The Chinese must go" and asking them if they were "ready to march down to the wharf and stop the leprous Chinese from landing." "The law of Judge Lynch" was threatened on any white employer who did not fire Asian workers. It was the capitalists and monopolists who had thrust these subhuman curs on good white workers, Kearney raved, and they would pay for that crime. "The dignity of labor must be sustained even if we have to kill every wretch that opposes us." Inside the hotel, pressed by Bee, Charlie Crocker was brave enough to claim that if the matter was put "calmly and deliberately" before the people of San Francisco, he believed 80 percent of them would want the Chinese to stay. Anyone, he said, could whip a storm of ugly fury. But Bee and Crocker were too generous about the sympathies of their fellow Californians. The reality was that 80 percent of the people of California were voting Democrat with exactly the opposite view in mind. When Crocker was asked by one of the senators whether he believed Chinese civilization was inferior to Western, he said he thought it was actually rather superior. Frank Pixley, a former state attorney general, was more representative of inflamed public opinion when he said that he could not wait for the day when he could stand on Telegraph Hill and see Chinese bodies hanging, with the rest leaving town.

As Jean Pfaelzer and Alexander Saxton have documented in moving detail, what then followed was an epidemic of American round-ups, mass expulsions, burnings, and murders, spreading from California to Denver in Colorado, Tacoma and Seattle in Washington, and Rock Springs, Wyoming. In most places an ultimatum would be issued to the Chinese communities to leave within a few hours or at most a day. To speed up the process, a few houses and stores would be torched, sometimes with people inside. Once the terrified population was on the march, the job would be finished by burning Chinatown down, top to bottom. In other places, like the town of Truckee, not far from Crocker's ranch and Bee's home in Placerville, the tactics, in response to the complaints of lawlessness, were slightly more subtle. There, Charles McGlashan organized a boycott of any employers hiring Chinese workers, which slowly threatened to strangle the entire economy of the town, forcing mass dismissals and evictions, but allowing McGlashan and the "Anti-Coolie" forces in Truckee to claim the Chinese had left of their own free will.

The response of local and federal authorities was mostly to turn a blind eye to all this, or worse, to ride the fury to power. In 1882, President Chester Arthur signed the Chinese Exclusion Act, which was supposed to assuage the movement. Chinese immigration, other than for "merchants, diplomats, students and travellers," was stopped for ten years, and the principle, already codified in the Naturalization Act of 1870, that no Chinese immigrant could ever be qualified for citizenship was reaffirmed. It would be renewed decade after decade and repealed only in 1943 after Pearl Harbor, when Kuomintang China suddenly became America's ally against the Japanese.

In the climate of such panicky hatred and violence, the courageous decency of the few Americans who clung to an older, less paranoid sense of E pluribus unum, and who even saw no reason why Asians should not one day be American, needs acknowledging. There was no reason why a respectable, late middle-aged Civil War veteran and businessman like Fred Bee should have undertaken to be the white knight of the Chinese community, much less their consul. He was paid for his services in court, but he was also guaranteed death threats. But Fred Bee went about it not just with more than the standard allotment of civic honor, but with the satisfaction an American citizen could get from making the law do what it was supposed to do: protect all those for whom it had responsibility. Bee targeted local mayors and governors whom he despised for betraying their public trust, and those much farther afield: the spineless politicians in Washington who were prepared to condone or even instigate mob rule the better to link their own fortunes to public rage, however ignorant and cruel. Bee took the opportunity of reminding Congress and the president that the United States government had been compensated with $700,000 for the burning of Christian missions in China. If local and state authorities continued to be indifferent to lawlessness, and the federal government refused to restrain them, the Chinese might well take it into their heads to inflict damage on Americans in their country, and it might not be so easy to seek redress. Some officers in the U.S. Attorney General's office were paying attention, and in something of a pyrrhic victory, National Guardsmen were sent to Rock Springs to keep order and escort to the coal face any Chinese miners who wanted to work in the pits. Guardsmen and federal troops continued to be a presence in the town until 1898.

Bee knew the criminal courts were rigged against convictions of any of those who had actually committed murder. A judge of the California Supreme Court, Hugh Murray, had handed down an opinion that since the Chinese had ancestrally crossed the Bering Straits, and over the centuries had become, in effect, Indians, the constitutional provision that disbarred Native Americans from giving evidence against citizens applied to them too. But these Chinese/Indians were of course often the only firsthand witnesses. So Bee changed tactics and did something outrageous. On behalf of the Six Companies he sued entire cities for losses of property during the riots and forced marches. He called them reparations. Though Bee seldom won any substantial damages, his persistence rattled local authorities, who saw themselves having to impose taxes on their citizenry if the plaintiffs were successful. The spirited determination of Bee and his partner Benjamin Brooks to use the Burlingame Treaty and the Constitution to establish basic legal decencies emboldened the Chinese themselves to think one day they might be treated with a modicum of human respect. When in 1892 a Californian congressman, Thomas Geary, introduced an act of Congress requiring all Chinese to carry photo IDs and President Benjamin Harrison—too timid to defy public prejudice in an election year—signed it into law, the Six Companies ordered over 100,000 of their people to defy the law and refuse to carry the degrading cards. In their official statement, probably drafted by Fred Bee, they actually dared to presume that "as residents of the United States we claim a common manhood with all other nationalities." Despite another economic panic in 1893 scapegoating them, the Chinese community asked for "an equal chance in the race of life, in this, our adopted home."

There were many in positions of authority who thought "over our dead bodies." One of them was Terence Powderly, who had led the Knights of Labor that had been in the forefront of the anti-Chinese movement in the 1870s and 1880s and who from 1897 to 1902 was commissioner general of immigration. Ensuring that Angel Island, the holding center in San Francisco Bay, was designed to keep out as many Chinese as possible, Powderly declared with a flourish that set the tone for generations of immigration officials to come, "I am no bigot but I am an American and believe that self-preservation is the first law of nations as well as nature." Self-preservation decreed that almost no Chinese women be admitted since they either were prostitutes, or if apparently legally married, would seal the fate of the United States by breeding generation after generation of heathen Chinee.

But the history of just one of those young Chinese Americans born in the United States pointed the way to a less paranoid future. Wong Kim Ark, the twenty-three-year-old son of a San Francisco merchant family, had been visiting family in China in 1895. His papers were straightforward, but the famously prejudiced Collector of Customs John H. Wise, responsible for West Coast immigration, denied him entry on the grounds that he was not a citizen and thus was barred by the Exclusion Act. Wong had in fact been allowed back to California after an earlier trip in 1890, and when he was detained on board a ship in the bay hired an attorney to file a writ of habeas corpus. The Fourteenth Amendment, which specified that all those born in the United States were entitled to citizenship, held as much for the children of ineligibles like his parents as anyone else. The U.S. district attorney argued that for people of ethnic groups deemed unassimilable, birth was not enough to give rights of citizenship and painted a picture of national self-destruction should Wong's claim be upheld: America at the mercy of "persons who must necessarily be a menace to the welfare of our Country." Happily, as Erika Lee records in her fine account of the case, the presiding judge thought the matter much simpler: "It is enough that he is born here whatever the status of his parents." Only criminal acts could waive this right guaranteed by the Fourteenth Amendment. Wong was released, and when the case was heard on appeal by the Supreme Court of the United States, the California judge's opinion was upheld. Though Chinese immigrants would be maltreated for many generations, the mere idea of an Asian American was no longer a contradiction in terms.

And for those who, unlike Wong Kim Ark, had not been born in the United States, there was another option in the decades after exclusion. They could pretend to be Mexicans. Because the Mexican government was more hospitable to Chinese immigration and because controls at the frontier were more lax, the need for temporary Mexican labor in the farms and orchards of Southern California being acute, the first generation of *coyote* smugglers could ship the Chinese, dressed in serapes and sombreros, queues cut off, over the frontier. Often they would arrive in San Francisco Bay, switch ships to steamers heading for Mexico, and then be taken in boxcars, or sometimes (if the disguise was good enough) by mule train or even foot, across the border between Sonora and Arizona, or between Baja California and San Diego. The routes were exactly what they are now; the business was as lucrative as now; the businessmen were sometimes Chinese-Mexican like Jose Chang; pure Chinese like Lee Quong "the Jew"; sometimes American operators like B. C. Springstein or Curly Edwards. And already, in the early decades of the twentieth century (especially during Prohibition) the profits of the human contraband were enriched by having the illegals take drugs and drink (opium and whiskey) with them. The border was 300 miles long; there were never enough patrolmen, or "in-line" riders on the freight trains, and the industry in Sonora forging residence certificates for the "Mexicans" was brilliantly professional. Besides, as one of the immigration inspectors said, it was hard telling Mexicans and Chinamen apart. At least 17,000 undocumented Chinese entered the United States this way between 1882 and 1920, a drop in the bucket of what was to come. But the peoples for whom the Crèvecoeur promise had been most bitterly betrayed were finding their own way to make it come true.

31. Grace under pressure

Peering out at the fog and rain, Frédéric-Auguste Bartholdi stood in the head he had designed: the head of Liberty Enlightening the World. The day, 28 October 1886, had gone well despite the weather. A million had watched the parade from City Hall Park down Broadway. The bands had been properly rehearsed; the flotilla of tugs and steamboats in the harbor, a happy cacophony of horns and whistles. Even the poem written by one Sidney Herbert Pierson had suited the occasion: "Today the slaves of ancient scorn and hate / Behold across the waters . . . / Her blazing torch flame through ocean's gate." Around four o'clock with the light fading, Bartholdi listened intently for the end of Secretary Evarts's speech, the signal for him to unveil the statue. A burst of applause came from the 2,000 dignitaries seated before the pedestal. Bartholdi tugged at the ripcord, and, with the precision he had prayed for, the great tricolor veil fell from the face of the colossus. A roar went up from the audience and a mighty tooting from the tugboats. But then, when the sounds eventually died away, Secretary Evarts went on with his unfinished speech. Grover Cleveland's face (which liked a good prank) was a mask of attentive self-control even though the temptation to chuckle must have been gut-busting. Instead, in turn he rose to his feet, sonorous and apt as usual, to accept the gift of the statue from the sister republic of France. "We are not here today to bow before the representation of a fierce warlike god filled with wrath and vengeance but we joyously contemplate instead our own deity keeping watch and ward over the open gates of America."

They were indeed still open. In the next six months a quarter of a million immigrants saw the upraised copper arm with its beacon of liberty as their ships sailed into the harbor and onto processing sheds at Castle Gardens. On 11 May 1887, thirteen steamers, coming from Liverpool (the *Wyoming, Helvetia,* and *Baltic*), Antwerp, Glasgow, Bremen, Hamburg, Marseilles, Le Havre, and Bordeaux (the *Chateau d'Yquem!*), unloaded just short of 10,000 on a single day. And the *New York Times* had had enough of the spirit of hospitality. "Shall we take Europe's paupers, her criminals, her lunatics, her crazy revolutionaries, her vagabonds?" the paper asked. These were laborers "who lived on garbage" and were a "standing menace to the city's health." Another editorial (for the *Times* sounded off regularly on the subject) opined that "in every Anarchist meeting, every official statement concerning the condition of labor or the inmates of our almshouses and asylums for the insane, every report relating to plague spots in the slums of our great cities may be felt something to remind the people of the United States that immigration under restrictions now provided is not a blessing."

Seven years later, in 1894, the Immigration Restriction League was duly founded to combat the irresponsible, sentimental universalism (as it saw it) of those who looked upon the torch of Liberty in New York Harbor and wiped a tear from their eye. The men who created the league were dry-eyed when it came to the fate of the tempest-tost. If they were not sentimentalists, they were also not street shouters like Dennis Kearney, or labor tub-thumpers like Terence Powderly. They were from the cream of the eastern patriciate; those who flattered themselves as belonging to its intellectual as well as social aristocracy, and a disgraceful number of them were professors. Not any professors either, but the founding fathers of the social sciences in the United States: statisticians, eugenicists, biologists, economists, and ecologists. Sometimes, like Madison Grant, the author of *The Passing of the Great Race* (1916), they were a combination of all those scientific endeavors, for Grant published his apprehensions about the vanishing moose and caribou before declaring that white America was committing "race suicide" by allowing the biologically degraded to take so many jobs that those in a more exalted tier had no option but to limit the size of their families.

They were not, then, xenophobic crackpots, the restrictionists.

Princeton and Yale were prominent among their alma maters. Their most strenuous mind, arguably, was Francis A. Walker, the dean of American statisticians and the president of the Massachusetts Institute of Technology. And the league was itself born upstream on the Charles in the sacred halls of Harvard by three graduates whose names are purest Brahmin: Prescott Farnsworth Hall (who would serve as secretary of a national organization of immigration restriction societies into the 1920s when their policy had become law), Robert DeCourcy Ward (the first professor of meteorology at Harvard), and Charles Warren, whose name still adorns the graduate center for American history at that university. The aim stated in their constitution was "to arouse public opinion to the necessity of a further exclusion of elements undesirable for citizenship or injurious to our national character." They used the already formidable network of Harvard alumni to spread the word; to reach powerful politicians like the Massachusetts senator Henry Cabot Lodge and his close friend and fellow alumnus Theodore Roosevelt. With over five million immigrants arriving between 1880 and 1890, they believed the American future at stake. The nation's virtues had been inherited from "sturdy" (a word they liked to repeat) stock of the English, Scots, and (even) Irish along with a decent Nordic smattering of Scandinavians and Germans. That inbred pedigree of resolute will, toughness, and beauty, the product of generations of trial, was now under siege from the polluting under-races pouring through New York from southern and eastern Europe: Italians, "Slavs" (Poles, Ruthenians, Lithuanians), Hungarians and Rumanians, Armenians and Syrians and most abominable of all, "the Hebrews."

From their faculty houses and gentleman's clubs (no Hebrews need apply), the professors and the patricians could smell the reek of cooking onions and grimy underthings; they could see the dirt-clogged nails of the sweat workers in the tenement garment shops, and they trembled for America as they pressed their lawn handkerchiefs to their noses. They were all well traveled. They all adored Europe; but it was the Europe of Michelangelo, of countless Hotel Bristols, not the chicken-gizzard slums and the greasy brothels. Now the very worst of Europe was invading the American shore, dispatching its diseased madmen, tubercular paupers, and sinister agitators. Only they who understood, as they kept on saying, the *scientific* basis of the threat stood between America and death by subhuman infestation.

Eighteen ninety-three was the perfect year to begin the campaign that would culminate in the establishment of the Immigration Restriction League. The country was in the midst of another of its economic meltdowns: failing banks; massive unemployment. In an attempt at rallying the national spirit, the Columbian Exposition had opened in Chicago, and had proven an electric-lit wonder. But even there the lecture delivered by the Wisconsin professor of history Frederick Jackson Turner, attributing the triumphant expansion of democracy to the moving continental frontier, had a valedictory ring to it when Turner declared that frontier closed. Ideological claustrophobia bred paranoia. Now that the invasion of the Inferior Races had penetrated the interior of the United States, there was nowhere to flee (except to their elegant summer homes in Maine and Long Island). Had they managed to shut the door on the Chinese in the West only to succumb to what the *Times* called "the Chinese of the eastern cities"? Lengthening unemployment lines and a fierce scramble for jobs recruited the forces of organized labor to the cause of restriction. In the rural South and parts of the Midwest, the sense of a capitalist plot to swamp America with what the populist politician Tom Watson called "the scum of creation," at the same time as they upheld the gold standard to make credit harder for regular folks, aggravated the resentment. It was, after all, the United States Chamber of Commerce and manufacturers' associations who were resisting immigration restrictions in the name of cheaper labor costs. In the meantime honest white workers were left to cope as best they could.

In June 1896, MIT's Francis Walker published his own arguments for restriction in the *Atlantic Monthly*. The fact that he had earned respect as Civil War soldier, commissioner of Indian affairs (presiding of course over the golden age of their liquidation in the 1870s), and as the founder of national associations of both economists and statisticians, meant that Walker's adherence to the restrictionist cause gave it priceless intellectual respectability. In the article he acknowledged that America had been built on the open hospitality of the Founding Fathers, but that did not necessarily mean their word should be law forever. They had cleared forests with abandon; now it was thought prudent to conserve them. So while our "fathers were right . . . yet the patriotic American may properly shrink in terror in contemplation of the vast hordes of ignorant and brutalised peasants who throng to our shores." Immigration had once been a test of will and fiber; now it was "pipeline immigration" run by unscrupulous agents in central and eastern Europe who locked their victims in boxcars, disgorged them on Ellis Island, and then drove them to the coal face in Pennsylvania and the Appalachians. To those who said "they do the jobs we do not wish to perform," Walker wondered whether that was a good thing seeing as there had been no jobs the generation of Andrew Jackson and Ralph Waldo Emerson had thought beneath them. If the Irish now liked Italians doing the menial work once allotted to them, perhaps if Baron Hirsch sent two million Jews (the fear of Jews was always counted in millions), Italians could stand aside from work they judged demeaning, but at what cost to the republic? Walker, who when he chose to turn it on could wax Gothic in his lurid, comic-book horrors, summoned up what America would become if nothing was done: a nocturnal vision of "police driving from the garbage dumps the miserable beings who try to burrow in those unutterable depths of filth and slime in order that they sleep there. Was it in cement like this that the foundations of our republic were laid?"

The restrictionists knew how to seem reasonable, demanding at the beginning a literacy test. Was it not common sense to require immigrants to be able to read fifty or so words *in any language*? (This usually meant the official language of their nation of origin, which would have barred the Jews of the Pale of Settlement who for the most part knew only Hebrew and Yiddish; or Czechs of the Habsburg Empire who didn't care to speak German.) But pressure mounted on Congress, which heard Henry Cabot Lodge's speeches on the subject, and a law went through both houses only for President Cleveland (in his second term) to veto it and to do so with an eloquent restatement of the classic Crèvecoeur–Paine case for the uniqueness of the American experiment. Perhaps he remembered that rainy day in October 1886. Such a law, the president said, would be "a radical departure from our national policy relating to immigrants. Heretofore we have welcomed all who come here from other lands except those whose moral or physical condition . . . threatened danger to our national welfare and safety. We have encouraged those coming from foreign countries to cast their lot with us and join in the development of our vast domain, securing in return a share in the blessings of citizenship." In repudi-ation of the restrictionist case that immigration meant economic damage, Cleveland went on, "This country's stupendous growth, largely due to the assimilation of millions of sturdy adopted patriotic citizens, attests the success of this generous and free-handed policy." Similar proposals would be brought to the desks of Presidents Taft and Wilson, and each would apply the veto once more.

A war was looming, in the ranks of social scientists as well as in Serbia and Belgium. In 1914 Edward Alsworth Ross, another of social science's most revered patriarchs, fired from Stanford in 1900 by Leland's widow for injudicious remarks about silver-backed currency and support for Asian exclusion, published *The Old World in the New*. Ross's book, the most influential in the whole debate before Madison Grant's racist bible, is the familiar litany of evils said to have been brought by the "inferior races" of the new immigration. And like many in the genre, under the guise of science it actually drove home its fears in deranged hyperbole. With Polish women producing seven children in fourteen years, "the Middle Ages" had been brought to America. The Hebrew mind was calculating and "combinative," fit for anticipating stock prices, in contrast to "the free poetic fancy of the Celts." The most eugenically minded chapter spoke of how the "blood now being injected into the veins of our people is sub-common." Look at the crowd coming down the gangplank, Ross wrote, and you will see "hirsute, low-browed, big-faced persons of obviously low mentality [who] clearly belong in skins in wattled huts at the close of the Great Ice Age." (Many of the restrictionists were associated with the natural history and zoological societies and designed their displays.) "Ugliness," Ross goes on, is both symptom and eugenic threat for "in every face there was something wrong: lips thick, mouths coarse, upper lip too long, chin poorly formed, bridge of nose hollowed . . . there were sugarloaf heads; moon faces, slit mouths, lantern jaws, goose bills that one might imagine a malicious djinn amused himself by casting human beings from a set of skew molds discarded by the Creator." This was the sort of stuff that would get a hearty roar of approval from Nazis like Alfred Rosenberg, not to mention his Leader.

But the dominant social-science paradigm did not go completely uncontested. The great Columbia anthropologist Franz Boas, the grandchild of Orthodox Jews as both his admirers and demonizers like to recall, devoted a life to attacking the social Darwinism of Herbert Spencer; the pseudo-biology of racial norms. Cultures, Boas argued, were certainly different but not to be arranged in some sort of hierarchy of mental and physical capacity. At the end of his *The Mind of Primitive Man*, Boas hoped that his work might "teach us a greater tolerance of forms of civilizations different from our own and that we should learn to look on foreign races with greater sympathy."

The presumptuousness of American university presidents and the grandest of their faculty to speak for their institutions seems only to have provoked dissenters to articulate a challenging view. Boas was determined to act as a "public intellectual" so as to deny Columbia's president, Nicholas Murray Butler, a restrictionist, the right to speak for the college. And there were some sons of Mother Harvard—and students of the philosophers George Santayana and the pragmatist William James—who begged to differ with their president Lowell, another Boston Brahmin crusader for the superior race. In the wake of attacks in the press on "hyphenated Americans" in 1915, Horace Kallen published an article in the *Nation* giving a subtler view of immigrant adjustment to American life. Kallen believed that following the initial urge to assimilate, immigrants often revisited their cultural traditions and language without any sense that they were compromising American allegiance, a step that Kallen called "dissimilation." His particular target was his colleague at the University of Wisconsin where they both taught: Edward Alsworth Ross. Why, he wondered, was Ross so attached to the white-bread insipidity of one version of American identity, and why so terrified of "difference"—the first time, I think, that that word was used to validate cultural character. Kallen proposed replacing the obligation of homogeneity by "harmony" which he then extended to seeing America as an "orchestra of mankind," each section with its own tone and musical texture; yet each a part of a miraculously bound whole.

It is too soon to say whether the founders of American cultural pluralism—the likes of Boas, Kallen, and Randolph Bourne, who in 1916 praised what the new immigrants had brought to America's stagnation—have won the war. Perhaps it will always be too soon. Ross's descendants like Samuel P. Huntingdon of Harvard, exercised about a war of civilizations fought out on the Rio Grande border, are still very much with us. And it was the restrictionists who won the immediate battle where it mattered, in the halls of authority and power. University presidents Lowell and Butler managed to establish quotas for Columbia and Harvard (Yale and Princeton were no better) that sharply reduced the numbers of Jews admitted after World War I, and a more serious quota system was instituted at the same time based on a ranking of ethnic and cultural desirability. It was that policy that shut the door on immigrants who desperately needed Crèvecoeur's refuge and instead perished in their millions in the camps of the Final Solution. But Madison Grant sleeps in his tomb in Sleepy Hollow, New York, along with the patriots of the American Revolution.

It was not until 1965 that Lyndon Johnson, building somewhat on the head of steam supplied by Kennedy's purported authorship of *A Nation of Immigrants*, succeeded in abolishing the quota system. But what helped more than the assassinated president's Irish pride was a different tradition of understanding the immigrant experience, one that had proceeded alongside all the noisy jeremiads about the damage they were doing to the cultural and social essence of the national character. That work was empirical and practical rather than common-room grandstanding, and it was done by an entirely different class of social workers from people like Madison Grant and Edward Ross. Is it surprising that the professors and the patricians holding their noses at the tenements were all men, but those who actually went into them— who listened to the stories, mopped sickly brows, and who actually bothered to travel to the remote regions in Ruthenia and Poland where the immigrants *came from*—were women?

And what women! The most often celebrated has been Jane Addams, who founded the first of the city settlement houses, Hull House on Halsted Street in Chicago. But it was her brilliant and tireless protégée Grace Abbott who, three years after Edward Ross's farrago of paranoid myths masquerading as social science had been published, wrote the first sympathetic work on *The Immigrant and the Community*. Nineteen seventeen was the year in which the United States entered the World War, and both of the parties had whipped up a froth of patriotic fury, Democrats disgruntled with Wilson's internationalism called for "America First," while Republicans trying to outbid them demanded "Undiluted Americanism." If there must be a way, immigrants might constitute a fifth column especially if they were mere "hyphenated Americans." Grace Abbott wanted to refute, systematically and statistically, each of the truisms recycled by Ross about the new immigration, and she showed conclusively, for example, that born Americans convicted of crimes constituted a much higher proportion of the population than the foreign-born. In northeast America, where immigrants were principally concentrated, white natives receiving poor assistance were also a higher proportion than among the "new immigrants."

But what gives Grace Abbott's book its enduring value is not its counterpunch at a pseudo-sociology; rather it is its narrative power: the portrait of the immigrant experience itself, seen through the many life histories to which the author had sympathetically listened. Moving through her pages are her own hours spent in the tramcars and sweatshops, the police courts, the ward boss saloons, and the huckster "immigrant banks." *The Immigrant and the Community* is, at the same time, the story of the tribulations and endurance of Polish girls without a word of English thrown into farm labor or domestic service, of the Italian railroad worker, of the Jewish seamstress; a handbook for their survival and a prescription to public authorities about how to help these multitudes make it in America. Abbott was no longer interested in debating restrictions. She assumed she and her more liberal kind had, for the time being, lost the argument; that there would be strict criteria of admission in place; but she wanted to do everything in her power to bring the immigrants who passed muster into the stream of American life. Like Kallen, she understood American life not as some imaginary and dull homogeneity, but as a rich adjacency of cultural neighborhoods. That, Grace Abbott supposed, was America's unique glory.

But then Grace Abbott was raised to independent thinking. Her father, a Civil War hero, became a reform-minded governor of Nebraska; her mother was an ardent early feminist and suffragist. Her older sister Edith was a recruit to Jane Addams's Hull House who went on to be dean of the University of Chicago's School of Civics and Philanthropy (later Social Service Administration). Edith's success in Chicago called Grace there too, in pursuit of a doctorate. Once acquired, she knew exactly what she wanted: to be of use in the uproar of the great metropolis. She moved into Hull House, was spotted as exceptional by Addams, and in 1908 became the first director of the Immigrants Protective League, which she ran alongside the equally remarkable Kentuckian Sophonisba Breckinridge.

It was as though Grace figured out, early, what had been wrong with all the grand theorizing about immigrants and American life: that it had been done by people who had never gone near them, who had no clue where Ruthenia was or what it was, and who would have reached for the smelling salts at the mere suggestion that publicly paid interpreters would be a good idea to help the immigrants in their resettlement. Instead of some lordly overview, Abbott simply went along, as far as she could, for the ride. What she noticed right away was the extraordinary number of unmarried and unaccompanied girls and young women. Between 1909 and 1914 there were a half a million of them aged between fourteen and twenty-nine; 84,000 Polish girls alone, 23,000 from Galician Ruthenia, 65,000 Russian and Polish Jewish girls; all most obviously at the mercy of an entire industry waiting to exploit, cheat, and indebt them in any way it could. Grace listened to stories of girls who had been sent to uncles, who took them in for a night or two in the district around the stockyards, perhaps found them poor lodgings, and then left them to sink or swim. Some who had expected to be met at the railway station never were, and stood with their pathetic little bag without any English on the platform until they were approached by a local vulture who would slip their arm in his and take them off to a saloon. Many were cheated even before they got on the train at the port of entry as steamboat companies who supplied the onward journey rail tickets made a killing from absurdly circuitous journeys; taking passengers, for example, from New York to Chicago via Norfolk, Virginia.

There was not a lot Grace Abbott could do about those frauds except publicize them, but she set up a system of greeting and reception for newcomers, especially females, by taking premises right opposite the stations, staffed by women in particular who spoke the relevant languages. Prominent signs in those same languages were posted on platforms together with staff meeting the trains even when they arrived, as they often did, between midnight and 6 a.m. If they were going on to a farther destination accommodation was found for the night and the next stage of the journey clearly explained, avoiding the fate of the Norwegian girl bound for Iowa who was taken off a westward train from Chicago by men claiming she needed to change trains, robbed, and abused. In 1913 Congress and President Wilson authorized the secretary of labor to make Abbott's receiving stations official; all the more needed because she complained that invariably the administration of the immigration laws was left in the hands of unsympathetic men. They were especially heartless if faced with pregnancies out of wedlock. Abbott recalled the prewar case of a young Austrian couple from Galicia: the man, prevented by military service from emigrating, finally arrived with his girl in a state of advanced pregnancy. He was admitted, and she denied entry on the moral grounds available to the immigration authorities. Horrified by the separation, Abbott whipped up a campaign among her Chicago women, who mounted enough pressure to get the decision reversed, and the couple were married on the day of the girl's arrival.

Much of the book is a compendium of advice about what to avoid: fake employment agencies that were often a conduit to prostitution; immigrant "banks" that were mostly used to remit earnings back home and that were often swindles managed by Russians or Hungarians who then disappeared along with the deposits; untrained midwives with filthy equipment capable of causing postnatal blindness or worse. (Abbott was hugely in favor of midwifery, not least because women themselves preferred home births to hospital deliveries, but wanted properly regulated training and licensing in whatever languages were appropriate.) Where she couldn't do a whole lot, at least she hoped to educate the defenseless in what to expect: the wretched conditions of seasonal workers in the Maine lumber camps or the Dakota wheat fields, or worst of all the railroad hobos who were made to sleep in freight-car bunks and forced to pay the company four or five dollars out of their pitiful earnings for foul food, liquor, tobacco, and gloves. Nothing good can be expected of the connections between the ward bosses and the police, she warned. Wise up, in particular to the latter, who will expect bribes that may or may not preempt brutality. The Chicago police sometimes treated immigrants as sport. One man who failed to understand the shouted order to get off the garbage can on which he was sitting was shot and killed for his incomprehension. A fifteen-year-old Slovenian boy, playing dice in a house with his pals, ordered to put his hands up by armed police, was nonetheless shot and killed at point-blank range. When the police got news of a wildcat strike or a bomb going off, they were capable of making random arrests in the community without even a pretense of connecting the arrested to the crime, simply as a message to the Italians or the Poles to "behave."

It's when she gets to the closing passages of her book, though, that

Grace Abbott makes all the hyperventilation of the professors and the patricians seem remote and absurd. When they demand that children repudiate their parents' tradition and language, do they stop to consider that because they have some facility in English, children are the conduit through which fathers and mothers speak to the bosses, the police, the judges, the physicians? And that it is necessary *for the children's sake* to restore the natural authority of their father and mother? How could the wholesale repudiation of their native culture help that or indeed do any good for America? After all, many of the immigrants, Czechs and Lithuanians, had come from countries where the authorities, German or Russian, had attempted to stamp out their native tongues. Abandoning them now would be a betrayal. "In our zeal to teach patriotism we are often teaching disrespect for history and traditions that the ancestors of immigrant parents had a part in making."

"Americanism," Grace Abbott says, following Horace Kallen, is a shibboleth, a weak-minded convenience, and she quotes the anthropologist William Sumner approvingly when he commented that what it often amounts to is a "duty to applaud, follow and obey whatever ruling clique of newspapers or politicians command us to say or do." What is an American? she (more or less) asks, and gives an answer richer and subtler than Crèvecoeur's, an answer for America's modern age: "We are many nationalities scattered over a continent with all the differences and interest that climate brings. But instead of being ashamed of this . . . we should recognize the particular opportunity for the world's service. If English, Irish, Polish, German, Scandinavian, Russian, Magyar, Lithuanian and all the other races on earth can live together each making his own distinctive contribution to our common life, if we can respect those differences that result from a different social and political environment and the common interests that unite all people, we shall meet the American opportunity. If instead we blindly follow Europe and cultivate national egotism we shall need to develop a contempt for others, to foster the national hatreds and jealousies that are necessary for aggressive nationalism."

After the war, would it not therefore be right for the United States to champion internationalism, and the cause of "oppressed nationalities . . . their cause should be our cause?" So it probably came as no surprise to Abbott when Henry Cabot Lodge, the restrictionist par excellence, spoke bitterly and successfully against American membership in the League of Nations.

Then again, you feel, reading her lovely pages, that none of this matters if she can just go out onto the sidewalk, into the raucous din of Halsted Street with its clatter of tramcars, newsboys, and street vendors, ragtime piano coming from the saloons, and see something that makes her feel all over again this faith in this America of neighborly difference, feel it deep in her Nebraska bones. On Greek Good Friday (for the Greeks are the closest community to the settlement house) she does just this, coming upon a procession of dark yellow candles wending its way down the street, priests and boys chanting the low anthems of the Aegean centuries. If someone came upon such a scene, she writes, they would never suppose they were in the United States of America. But "after a moment's reflection," as they notice the Irish cops who clear the way for the procession and the lines of Jews and Poles and Lithuanians looking on with a mixture of reverence and curiosity, they realize "that this panorama could only be enacted in an American city."

32. Jefferson's Koran

All around town unsellable Expeditions and Explorers were lined up in dealers' lots like cattle waiting for the abattoir truck. As the skies lowered over Dearborn, gray and sultry, nervous corporate accountants were trembling over their abacus, which by the end of the spring quarter would post an $8.7 billion loss. But if this was the beginning of the end of Ford Motor Company, you would never know it at Fair Lane, Henry Ford's prairie-style urban ranch, built in his manorial years just two miles from the farm where he was born. The heavily horizontal Fair Lane is built in rusticated Marblehead limestone and has the baronial dimness that the great captains of industry often required, as if a wash of light might somehow distract even the weekend guest from a properly Calvinist appreciation of the relationship between toil and triumph. Halfway up the oak-paneled stairway are stained-glass windows with Fordian heraldic supporters—dairy cattle and wheat sheaves, reminding the admirer of his humble rustic origin, together with homilies that repeat in the subterranean den: "He who chops his own wood is twice warmed."

In so many ways Henry Ford was the opposite of the freebooting, free-market capitalist of classical American economic theory. All his life he was obsessed with the damage that unregulated industrial life was having on an older agrarian America, whose true son he imagined himself to have been (even if there was an equally strong part of him that could not wait to get the smell of cow off his dungarees). It would not be wholly absurd to see him as the last Jeffersonian rather than the personification of Hamiltonian big business, for he prized yeoman self-sufficiency in the worker as much as it had been traditionally cherished in the farmer. It's often forgotten that he built power-driven farm machinery and tractors before the Model T, and he always thought of his cars as liberators of those prairie-dwelling folk who were otherwise imprisoned by slogging labor and immense distance. From beginning to end he was the farmer's friend.

So once Ford had charge of the livelihoods of thousands, he felt the same kind of seigneurial responsibility toward their welfare that Jefferson had felt for his slaves at Monticello. He was perfectly willing to incur the wrath of the *Wall Street Journal* and his fellow captains of industry in 1914 by doubling the wage of the workers on his production lines at Dearborn to an unheard-of five dollars a day. But there were strings attached to the generosity. Ford had been reading social science and understood the net output of his workers to be conditional on their social experience and character, rather than simply a calculus of men, machines, and capital. Hence the need to control that social character. And given that so many thousands of his workers at Dearborn were new immigrants, he introduced the company's "Sociological Department" to take care of, and police, their daily life. Sobriety and conjugal virtue were not merely encouraged but ordered. But the most imperative need of all, if they were to prove themselves true American workers, was English. So a special English School was established for adult education. Attendance was not optional if the workers wanted to keep their jobs. The first phrase those who had come from Hungary or Poland, Syria or Sicily, were taught was "I am an American."

On the first graduation day of the English School in 1914, at a Dearborn ballpark, a stage had been erected with a painted backdrop representing an Atlantic steamer named—what else?—*E pluribus unum*.

In front of it was an enormous wood and pasteboard cookpot with handles like an item from a Brobdingnagian kitchen and FORD ENGLISH SCHOOL painted on it in large white lettering. Down from a gangplank marched the graduating students in the costume of their ethnic origin—Hungarian, Polish, German, Italian—into the "Melting Pot," emerging in suits and academic gowns, holding American flags as the works band played "My Country 'Tis of Thee." Samuel Marquis, a former Episcopalian minister and the presiding genius of both the Sociological Department and the English School, had orchestrated the ceremony so that each month of the nine it had taken the students to master English was represented by a minute of their graduation ceremony. In following years both the students and alumni (whose organization was called "the American Club") participated in Americanization Day festivities held on the Fourth of July, in which thousands marched to City Hall in Detroit, a city where around three-quarters of the white population were foreign-born.

It seems unlikely that Henry Ford would have known that the symbol of the melting pot for the assimilating transformation of prospective citizens had been the invention of Israel Zangwill, a British Jew and thus the member of the one race that Ford believed utterly unassimilable to America; a message hammered relentlessly in his *Dearborn Independent* and in his book *The International Jew: The World's Foremost Problem*, a work the young Hitler found so instructive that he kept a portrait of his hero Ford close by in his study.

Zangwill's four-act play *The Melting Pot*, first performed in 1908, features an array of stereotypes: the moody, sensitive violin teacher, David Quixano (a peculiar choice of name for a Jew from the Pale), the sole survivor of the Kishineff pogrom in which he had seen his family, including a small sister, murdered before his eyes. Playing opposite David the fiddler is the aristocratic settlement worker Vera, also of course from Russia. On a visit to thank him for playing to the settlement children, Vera dreamily discovers his heavily thumbed scores of Mendelssohn and Bach's Chaconne and falls for David hook, line, and sinker, for, as she murmurs, was not King David a harpist? She yearns to hear the great symphonic work on which he is laboring entitled *The Crucible*: "America is God's Crucible," he explains to the wide-eyed, fine-boned gentile, "where all the races of Europe are melting and re-forming . . . your fifty languages, your fifty bloody hatreds and rivalries, your fifty feuds and vendettas, into the Crucible with you all. God is making The American." But WAIT! A small, dark cloud lurks on the horizon; a Terrible Truth. Vera is herself the daughter of the anti-Semitic Russian baron who had led the pogrom. When he encounters in the evil baron (visiting New York, and why not?) his nemesis, All Seems Lost, even though Vera is a good sort, willing to convert. Months pass, and there is a shy and painful reunion that ends with the two in each other's arms. Behind them, the sun slowly sets on New York Harbor as the ecstatic David exclaims, "Ah Vera, what is the glory of Rome and Jerusalem where all nations and races came to worship and look back compared with the glory of America where all races and nations come to labor and look forward?" Suddenly a shaft of gold illuminates the outstretched arm and torch of the Statue of Liberty as the curtain very slowly falls.

It may not be Crèvecoeur, but there was nothing in Zangwill's *The Melting Pot* to give the baron of Dearborn heartburn, even if it had been written by a ravening Semite. But how would he have felt about the other Semites who gather these days at the American Islamic Center on Ford Road? For the enormous mosque with its three golden domes and two minarets, sandwiched—at a becoming distance—between the Greek Orthodox church and the Lutheran chapel on the same strip, represents the fruit of Horace Kallen's and Grace Abbott's philosophy rather than Ross's and Ford's.

"I'm not hyphenated," says Chuck (Khalil) Alaman to me, rather adamantly, as we sit inside the mosque before Friday prayers. "I'm an American who happens to have some Lebanese blood, some French Canadian blood, and who is a Muslim."

Chuck is a retired civil engineer in his seventies, silver-haired with an easy, elegant manner. If Muslims like Chuck have Americans losing sleep, then the country is in more trouble than it would seem, what with one of the major parties nominating someone whose middle name is Hussein. For Chuck, who served in the military during the Korean War, is a deeply passionate patriot who went through a very bad time after 9/11, when to be a patriotic Muslim American was looked on as an impossibility. "That was a bad time, the worst. *Asalaam aleykum*," he greets an incoming worshipper, following it with "I like to say I'm American as apple pie."

Or *lahme*, I want to say but don't. Muslims from Lebanon and Syria have been in Dearborn for more than a century, part of the same emigration out of the relics of the Ottoman Empire. I know about this, given the route map of the Schamas, and am momentarily overcome by a craving for mint tea. People from his part of central Lebanon came to the town, of course, for work in the Ford factories. In his repellent catalog of stereotypes Edward Ross included the Syrians, whom he classified as "strangers to the truth," shifty, oily Levantines trained in deviously unreliable ingratiation. This could not have been further from the Lebanese who came off their tough shift on the line and made a beeline for Chuck's father's store. There they could inhale and buy the dried savors of home: herb teas of all kinds, pistachios, figs and grapes, many kinds of rice, lentils, sesame, sumac; and notwithstanding those who might have been through Ford's Melting Pot, speak the softly beautiful Arabic of their old home.

Two more migrations happened: one after World War II when the French Mandate ended, Israel was established, and the Palestinians came in multitudes to Lebanon; a second in 1976 during the Lebanese civil war. Through all that time there had only been an informal small mosque and no halal butcher. The Dearborn Lebanese, perhaps about 500 families, bought their meat from kosher butchers, and growing up in that world, Chuck says, "I never felt burdened by either ethnicity or religion. We were known as Mohammedans then, and the kids at school didn't know much about us; we were just kids who played touch football in the park with everyone else. I was, well I am, part of the mosaic." Chuck is indeed exactly the kind of American Grace Abbott had seen as a hope for the future: one whose faith and culture were a tribute to America's capacity for pluralism rather than a problem.

And then came 9/11 and the Iraq War, and life became a lot more difficult. Chuck remembers being in a store when it happened, staring in disbelief at the terrible images on the television screen and immediately praying silently that it not be Muslims who had done this. In Dearborn, he says, there was almost never a problem. No one turned their back or refused his hand; if anything they were more sympathetic, knowing the strength of his American heart. Outside the city, "Well . . . ," he says, not wanting to spell anything out but leaving me to guess. On the wall of the vestibule where we're sitting is a code of honor, unexceptionable for the faithful but including a telling warning to them not to pay heed to "outside literature," meaning, it's quite clear, incendiary jihadi calls from Muslims who have no share in Dearborn traditions. And off camera Chuck confesses that the most recent immigration— from Iraq especially—has not, for the most obvious reasons, been as smooth as he would like. "They don't speak much English and they don't want the Koran translated. They keep themselves to themselves more than I'd like, I will say that." When I give Obama a hard time for hastily removing two women in headscarves from the photo op at one of his campaign rallies, Chuck sighs but says he understands. He might have done the same, he says.

But right now Chuck Alaman is an optimist. The mere fact of the Obama candidacy, with roots in Muslim Kenya and Christian Kansas, is a source of marvel to him; a vindication of America, whatever the outcome of the presidential election. "Did you know," he asks, "that the first Muslim member of Congress, Keith Ellison [also an African American] from *Minnesota* mind, took the oath on Jefferson's Koran?" I confess I didn't even know that Jefferson had a Koran. But Chuck was right. The Virginian had bought the two-volume London translation of George Sale in 1765 and into the Monticello library it went, used, it is sometimes said, for his research in comparative law rather than theology. Though there is no reason why Jefferson should have had any less interest in Mohammed as a moral teacher than Jesus. When the news spread in March 2007 that a Congressman was indeed going to swear his loyalty oath on the Koran, it was no time before a fellow House member, one Virgil Goode, warned that it was the Beginning of the End of America and felt emboldened to add that "we will have millions more Muslims in the United States if we do not adopt strict immigration policies."

For most everyone else, the uncontroversial moment of Congressman Ellison's swearing-in suggests all that is right about the United States when much is wrong. Best of all, the congressman's mother, Cilida, when asked how much that day meant to her, offered the perfect reply. "I'm a Catholic. I go to Mass *every* day."

PART FOUR

I V : A M E R I C A N P L E N T Y

33. Running on empty?

I wasn't on the flatboard before the truck roared off; I lurched, a rider grabbed me, and I sat down. Somebody passed a bottle of rotgut, the bottom of it. I took a big swig in the wild, lyrical, drizzling air of Nebraska. "Whooee, here we go!" yelled a kid in a baseball cap, and they gunned up the truck to seventy and passed everybody on the road. "We been riding this sonofabitch since Des Moines. These guys never stop. Every now and then you have to yell for pisscall, otherwise you have to piss off the air, and hang on, brother, hang on."

I looked at the company. There were two young farmer boys from North Dakota in red baseball caps, which is the standard North Dakota farmer-boy hat, and they were headed for the harvests; their old men had given them leave to hit the road for a summer. There were two young city boys from Columbus, Ohio, high-school football players, chewing gum, winking, singing in the breeze, and they said they were hitchhiking around the United States for the summer. "We're going to LA!" they yelled.

"What are you going to do there?"

"Hell, we don't know. Who cares?"

Jack Kerouac, *On the Road* (1957)

In 1958, that was the way I saw America from afar, willing the number 226 to take a wrong turn down Cricklewood Lane and end up in Oklahoma: corn as high as an elephant's eye (I'd seen the show); a cloudless blue bowl of a sky; the whip-poor-wills doing whatever it was that whip-poor-wills did; prairie chickens ditto; a straight old empty road heading west to happiness. *Whooee!*

I was right. If you want one word to describe the American state of mind, it would be: "boundless"; the beckoning road trip, the shaking loose of fetters. Natural limits—mountains, rivers—have been there to be wondered at and then, in short order, crossed, forded, mapped, left behind. The country was invented to slip the bounds of parish, manor, estate; all the ancient jurisdictions of the Old World that cramped free movement. (In eighteenth-century France, for example, seasonal migrants needed papers signed by their parish priest to avoid the attention of the police.) America was about casting off the security of a nervously watchful church and state. Beginning with the moment you stepped on board and stared at the gray seawater out beyond the harbor, America told you to embrace the peril of unbounded space for the chance of starting over. In the Old World you knew your place; in the New World you made it. So American liberty has always been the liberty to move on. Whatever ails you, whatever has failed; whenever calamity dogs your heels or your allotted patch feels too small for your dreams, there's always the wide blue yonder, the prairie just over the next hill, waiting for your cattle or your hoe. Say howdy, give it a good poke, and up will pop your very own piece of plenty: a crop of corn, a magic glint in the stream, a gush of black gold. Come sundown you can rock on the porch and survey your little kingdom, the kingdom of the common man; your heart's content.

It was when eighteenth-century British governments decided that the line of the Allegheny Mountains would be the western limit of their American empire, any greater extent being expensively indefensible, that it doomed itself, irreversibly, to destruction. America has always been about the forward propulsion that will beat the confinement of the regulating state every time. Benjamin Franklin did his best to explain to friends and Parliament that the sovereign fact about America was territorial magnitude. It was, he wrote to his philosophical Scottish friend Lord Kames, "an immense country, favoured by Nature with all the advantages of climate, soil, great navigable rivers and lakes etc. and *must* become a great country, populous and mighty and will in less time than generally conceived, be able to shake off the shackles imposed on her." "There appears everywhere an unaccountable penchant in all our people to move westward," he wrote on another occasion, seeing in his mind's eye, even as he sat in his London house, the ax and the hoe, trees falling; deep woods cleared; the land put under the plow;

river valleys opening for traffic. But the lords of mercantile empire in
London could not see it. For theirs was a calculus of national profits
divided by the costs of defense and revenue collection. They were not
much interested in the settlement business, except as a source of raw
materials and a mart for British manufactures; and perhaps a sponge
to draw up the "viler sort" of their own islands.

Continental North America was, then, a convenience (as distinct
from the sugar islands of the Caribbean, which were a necessity). But if
its value was not to be made worthless by the costs of perpetual warfare,
it needed to stand on a fiscally defensible frontier. Nervousness about
overreach elsewhere in the British Empire—in India in particular—
affected decisions about America. In the official mind, pioneering was
another name for strategic irresponsibility. But even as it tried to hold
the line, the far-off government had no choice but to encourage some
settlement on the western frontier, if only to preempt the French and
protect the borders of British America from their soldiers and Indian
allies. The snag was that there was always another interior line—the
Mississippi, for example—which if taken and fortified by the enemy
might yet put a chokehold on British America.

So British policy on the western frontier wavered between confinement and permissiveness. But in the end, American perception of
territorial conservatism triggered the first revolution of disaffected
real-estate agents who believed that the market for prime land must
overcome geopolitical timidity. To map the backcountry had been to
spin an investment. Washington's first career was as a land surveyor,
and he had an almost mystical faith in the Ohio Valley as the crucible
of American continental empire. For Franklin the Ohio Valley meant
men and money, lots of both, and a nice cut for him.

In 1782, after the fighting had ended and backcountry land was open
for sale, Franklin published *Information to Those Who Would Remove to
America*, in which the cautionary note surrendered to the shameless
come-on. After grandly disabusing potential immigrants of the easy
availability of "profitable offices" and the myth, apparently wide-
spread, that land and Negroes were given away free, Franklin made
his beguiling pitch. "What are the advantages they may reasonably
expect?" he asks. The first is that "Land being cheap in that Country,
from the vast Forests still void of inhabitants, and not likely to be occu-

pied in an Age to come, insomuch that the Propriety of an hundred acres of fertile soil full of Wood may be obtained near the Frontiers, in many places for Eight or Ten Guineas, hearty young Labouring Men who understand the Husbandry of Corn and Cattle . . . may easily establish themselves there. A little Money sav'd of the good Wages they receive there, while they work for others, enables them to buy the Land and begin their Plantation in which they are assisted by the Good Will of their Neighbours, and some Credit. Multitudes of poor People from England, Ireland, Scotland and Germany, have by this means in a few years become wealthy Farmers who, in their own Countries, where all the Lands are fully occupied . . . could never have emerged from the poor Condition where they were born." To move west, Franklin implied, is to make money fast. He "personally knew" several people who bought large tracts of land on the western frontier of Pennsylvania for £10 per hundred acres and who, as the farmland boundary pushed west, sold the same land for £3 an acre: an American killing!

A generation later, in the 1830s, Alexis de Tocqueville judged this "restless spirit" the great American peculiarity. No matter how prosperous the times, or how well the citizen might be doing, standing pat on what one had was out of the question for a true American. Dreading the possibility of loss, Americans were "forever brooding over advantages they do not possess," compelled to find an expeditious way to something still better. Passive contentment was apparently not an American option. Tocqueville saw with his usual astuteness the tension in American life between settlement and stir-crazy impatience to be Getting On. They were, and are, two impulses in contention. On the one hand, there was the direction towards which the wagons rolled, the log cabin in the clearing; which as the land was opened and tilled, would give way to a picket-fenced yard and farmhouse. On the other hand, there was the irrepressible itch to be up and Improving. From the unresolved tension between the two instincts, Tocqueville thought, could come social madness: happiness as a malevolent will-o'-the-wisp, forever capering before the breathlessly pursuing Americans who were trying to catch it. How else to explain irrational habits that barely raised an eyebrow in America? "In the United States a man builds a house to spend his latter years in it and sells it before the roof is on and lets it just as the trees are coming into bearing . . . he settles in a place which he soon afterwards leaves, to carry his changeable longings elsewhere. If his private affairs leave him any leisure he instantly plunges into the vortex of politics; and if at the end of a year of unremitting labor he finds he has a few days' vacation, his eager curiosity whirls him over the vast expanse of the United States and he will travel 1,500 miles in a few days to shake off his happiness. Death at length overtakes him, but it is before he is weary of his bootless chase of that complete felicity which is forever on the wing."

But America couldn't help itself. It was always moving toward Gatsby's "orgastic future," signaled by the green light at the end of the dock. "It eluded us then, but that's no matter—tomorrow we will run faster, stretch out our arms further . . . And one fine morning—" Fitzgerald's boat, in the end is "borne back ceaselessly into the past." But that end note of pessimism is tellingly misread by most of the high-school students assigned the book, at least according to the *New York Times*. In February 2008, the paper reported from a class of first-generation immigrant students at Boston Latin School, most of whom took the green light not as a tantalizing mirage, the glow that lit Gatsby's doom, but something like the opposite: a beacon of hope; their very own go signal. Jinghzao Wang, fourteen and a first-generation immigrant, told the *Times* that she had adopted Fitzgerald's green light as a symbol of her determination to get into Harvard.

You can see her point. No one has yet won an election in the United States by lecturing America about limits, even if common sense suggests such homilies may be overdue. In 1893, the Wisconsin historian Frederick Jackson Turner noted that the superintendent of the U.S. Census had declared that since the "unsettled" area of the country was so broken, there could hardly be said, for the purposes of the census, to be a frontier any longer. Turner took that as momentous, the end of the "first epoch" in American history. It was, he lamented, an end of repeated beginnings, since with each westward push, American society had had to start all over, thus giving the nation its bracing sense of perpetual youth. But all was not entirely lost. Three years later, in 1896, in an essay called "The Problem of the West," Turner prophesied that some sort of robust American response to the closing of the frontier would shape what he called a "new Americanism":

a "drastic assertion of national government and imperial expansion under a popular hero." The frontier had given birth to democracy. Would its closing give birth to something more startling? "The forces of reorganization are turbulent," he wrote darkly, "and the nation seems like a witches' kettle."

Each time the United States has experienced an unaccustomed sense of claustrophobia, new versions of frontier reinvigoration have been sold to the electors as national tonic. In the 1890s, Teddy Roosevelt's muscular imperialism, the answer to Turner's prayers, was meant to get the country out of its end-of-frontier funk. And if the moving frontier could no longer generate democracy in, say, Montana or New Mexico, then perhaps it could do so in Slovakia, Latvia, or Cuba. That at any rate was the hope of Wilsonian internationalism; the irrepressible urge to pioneer through politics. In 1960, in the midst of the cold war, experienced once more as a constriction of national energy, Kennedy's inaugural speech promised another breakout. America's back was no longer against any sort of wall, Berlin or otherwise. It was Up and Doing; it would "go anywhere" and "pay any price" to defend freedom. Anywhere came to include the Newest Frontier of the moon, on which the astronauts of Apollos 15 to 17 took the ultimately meaningless road trip, cruising about the lunar surface in their big-wheeled buggy looking for cool rocks.

The great exception to the obligations of optimism was Jimmy Carter. His fate in the election of 1980 against the unshakably sunny Ronald Reagan became an object lesson in the penalties of candor. Carter gave no fewer than four television speeches on the subject of energy, which, prophetically, he saw as the arbiter of security. From April 1977 to his most dramatic speech in mid-July 1979 setting out a national energy policy designed to reduce dependence on foreign oil by a quota on imports, conservation, and tax incentives for investment in alternative fuels, his television audience went south (eighty million to thirty million). Press reaction to the extraordinary speech of 15 July, in which the born-again Baptist read admonitions to himself from perplexed citizens before he asked America to face the facts about oil dependence, was hostile. The LA Times took the president to task for "scolding his fellow citizens like a pastor his profligate flock." William Buckley's National Review drolly confessed its surprise to discover that God was a member of the Carter Cabinet. On the other hand, popular reaction to the speech as measured by opinion polls was positive, the president's ratings rising by 11 percent. The several disasters that overwhelmed the Carter presidency—not least the Iran hostage crisis, severe inflation, the gradual but unmistakable collapse of the president's own capacity to call on the country in ways that would balance frankness with optimism—all put the prescient courage and clarity of his energy policy under a cloud. In retrospect it became just another item on the preacher's sin list for which the solution was collective repentance.

Ronald Reagan, on the other hand, was no breast-beater. His solution was to displace the responsibility for America's economic predicament from the shortsighted habits of business and consumers that had troubled Carter onto the lumbering leviathan of government: the "problem" not the "answer," as he memorably said. Instead of Carter's interventionism in the name of the common good, what was needed, Reagan proclaimed, was for government to get out of the way; to deregulate the energy industry and get as much hydrocarbon fuel out of the ground as possible, as quickly as may be. America would deliver; America always had. The people ate it up. Watching the two together on television was like beholding a happy elderly parrot, crest cocked to one side, confront a gloomy creature of the deep, the fish lips parting occasionally to reveal a frightening grin. A parrot-fish of course is what Americans needed but, obliged to choose, they had little hesitation. Just six states plus the District of Columbia withstood the landslide to fall in Carter's column.

The next time that a candidate was brave or foolhardy enough to suggest that perhaps the United States was getting to the end of plenty, and that there was a price to be paid for both the use and the depletion of fossil fuels, he too paid a price at the polls. The contrast could not have been more glaring. Both George Bush and Dick Cheney came out of the oil industry; both were skeptics of global-warming science; both thought there was no energy shortage, even in the age of the SUV, that could not be put right by getting rid of George H. W. Bush's ban on drilling offshore and in environmentally sensitive regions, and developing nuclear power as fast as possible (an option Carter too had recommended). Al Gore, on the other hand, was cast by his opponents as a false prophet of energy apocalypse, warning of the wrath to come if America did not mend its profligate ways. The

more strenuously "ozone man" lectured the country, the funnier the
joke was supposed to be. And once the Bush–Cheney administration was installed in power in 2001, it lost no time in repudiating the
Kyoto accords on measures to combat climate change, and deregu-
lating the energy industry as completely as possible. A task force on
energy policy was convened under the chairmanship of Vice President
Cheney. Its proceedings remained secret, and as soon as the suspicion was raised in the press that many of its personnel and those it
consulted were heavily skewed toward executives from the industry,
a decision was made to keep their identity private. Invoking executive
privilege, Cheney resisted all demands from Congress and suits from
environmental organizations to disclose names. When, in 2007, the
Washington Post revealed the identity of forty of those consulted, few
were surprised to discover that they included executives from Exxon
and Mobil, a drilling friend of Cheney's from Wyoming, and Kenneth
Lay, the CEO of Enron.

Lately, though, George Bush has begun to sound like Jimmy Carter,
with a smidgin of Al Gore's concern for climate change. In December
2007, the Act for Energy Independence and Security was signed,
mandating fuel efficiency standards for automobiles of at least thirty-
five miles to the gallon by 2010 and increasing the supply of biofuels to
36 billion gallons. The modest commitment to higher fuel efficiency was
immediately denounced by Grover Norquist, the director of Americans
for Tax Reform, as a measure that would kill Americans by forcing
them to drive smaller, more vulnerable cars.

Norquist's quaint anxiety that America's national character, not
to mention its life and limb, were being imperiled by environmental
fanatics has been overtaken by the four-dollar gallon, a popular rush
back to public transport, and the abandonment of the SUVs and minivans that have been the mainstay of the automobile industry's profits.
Caught wrong-footed by a sudden and massive shift in demand, the
Big Three manufacturers are buried in redundant back inventory.
But the bigger issue for the present campaign and for the future is
whether an America of limits can actually be sold to the electorate.
John McCain, who for years subscribed to environmental pessimism on
global warming and ecological damage, understands the bewilderment
of Americans who have had to abandon plans for a driving vacation

and have a van in their garage they can't use, and has had a change of heart. Now his position is closer to the unreconstructed George Bush than the reformed version. He too wants to abandon the elder Bush's ban on drilling in environmentally sensitive regions. Open the Arctic National Wildlife Refuge, bring in the rigs, and the price of gasoline will magically return to the proper American level of around two bucks a gallon. It was McCain's bad luck that on the day he was due to offer this remedy, a hurricane prevented him from landing on an offshore platform and an oil tanker collided with a barge in the Mississippi Delta, spilling crude and producing a twelve-mile oil slick.

McCain is betting that environmental optimism is hard-wired into the American character. He may yet be right. Even in their tight spot, not many want to hear that the country has finally come to the end of its providential allotment of inexhaustible plenty. Public support for an end to drilling restrictions has risen sharply even while consumers try to trade down to more fuel-efficient smaller cars of the kind that Grover Norquist condemned as unpatriotically hazardous. But there is equally a sense that if nature comes up short, that other infinite resource, American know-how, can make up the difference. It was native ingenuity, planted in the most unforgiving soil, that could deliver a yield. Taming the untamable Colorado River with the stupendous Hoover Dam produced a water supply in the arid western desert copious enough to supply populous cities and intensive farming. In Imperial Valley, California, one side of the All-American Canal is a dunescape so barren that it could (and often does) serve as Hollywood's version of the Sahara; on the other are fields so abundant they produce green beans, asparagus, and strawberries for the supermarkets, and alfalfa for the cattle feedlots, all year round. Never mind that the reservoir of Lake Mead that delivers water to those cities, and farther downstream, the farms, is at 50 percent of capacity; all will somehow be well. The taps of Los Angeles running dry? The solution, a farmer, waxing indignant at the thought of selling some of his surplus water to the cities of Nevada and California, told me, is right on "their" doorstep: desalinate the Pacific!

It's not over, then, the American sense of a national entitlement to plenty, in which no one gets shortchanged and the next generation is always better off than the last. But then the dream of cost-free abun-

dance, a mirage of America as a perennially fruitful garden, goes back
a very long way into the national past.

34. Strawberry fields, 1775

It was when the hooves and fetlocks of his horse were dyed scarlet
from the crush of strawberries that Billy Bartram reckoned he must
have arrived in the Elysian Fields. He had just not expected them to
be located in the northwest hill country of Georgia, nor their inhabitants to be Cherokee. Bartram was botanizing in the South for Dr.
John Fothergill, the London Quaker whose collection of American
flora was second only to Kew. The vocation came naturally, as his
Philadelphian father, John, had been recognized as botanist to the
king. No one had been more assiduously encyclopedic in his mission
to spread abroad the reputation of *flora americana* than John Bartram,
yet he had not been especially content that his son should follow in
his footsteps. He would rather he had pursued some more lucrative
profession, and was dismayed by Billy's (ultimately unsuccessful) efforts
to turn indigo planter. But reluctantly or not John Bartram recognized
in his son someone who could not help warming to the discovery of
a new variety of *Robinia* or *Philadelphus*, and so he reconciled himself
to his son's path.

That had led William south into the Carolinas and Florida, making
copious notes and sketches for Dr. Fothergill, and recording his impressions of scenery and people as he went. The difference between father
and son was one of cultural generations as well as personal temper.
Bartram Sr. was an Enlightenment rationalist, for whom scientific
information, scrupulously recorded, was wonder enough. William, on
the other hand, while priding himself on exact detail, was a botanical
romantic, for whom around every turn of a hilly trail, a living miracle lay
waiting. Thus a singularly beautiful species of *Aesculus pavia*, growing
six feet tall on the crests of a Georgian hill, had limbs that terminated
"with a heavy cluster or thrysis of rose or pink-colored flowers, speckled
or variegated with crimson." But what Bartram wanted to be seen in
his hot prose was "these heavy spikes of flowers, charged with the
morning dews, [that] bend the flexile stems to the ground."

Compared to this kind of excitement, of what import was the news, picked up at Charleston, of the fight between the British and Patriot minutemen at Lexington, some weeks earlier? As Bartram progressed up the Savannah River, away from the flooded rice fields of the Low Country, and reached undulating hills, his excitement climbed along with the topography. Between Augusta and Fort James he rode past "heaps of white, gnawed bones, of ancient buffalo, elk and deer, indiscriminately mixed with those of men, half grown over with moss." Billy was entering his own Gothic romance. "How harmonious and sweetly murmur the purling rills and fleeting brooks, roving along the shadowy vales, passing through dark, subterranean caverns, or dashing over steep rocky precipices, their cold humid banks condensing the volatile vapors which falling, coalesce in crystalline drops on the leaves and elastic twigs of the aromatic shrubs and incarnate flowers!" Steady on, the reader protests, but it's too late; Bartram is already lost in the Vale of Tempe, deep inside the "New Purchase," which meant the land torn from the Cherokee after a defeat. No romance would be complete without mysterious ruins, which Bartram duly beholds in the form of ancient Indian tumuli, flattened at the top like the Wiltshire mounds but here overspread at their summit with red cedar, the sides indented with "lookouts" or sentry places, and terraces below planted with corn. "It is reasonable to suppose," Bartram speculates, that "they were to serve some important purpose in those days as they were public works and would have required the united labor and attention of a whole nation."

The stock romantic image of the American Indian is of a solitary, nobly brooding savage, imagined in forest glades, or bow-hunting deer and elk; far less familiar is an idea of tribes like the Cherokee dwelling in towns and large villages, building public works, constituting a complex society. But that is exactly what Bartram would encounter, the deeper he penetrates the interior of the Cherokee world. At Fort Prince George, he waits for three days for the Indian who was deputed to be his guide, before deciding to set off on his own under thundery skies. Climbing to the crest of the Oconee range, he looks down on the mountain wilderness "undulated as the great ocean after a tempest, the undulations gradually depressing, yet perfectly regular as the squama of fish or imbrications of tile on a roof." (It is beneath the poetic-scientific dignity of Billy to call them scales.) He descends by the banks of glittering rivers, names a mountain (Magnolia) and throbs to the spectacle of "roving beauties" (*Calycanthus floridus, Philadelphus inodorus, Convallaria majalis, Leontice thalictroides, Anemone hepatica* . . .) that "stroll over the mossy shelving, humid rocks." He finds a deserted Indian hunting lodge in which to shelter for the night from an immense electrical storm, dines on dried beef and biscuit, and listens to the whip-poor-wills. In the morning before him are spread, in their gaudy splendor, "the painted beds" of wild strawberries. Before long, he is entertained by the chief of Watauga and his two sons at their house (a log structure, plastered with clay inside and out), served "sodden venison" and hot corn cakes with milk and hominy pudding, afterward taking a pull from a mighty four-foot pipe, wrapped in "speckled snakeskin." "During my continuance here, about half an hour, I experienced the most perfect and agreeable hospitality conferred on me by these happy people . . . I mean happy in their dispositions, in their apprehensions of rectitude with regard to our social or moral conduct."

With his romantic projection in overdrive, Billy Bartram is now enclosed within the American terrestrial paradise. He rides through a forest glade and emerges to look down on "a vast expanse of green meadow . . . a meandering river, saluting in its various turnings the swelling green turfy knolls, embellished with parterres of flowers and [of course] strawberry fields; flocks of turkeys strolling about them, herds of deer prancing in the meads or bounding over the hills," and, most enticingly of all, "companies of young, innocent Cherokee virgins" filling baskets with berries with which they will stain their lips and cheeks. Stirred by the strawberry girls, Bartram and his trader friend spy on them, "although we meant no other than an innocent frolic with this gay assembly of hamadryades, we shall leave it to the person of feeling and sensibility to form an idea to what lengths our passions might have hurried us, thus warmed and excited, had it not been for the vigilance and care of some envious matrons who lay in ambush and espying us, gave the alarm, time enough for the nymphs to rally and assemble together."

After eros, power. Riding on alone into the Overhill towns, Bartram encounters the "caravan" of Little Carpenter, Ata-cul-culla, whom he calls "emperor or grand chief" of the Cherokee. He has enough self-possession that, when the chief asks him if he knows his name, Bartram replies that he certainly does and introduces himself as from the tribe of "white men, of Pennsylvania, who esteem themselves brothers and friends to the red men, but particularly so to the Cherokees and that notwithstanding we dwell at so great a distance we are united in love and friendship and that the name of Ata-cul-culla is dear to his white brothers of Pennsylvania."

The chief bids Bartram welcome but goes on his way down country to the meeting near Charleston, the outcome of which would bring disaster to the Cherokee for it would ally them with the king rather than the revolution. (With the exception of 1812 they had a record of choosing the wrong side in American wars.) But although Bartram published his *Travels* in 1791, he refuses to cloud the moment of instinctive human fraternity with anything ominous. In fact the friendship between Quaker botanist and Cherokee deepens when, on his way back to the coast, he visits the town of Cowee, where about a hundred houses were built around a grandly pillared circular "Council House" or "rotunda" where townsmen could meet to discuss tribal business or the state of the harvest. The Cherokee honor Bartram with an elaborate feast where he listens to "orations" from elders and then sees the ballplay dances (like lacrosse, only played with two racquets), whooping young male dancers "ornamented with silver bracelets, gorgets and wampum"dancing in a semicircle before a line of singing be-ribboned girls. Bartram is, as usual, spellbound.

Though he understands the enmity between Cherokee and Creek, and both of those tribes and the Choctaw, Bartram is innocent enough to idealize what he sees in the wooded mountains, the brilliant meadows and the strawberry fields as the perfectly self-sustaining American society. "Physically tall and graceful, the Cherokee are fond of their children. The men behave well to their women and they cherish their aged. The tribe hunts but it also cultivates, so that corn, melon, beans, pumpkin and squash are raised in a common garden." Plainly they are in no need of a white "civilization" whose incursions have brought them only trouble and rum. And they don't need Improvement. It is something of a wonder, Bartram muses, that they have been able to resist the corruption of the white world for so long, and he fears the magic of their world may not much longer survive the hordes of land-hungry settlers pressing on their territory.

But, Bartram hopes, in a country so abundantly blessed, is there not plenty for all to go around?

35. White Path, 1801–23

There were days when the agent thought his commission impossible to fulfill and wondered why, in his sixties, he had accepted so thankless an office. But then Return Jonathan Meigs had never been one to shrink from a challenge. Portrait engravings of the colonel show a tough old turkeycock of a man, beady-eyed and bony. So when his old comrade from the Quebec campaign of heroic and disastrous memory, Henry Dearborn, now President Jefferson's secretary of war, had inquired if Meigs might go to Tennessee to be United States agent among the Cherokee, he had not hesitated. Everyone seemed to go to Tennessee sooner or later. Besides, his son, Return Jonathan Jr., could now be left safely in a place of authority and eminence in Ohio, with the brightest prospects in the world, needing no paternal eye for his further advancement. Meigs had heard fine things of the Smoky Mountains, of the high country of north Georgia, and the old adventuring passion that had sent him in 1788 into the wildest regions of western Ohio had not yet died in his old body. It had been on the Ohio frontier that he had seen the Indians in full fury, but when the time came to speak with them about the return of captives, he found they were men like himself; men whose cast of mind he believed he understood. Could the Cherokee be much different?

But those Indians in the Ohio country had been braves and warriors. Now, as he understood his appointment, he was being asked to do something quite different, something about which Return Jonathan, from the beginning, had decidedly mixed feelings. The policy of the government toward the Cherokee, as it was to the other tribes of the Southeast, was to uphold them as proprietors of their land, and protect those rights against white frontiersmen who sought to dispossess them by simply squatting and daring the law to come and evict them. Neither Washington nor his secretary of war, General Henry Knox, imagined the Indians could be protected forever. But neither did they wish to have endless wars on the frontier. So the policy was to be one of social rather than military pacification. The Cherokee (and for that matter the Creek, Chickasaw, and the Choctaw) were to be turned into true Americans, which meant farmers. In any case, their old hunting grounds had been invaded by whites and depleted of game; and (so it was said) their communal gardens never yielded enough crops to see them through from year to year. Bartram's Cherokee Eden had been a fantasy. But if the Indians could somehow be persuaded to adopt a civilized life, the tomahawk replaced by the plow, each family with its own, the women supplementing food crops with cotton (for the Cherokee were slave owners), they could card and spin, and no longer be a threat. To the tidy, economical American mind, hunting grounds were a shocking waste of good land that could be made productive. Since there were at most 16,000 Cherokee claiming to occupy millions of acres, the abandonment of the old life would liberate a great portion of the land for sale and tillage. Perhaps the Cherokee could be induced to part with it in exchange for the settlement of the exorbitant debts they seemed to run up to white traders.

In one of the many meetings he had with Indian chiefs in Washington, Thomas Jefferson put the policy most clearly and generously, so Meigs thought: "Let me entreat you, on the lands now given to you, to begin to give every man a farm; let him enclose it, cultivate it, build a warm house on it, and when he dies let it belong to his wife and children after him. Nothing is so easy as to learn to cultivate the earth; all your women understand it and to make it easier we are always ready to teach you how to make plows, hoes, and necessary utensils. If the men will take the labor of the earth from the women, they will learn to spin and weave and clothe their families . . . When once you have property you will want laws and magistrates to protect your property and person . . . you will find our laws are good for this purpose . . . you will unite yourselves with us, join with us in our great councils and form one people with us, and we shall all be Americans; you will mix with us by marriage, your blood will run in our veins and will spread with us over this great continent."

The idea may have been noble, and it would inspire an entire generation of Cherokees to take the Jeffersonian dream seriously. But the president was playing a double game. In 1802 he let it be known to the state of Georgia that one day Cherokee rights would be retro-ceded to it. But his own officers, especially Colonel Meigs, believed his grander intentions. The first federal agent to the tribe, Leonard Shaw, had been sympathetic to the fusion of Indian and American and had married a full-blood Cherokee, instantly earning the suspicious hatred of the frontiersmen. Return Jonathan came to know the Cherokee well enough to understand that there could never be any question of their easy mass conversion to horse-and-plow farming; that hunting was ingrained in their culture; that it gave them food and clothing, and that attached to the division of labor between the sexes was an entire cultural calendar. Their religion, their dances, their food and tobacco were all unthinkable without this union of opposites: forest and garden.

But Meigs also knew that the Cherokee world had already been badly hurt by history. Choosing the British in the Revolutionary War had forced punitive land cessions out of them so that only about a third of the territory they thought of as ancestrally tribal was actually now theirs. Pressure from white settlers was relentless and supported by the Georgia politicians who wanted their votes. The whites, often ex-Patriot militiamen who had fought against the Indians, could not have been further from the high-minded paternalism of Washington, Knox, Jefferson, and Meigs himself, whom the Cherokee honored with the name of White Path. The settlers thought the Indians heathen savages who needed to be cleared out of the way or exterminated so that decent white Christian people, who understood what a hoe was, could make a go of it and make the wilderness bloom. So if for some incomprehensible reason the federal government was tender to the Indians, they would do their best to give the red men good reason to leave, making them understand there could be no cozy living together in Georgia and Tennessee.

So even as Meigs labored to persuade those among the Cherokee themselves (generally the older chiefs and the younger braves) who were skeptical of the good faith of the government, that the White Father meant what he said, the sorry record of casual theft, knife attacks, and murders, with American culprits going scot-free, undermined his best efforts. In 1812, after a series of eight murders of Cherokee for which no one was brought to justice, a furious Meigs wrote that "the Indians are condemned and executed on the testimony of any white citizen of common character and understanding, when at the same time a white man can kill an Indian in the presence of a hundred Indians and the testimony of these hundred Indians means nothing and the man will be acquitted."

To add insult to injury, the federal government, distressed by the impossibility of offering true justice to Cherokee victims, offered cash instead. Secretary Dearborn thought $200–300 for each murdered man or women would be about right. Knowing how abhorrent this was, Meigs decided to offer it anyway as the only form of reparation the Cherokee would get. Initially the chiefs were horrified, but there were so many cases of suffering, that after 1803 they accepted some payment while keeping, as William McLoughlin writes in his extraordinary work *Cherokee Renascence in the New Republic*, a different account in their own books of the lives owed to them. A deadly pattern established itself. The frontier settlers stole land and attacked Cherokee and could not be held to account. The Cherokee retaliated by stealing horses, which gave the settlers further reason to treat them as red outlaws.

The inability of the United States government to deliver on its promises stirred up a faction among the Cherokee who saw no reason why they should meekly discard their traditional way of life. They also suspected (correctly) that at least part of the motive for turning them into model farmers was so that millions of acres of their land could be ceded and sold. To be satisfied with cotton and corn and be surrounded by a world of hostile, brutal whites who wished them to be gone was to die the death by a thousand cuts. The colonel, when he was honest with himself (as he often was), knew they had a point and believed, incorrectly, that the Cherokee heart could never be in sedentary agriculture. But he was conscientious in doing what he could to realize Jefferson's dream of Indian Progress. He distributed farming tools, mattocks, and plows, as well as carding machines and spinning wheels for the cotton, which he noted, somewhat to his surprise, the Cherokee women had already made a success of cultivating. He also believed that the true salvation of the Indians would be in education (a Meigs dynasty trait, this) and encouraged the cession of land in return for sums of money that were applied to the creation of schools and the payment of teachers.

Caught between lawless white encroachment on the one hand, and hard-line Indian resistance on the other, Meigs believed he had no alternative but to cultivate chiefs who were inclined to sell and settle.

The most prominent was Doublehead, and the colonel knew well that he had adopted the new way as a means of enriching himself as much as his village. But was not that, after all, the American way? When Meigs and Doublehead together embarked on a campaign for the sale of old hunting grounds, the consequence was predictable. After a ballplay in August 1807, Doublehead was killed by a group of young chiefs incensed at his betrayal of their homeland.

Meigs ought to have seen this coming. He had himself written to Dearborn that "they have long resolved not to part with any more land. There is not a man in the nation who dares advocate it." And yet the pressure for land was unrelenting. Iron ore deposits long known about but unexploited became yet another reason for dispossession. The first decades of the nineteenth century were a time of heroic road building in the United States. To connect the interior with the coast, the federal government as well as the states wanted a route that linked Nashville and Chattanooga with Augusta and the eastern seaboard, a line that went straight through Cherokee territory.

A gloomy fatalism began to make its way into Return Jonathan's canny old head. He had just buried his second wife, Grace, and with every day of trouble that passed he had second thoughts about the eventual fate of the Cherokee. If sedentary agriculture was not going to work, either through white aggression or Indian reaction, what would? Around 1808–9, he began to entertain an idea that Jefferson himself had raised five years earlier, that of an "exchange" of land; hundreds of thousands of acres of territory west of the Mississippi in return for abandonment of their present territory. Put another way, this was a policy of "removal," for the moment voluntary. Jefferson rationalized what he was proposing in terms of treating the Indians *more* like white pioneers rather than less. Why would they *not* want to remove west, where they would be rid of white marauders and squatters and where game was surely plentiful, if they insisted on keeping to their older way of life? For those who wished to farm, the money paid for their land in the East could be enough to buy western acres. They would be native homesteaders. Why would they not jump at such an opportunity? The answer of course was that this plan represented the abandonment of Jefferson's fine promise that Indians would share American destiny and abundance and the two races march forward together as one farming people. Besides, as the

president who clung through thick and thin to his Virginia hillside
knew perfectly well, land was more than territorial inventory. Land
was a place impregnated with poetic or even mystical qualities, and
the specific place that the Cherokee called home were those hills
draped in wild azalea that had so elated William Bartram. Home
was the strawberry fields. In this sense the Cherokee were not, by
Tocqueville's standards, true Americans at all. Moving was not an
opportunity; it was a calamity. Besides, the Cherokee knew nothing
of what lay beyond the Mississippi except that some white men called
it "the Great American Desert."

And yet, for all their forebodings, a few hundred of the Cherokee
who felt most under pressure or who put faith in White Path's reas-
surances did indeed make the journey west to Arkansas. Many of the
chiefs who opted for transplantation did so armed with the reassurance,
given by Meigs, that they would "take their land with them"—that an
acre of Georgia or Tennessee would entitle them to precisely the same
allotment in their new home. By far the greater part of the Cherokee
opted to stay and pinned their faith in the security of their land title
promised by successive treaties. Encouraged by Meigs, they joined, for
once, the right side in an American war, making up a volunteer force
in 1812–14 that served under Andrew Jackson, Indian fighter and land
speculator. Their neighbors, the Creeks, made the wrong choice, siding
with the British, and were duly punished. But to their horror, part of
the land cessions imposed on the Creeks included a cool two-million-
plus acres belonging to the Cherokee, America's ally! President Madison
had actually decorated Cherokee warriors for bravery in the war but
had been unable to stop Jackson's white militia from a spree of killing
and destruction of Cherokee farms and livestock on their way back
from the battle. For the militia it seemed like sport. Warriors returned
home to sacked villages, the rotting carcasses of hogs, sheep, cattle,
and horses, houses in charred ruins. The idea, of course, was to use
the opportunity of wartime to terrorize the Cherokee into vacating
their lands as soon as possible. Aghast, Meigs computed the damages
for losses suffered by the Cherokee at the hands of the demobilized
marauders at nearly $21,000, a huge sum by the standards of the day,
so enormous that it made Andrew Jackson chuckle at the idea that
anyone would actually believe the word of an Indian, much less pay
out damages.

For a few months, Meigs thought he had prevailed on their behalf in Washington. Damages were authorized, and the 2 million acres that the colonel had proved indisputably belonged to the Cherokee were returned to them. But this only brought the wrath of the frontier, led by Jackson, down on the head of President Madison. His message to Meigs was: make them cough up some of it if they don't want to lose it all. Meigs tried but in vain. And now, for all that the Cherokee had suffered by way of violations of the treaties of earlier administrations, the colonel was beginning to think that their presumption of being treated as a sovereign nation was deluded if not doomed. He detected something unprecedentedly brutal in Jackson, and he was right. Years of famine in 1816 and 1817 only persuaded the colonel that his pessimism was justified. Perhaps the Cherokee would indeed be better off somewhere else. At Hiwassee, where Meigs was based, he heard Jackson spell out to the Cherokee chiefs their alternatives: removal to Arkansas, the government providing a gun and a blanket, or staying and abiding by the laws of Georgia or Tennessee. They must understand they did not have ancestral "territory"; they would get 640 acres each. The chiefs replied "we wish to remain on our land and hold it fast. We appeal to our father the president to do us justice." They wanted neither to go nor to be American citizens. They had supposed they already had a nation and that Presidents Washington and Jefferson had thought so too. If they were forced to go west, they would be reduced to a "savage way of life," and had not Washington and Jefferson wanted them to "remain on our lands and follow the pursuits of agriculture and civilisation"? Jackson told them their third way was no longer an option and they must choose between the two he offered. If they refused, they would be considered "unfriendly." Intimidated by the sinister threat, half—but only half—of the sixty-seven chiefs signed.

Meigs told his Indian friends he thought the move was for the best. If they stayed, they faced eventual "extinction," while in the west they might still be a nation. This was also the language Jackson used, but "extinction" was not something preordained by history, just because of the numbers of white immigrants. It was an actual policy determined by actual men. What Jackson meant was that they could no longer expect the federal government to protect them in the name of some old treaties signed in a sentimental generation. What was government supposed to do: establish forts to fire on their own kind? Really, the idea was ridiculous. And the sorriest part of the story is that White Path had begun to talk like this too. The most adventurous of the great Meigs dynasty had become a moral coward, rationalizing that he was doing the right thing by the Cherokee, when a large part of him must have known the opposite was true.

What made it worse was that seeds that he himself had planted among the Cherokee in the Jefferson years were now bearing fruit, belying Jackson's slander that the Indians would never be capable of modernizing themselves. They had a cotton culture (including, lest it be forgotten in the Indian romance, slaves); they spun and wove. Their schools were starting to produce Cherokee children literate in their own language and English, and the influence of fairly benevolent Moravian missionaries had even succeeded in some conversions.

But the success story was exactly what Andrew Jackson neither believed nor wanted to hear. And after the Battle of New Orleans (won after the war with Britain had ended), he was politically untouchable. Jackson was, many people knew, a president-in-waiting, and he made it clear that he was eager to end what he called the "farce" of treaties between the government and jumped-up savages who imagined they were little kings.

It was politics as much as racism (of which there was a deep streak in Jackson, for all his adoption of two Indian children) that moved him. His was the authentic voice of the democratic frontier, both speaking for and expecting the loyalty of all those who could not wait to get their hands on Indian land. The dream of American plenty for the ordinary man was born from Andrew Jackson's determination to evict tens of thousands of Indians—Chickasaw, Choctaw, Seminole, and Creek as well as Cherokee—from the only homelands they had ever known, because they happened to be in the way. It was absurd to the point of offensive, Jackson told himself, to pretend that Indian "nations" could ever be incorporated into the Union on their own terms as Indian states.

Return Jonathan Meigs died in 1823 at the age of eighty-two, and the manner of his going belied his grim conviction in his later years that transplantation—in effect the disentanglement of Indian and white lives that had become so braided together in the up-country— would be for the best. An elderly Cherokee chief had come to see him at the agency at Hiwassee to debate the wisdom of removal. Meigs saw that the old man was sick and offered him his lodge while he took a tent outside. But it was Meigs whose head cold turned into pneumonia, taking him away. The funeral for White Path was attended by a long file of Cherokee chiefs, braves, and women, all of whom had trusted another Meigs, perhaps too much, with their best interests. A Cherokee death chant went up over the mountains with a wreath of smoke.

Perhaps it was as well that Meigs did not live to see the denouement, which was more tragic than he could possibly have imagined. And it was made so because the Cherokee had listened all too well to their tutor White Path. By the 1820s they had a written language developed by George Gist, known as Sequoyah, based on the eighty-six syllables at the core of their spoken tongue. Once the language was established, a bilingual newspaper followed, the *Cherokee Phoenix*. In place of the old tribal councils meeting in the round houses, John Ross, the charismatic chief of the anti-emigrant majority, an eighth Cherokee and seven-eighths Scots, produced a written constitution, modeled after that of the United States. Elections were organized; law courts and magistrates, a militia, and a police force put into being. In less than ten years, with the threat of mass deportation always at their back, the Cherokee had become a true microstate comparable to many of those claiming independence in Europe.

It was as though Ross and his allies in Cherokee nationalism, many of whom like Major Ridge had fought in the Creek War and who had been embittered by Jackson's shocking indifference to his own Indian militia, now wanted to demonstrate to the shades of Jefferson and Meigs that they could indeed make a modern political and social culture. It was precisely this *striving* to be like America that had periodically depressed Meigs and made him wish that they return to elk hunts and spirit dances in whatever country they could be left in peace. And it was certainly the fact of their progress in agriculture, the cotton industry, and politics that spurred Andrew Jackson to get rid of the Cherokee sooner rather than later. For who could tell what fancy notions they might get if the experiment in Indian nation building was allowed to mature? They might actually *succeed* in cash crop farming and then be impossible to uproot! They might get dangerous ideas that others might be called on for their protection, those others usually marching under the Union Jack, the nation that Jackson most detested in the world. As the Cherokee became more organized and aggravatingly articulate, they turned from a nuisance into a menace.

And suddenly fortune played the Cherokee a wild card. Gold ore was discovered in the hills of west Georgia in an area Bartram had traversed, by one Benjamin Parks, setting off the first "rush" in American history. The Cherokee had known about specks of alluvial ore, and the nearest town to the strike was called Dahlonega, Cherokee for "yellow money," but they had not bothered to do anything systematic about it. Now, the mere thought that, enriched by gold, the Cherokee might be able to hold off their transplantation indefinitely provoked Jackson, elected president in 1828, to introduce the Indian Removal Act in Congress in 1830.

Under its terms not just the Cherokee but all the "Five Civilized Tribes"—Choctaw, Creek, Chickasaw, and Seminole—were to be removed from their lands east of the Mississippi and taken to an alternative settlement somewhere to the west of the river, plumb in the region straddling the 100th meridian, called by Stephen Long, and with good reason, "the Great American Desert" (now Oklahoma). What little was known of the area made it clear that it was, especially west of the 100th, arid and unsuitable for precisely the kind of settled agriculture the Indians had been told they were supposed to master. The Removal Act attempted to mask its mass deportations as a voluntary invitation to emigrate, and to make that act seem akin to the westward movement of white pioneers on the Oregon Trail. But Cherokee chiefs like Ross noticed that the Oregon emigrants traveled straight past the region to which they were being removed. Had it had any serious potential, it would already have surely attracted settlers. But the president made it clear that should any decide to remain, they would henceforth be subject to the laws of the states in which they resided. It was, in effect, a threat. Jackson would eliminate all federal protection for the Indians (in violation of guarantees made by Washington, Jefferson, and Adams) and would unleash the governments of Georgia and Tennessee on them, which would undoubtedly ride roughshod over any notion that the tribes owned much land at all, and certainly not any that white settlers and gold prospectors had their eyes on.

It was one of the most morally repugnant moments in American history, one that ought by rights to take its prime mover, the seventh president, off the currency of any self-respecting nation. But there he is with his imperious forelock on the face of the twenty-dollar bill, the ethnic cleanser of the first democratic age. Jackson made several speeches on the subject, in which he cast himself and his policy as the Savior of Indians who otherwise would be doomed to disappear, surrounded as they were by the arts and industries of a "superior race." They had neither the intelligence nor the inclination, he said, to transform themselves into a viable modern society, in the face of all the evidence to the contrary.

It is not as if the ferocious immorality of the Removal Act were not noticed at the time. It squeaked through with bare pluralities, in the House of Representatives by just five votes, and many of the most eloquent voices of the day spoke bitterly and lengthily against it; Senator Theodore Frelinghuysen of New Jersey took three days to go through all the solemn undertakings given by past governments to the Indians on the implicit understanding that they had title to the land from "immemorial possession." "If we abandon these aboriginal proprietors of our soil, these early allies and adopted children of our forefathers, how shall we justify this trespass to ourselves . . . Let us beware how, by oppressive encroachments on the sacred privileges of our Indian neighbors, we minister to the agonies of future remorse." The most unlikely opposition came from a legendary figure who embodied the frontier even more emblematically than Jackson: David Crockett, like the president, from Tennessee. Crockett had fought alongside Cherokee like John Ross at the Battle of Horseshoe Bend and had come to admire their tenacity and courage. He would pay at the polls for the temerity of his opposition, losing a bid for reelection to the House, though winning on a second try. Many of the greatest figures in Congress like Henry Clay and Daniel Webster spoke in indignation at the iniquity of the bill, but the orator who most grasped the enormity of what was being proposed, a people's destruction engineered from greed and covetousness, and who attempted to summon American moral tradition against it, was Edward Everett, Massachusetts congressman, later governor of that commonwealth and president of Harvard. "The evil," said Everett, "was enormous, the inevitable suffering incalculable. Do not stain the fair fame of the country . . . Nations of dependent Indians, under color of law, are driven from their homes into the wilderness. You cannot explain it; you cannot reason it away . . . Our friends will view this measure with sorrow and our enemies alone with joy. And we ourselves Sir, when the interests and passions of the day are past, shall look back upon it, I fear, with self-reproach and a regret as bitter as it is unavailing."

None of this had the slightest effect on Jackson's implacable determination to purge the Southeast of Indians. The more civilized they were, the more reason to uproot them. Nor could the Supreme Court make much of an impression. When Ross and the anti-removal party sued the state of Georgia for illegal trespass and having no standing in altering binding treaties concluded between the federal government and the Cherokee Nation, the court initially dismissed the argument. But the greatest advocate of the day, William Wirt, pleading for the Indians, had no trouble showing from the letter of past treaties that the Cherokee had indeed been treated as a nation and Chief Justice John Marshall's court upheld their case, disallowing Georgia from, among other presumptions, holding a lottery to sell off Indian land. The newspaper editor Horace Greeley reported Jackson as saying, contemptuously, "John Marshall has made his opinion. Now let him enforce it." Apocryphal or not, the contempt for the Constitution certainly sounds like authentic Jackson.

Jackson advertised the measure as the purest benevolence and its realization as "a happy consummation." In his second address to Congress, Jackson finally obliterated Jefferson's dream of being "all Americans" by boasting that removal would "separate the Indians from immediate contact with the whites" and "enable them to pursue happiness in their own way and under their own rude institutions . . . and perhaps cause them gradually to cast off their savage habits and become an interesting, civilised and Christian community . . . Humanity has often wept over the fate of the aborigines of this country . . . To follow to the tomb the last of his race and to tread on the graves of extinct nations excite melancholy reflection . . . [but] what good man would prefer a country covered with forests and ranged by a few thousand savages to our extensive Republic, studded with cities, towns and prosperous farms . . . and filled with all the blessings of liberty, civilisation and religion? . . . Doubtless it will be painful to leave the graves of their fathers; but what do they [do] more than our ancestors did or than our children are now doing? To better their condition in an unknown land our forefathers left all that was dear in earthly objects . . . How many thousands of our own people would gladly embrace the opportunity of removing to the West on such conditions!" All the Indians had to do to become truly American was to hitch their wagons to the religion of Moving Right On.

Everything about this was specious: the caricature of the Cherokee as still "savage," unable to maintain themselves against the white hordes, needing separation in their own best interests, and the same kind of arguments would be heard again in the future history of ethnic cleansing in and out of Europe. America's honor lies in the fierceness of those who opposed Jackson. They continued to do so, but were up against the most authoritarian president the United States had seen, dressed withal as a new kind of Democrat; someone who had no compunction about violating the Constitution, ignoring the Supreme Court, unilaterally rescinding treaties. When a "Treaty" was signed with a small but influential faction of the Cherokees moved to despair by Jackson's bullying, and passed through Congress by a single vote, Everett demanded that the president go to the Capitol steps and burn the actual treaties with Washington's, Adams's, and Jefferson's signatures written on them.

And for what was this monstrous crime committed? So that the myths of the homestead movement, of America's millions moving into empty space, the "garden" that a generous Providence had bestowed on them, could be sustained. But what the white settlers of west Georgia and east Tennessee were actually doing was taking someone else's plenty.

Ross tried everything, overplaying his hand in an interview with Jackson in 1834, that may have contributed to the relatively meager sum—$5 million—that the Cherokee were to be paid for leaving their millions of acres. The lottery went ahead, and when properties were distributed, Ross went home from negotiations in Washington to discover his had been taken and that he and his family had been summarily evicted. The lottery winner was already installed, and Ross had to leave his horse with him while he went on foot to search for his family. They moved into a two-room log cabin for the rest of their time in the territory, and no one complained.

Jackson and the Department of War wanted the removals to be

"voluntary" but Ross refused to budge from his policy of defiant resistance, urging the Cherokee at their council to stay united. So they
became united in misery. The "treaty" enacted in 1836 by the majority
of one provided for a two-year grace period before the removal was
made coercive. Only a bare handful of Cherokee took advantage of the
offer. The vast majority, listening to Ross, awaited their doom.

In the late spring of 1838, under the presidency of Martin Van
Buren, Jackson's successor, the deadline for voluntary removal having
expired, the Cherokee, the last of the "Five Civilized Tribes" to hold
out in the Southeast, were rounded up and hunted down by General
Winfield Scott's 7,000 troops assigned to the glorious mission. The
traumatized and humiliated Cherokee, many of whom were children
and elderly, were held in pens and corrals in conditions of the utmost
squalor and deprivation. So many became sick and died of maltreat-
ment that officers in charge of the operation resigned in abhorrence
of what they were being asked to do. Thousands of the Indians were
herded onto rickety boats for the first stage of the journey west. One
of those, on the Arkansas River, a hundred feet long and twenty wide
according to the Reverend Daniel Butrick who saw it, was so over-
packed with terrified Indians that "the timbers began to crack and give
way and the boat itself was on the point of sinking . . . Who would
think of crowding men, women and children, sick and well together
with little, if any more room . . . than would be allowed to swine
taken to market?" Overnight they were guarded like convicts. Others
were packed into railroad boxcars, the dead and dying thrown out
before the survivors were made to walk, in a file miles long, the rest
of the 800 miles into their new home in the Great American Desert.
Appalled at the army's treatment, Ross pleaded that he might take
over the command of the transportation. He was allowed this, only
to see fully a quarter of his 16,000 people perish.

In January 1839, a traveler from Maine, who published his account in
the New York Observer, saw a sad caravan of 2,000 Cherokee extending
for three miles, "the sick and feeble carried in wagons—a great many
ride on horseback and multitudes go on foot—even aged females
apparently nearly ready to drop into the grave, were traveling with
heavy burdens attached to the back—on the sometimes frozen ground
and sometimes muddy streets with no covering for the feet except
what nature had given them . . . we learned from the inhabitants on

the road where the Indians passed that they buried fourteen or fifteen at every stopping place."

Back in Georgia and Tennessee jubilant homesteaders moved into their allotted properties. Many of those who hoped to make colossal fortunes were deceived, for there was less gold than the first excited wave of prospectors had predicted. From the time that the last of the Cherokee trudged west to the Mississippi to the end of the Georgia gold rush was just five years.

36.   1893

In early March 1893, the Reading and Pennsylvania Railroad went into receivership. The rail companies that could make no mistake in calculating their part in the American future had overestimated demand. Some of the biggest names in the business followed, burdened with credit liabilities and sharply dropping receipts. The bankruptcy of the National Cordage Company, thought solid, was another shocking blow. By the end of the year 15,000 businesses had declared insolvency, more than 600 of them banks that closed their doors. When the United States Treasury announced that its gold reserves, required to shore up the currency, had fallen below the magic figure of 100 million, on 5 May, Wall Street panicked, writing off millions of dollars of stock. The steady rain of failures turned into a downpour. By the summer most industrial cities in America had unemployment rates of 18 to 20 percent. In some of the heavy manufacturing states of the Midwest as many as one in two adult men were out of work. With surviving banks calling in loans, small businessmen were hit especially hard. One of them who owned an ironware company in Massillon, Ohio, Jacob Coxey, demanded the government use currency unbacked by gold, to finance public works. Written off as a socialist fanatic, Jacob, along with his son Legal Tender Coxey, led "Coxey's Army," a procession of some hundreds of unemployed in a march on Washington.

But the president was not well. Grover Cleveland, in his second term, was so nervous of the public mood that when his physicians told him he was suffering from cancer of the mouth and jaw and would need surgery, he kept the matter secret lest it roil the markets further. On the yacht *Oneida*, surgeons cut away a large part of his jaw as the boat steamed gently out of New York Harbor.

All of which made the 600 acres of the Columbian Exposition in Chicago more, rather than less, necessary as a bright show of faith in the American future. If half the country was stricken, the other half went to Chicago. Twenty-seven million passed through the gates at Jackson Park between May and the end of October. Westinghouse (rather to Thomas Edison's annoyance) supplied electrical lighting; a White City with work of America's most showy and inventive architects—Charles McKim, Henry Cobb, and Louis Sullivan among them—adorned the center with Beaux Arts cupolas and gleaming halls. In the national and "ethnic" pavilions it was possible to see hula dancers from Hawaii (the islands Cleveland declined to annex) or lacemakers from Bruges. But having done their dutiful tour of elevated exhibits, most of the crowds headed for Midway Plaisance, Sol Bloom's pleasure grounds where they could giggle at the ethnically incorrect "Arabo-Egyptian" hootchy-kootchy, sail 200 feet into the sky on George Ferris's Big Wheel (the first in the country), watch dumbstruck as Eadweard Muybridge showed his images of Animal Locomotion, and sample the enticing array of snacking wonders available for the very first time: Juicy Fruit gum! CRACKERJACK! If the mood took them, they could actually have listened to the historian Frederick Jackson Turner declare that it was all over with frontier democracy, before going on to Buffalo Bill Cody's show to make sure they got a sighting of Sitting Bull and bronco busters before it was too late. But amid all the ragtime honky-tonk clamor (Scott Joplin was there, as well as John Philip Sousa) and electrically charged fun, millions upon millions of the visitors lined up to see one small, quiet emblem of the America that had been and, they hoped, always would be.

The Idaho Pavilion was at its heart a log cabin; admittedly the grandest and most richly decorated log cabin imaginable, glowing with richly woven rugs and tapestries, all with western and Indian themes, heavy turned tables and sideboards in the oddly separated Men's and Women's Reception rooms. Some of its designs in wood and metalwork inspired the beginnings of the American Arts and Crafts movement. But its message, appreciated by the millions, was that amid modern wonders and sumptuousness, the true America was still log cabin simplicity, and for all the daunting immensity of a New York or a Chicago, the heart of the country beat in the most rudimentary shelters. "The children who come here," one account said, "should see these rooms as they teach the lesson of the growth of our national civilization and how step by step men have made their way in life as well as something of the costs of the luxuries we now enjoy. When the foundations of a national civilization are bared to our eyes we are apt to judge them with greater acumen . . . the foundation is here in the cabin and the log hut." The sense that there was still something solid to hang on to while the American economy disintegrated around them could not have been more important for the millions who wondered whether there would be plenty in their own American future. Was the dream of the pioneering homestead, 160 acres of good earth for every family willing to bend their backs in tillage, lost forever?

The thousands lined up on the dusty empty plain of the Cherokee Outlet on the morning of 16 September certainly didn't think so. Right across the neck of land called the Cherokee Strip, reporters looking to pump up a nice round number for their readers back East claimed 100,000; certainly a whole lot more than the 42,000 parcels that were up for allotment that day. It was going to be, they said, the last great Run. There would be nothing left of the Indian Territory after the Strip had gone, so this was their last crack at plenty. If they were fast and lucky and got to make their claim at one of the land offices, they would move their family out to the plain, cut themselves a sod house, raise some corn, and stay maybe five, six years, just until they had enough to move on again. Who knew how long the good times on the prairie would last; when the next big drought would come?

The railroad companies had scooped up the best land as the government had sold off millions of acres and then had promoted westward migration and farm life on the plains just as hard as they could. Come to Kansas, their papers and flyers had said, the rainiest place in America, fertile and green. On came the greenhorns from the cities; anything to escape the tenements, and the railroad had them in its power. The farmers would need the railroad and the country banks they owned to advance them money for tools and seed and a team of oxen. They would need the company dry-goods stores for their provisions, and they would need the railroad itself to get their crops to mills and markets.

Here was one last chance to get a piece of land the railroad people hadn't put away. So they couldn't turn their back on the Run, everyone waiting, fidgeting at the line ready, once the cannon fired, to charge hell for leather. Lined up on the prairie was every kind of conveyance to get a man where he needed to go fast: wagon teams, buckboards, buggies, every kind of horse too, including the iron one, a train steaming on its tracks, loaded with men hanging out of the cattle cars, whooping away.

But no Indians. The Cherokee Strip Run would be another farewell, but this time they were not shedding tears. With demand, first for cattle pasture and then for tillage, the nation had leased and then gradually sold off most of its remaining acres on the Outlet, the ungenerous, tight-packed sod that had never done much for them anyway. The 12,000 who survived the Trail of Tears under the leadership of John Ross had their worst fears about the Great American Desert confirmed. The landscape could not have been more different from the hills and valleys they had left behind in Georgia and Tennessee. All there was here was an endless high plain, covered with tough, short buffalo grass. No trees broke the force of the winds that blew from the north, cutting them like knives in the winter, burning them in the summer. They could graze cattle and hunt bison, but tribes already there—the Osage in particular—saw that as their preserve and were none too happy with the arrival of a foreign people in their midst. All too soon mass slaughter by the Americans had finished off the buffalo herds anyway, their skeletons littering the prairie. The deer and beaver were nowhere to be seen. They became reduced to bow-hunting jackrabbits and prairie chickens. Most difficult of all, they had lost the streams and waterfalls of their old world, the water sent by the Great Spirit to make their crops grow. This place was dry. When it rained, which it seldom did, the water ran away against the hard, grass-packed dirt; an earth so tight that to dig it was to make war on it.

The years after 1840 were grim for the Cherokee. They went from the inconsolable loss of their uprooting to the impossibility of replanting in Oklahoma. The Ghost Dancers came among them again, and with their pipes they smoked dreams of the rain-swollen white clouds that hung over their lost green mountains of the East. From Washington they learned to expect nothing but merciless betrayal. So when in 1861 they had to make a choice, they chose for the Confederacy, for the one thing some of them still had were their slaves. The fateful decision compounded their calamities. In 1866 they ended up with neither slaves nor mastery of the acres they had been promised in perpetuity by Jackson for their removal. Now they had to make a treaty all over again, which gave the government the right to settle other Indians—how many were they bringing to this wilderness?—whenever they chose; and to insist that they relinquish title for sales should that need arise. They signed virtually at the point of a gun and continued to eat their bitter meal. One by one their councils were set aside until by the 1890s there was virtually nothing left of their old freedom; save their language, their religion, and their horses and dogs. But that little meant at least they were still Cherokee.

Indians like themselves, they tried to tell America, were people of the East. Over the generations they had learned enough about the land to know how different were East and West. East of the 100th meridian, they could expect rain enough to grow corn; west, they could never rely on it. Sometimes it came, more often it did not. So most of the Cherokee Nation began to live east of the 96th, leaving the wide-open tracts of what was called their Outlet to those who wanted to buy or rent it. Cattlemen did, because they needed pasture for their herds, uncountable hundreds of thousands of longhorns, raised to feed America's beef habit, driven up from Texas to the Kansas railheads. The Texans were prepared to pay handsomely for the lease of their land, and the Cherokee were happy to take it. But in 1886–87 there had been a winter so brutal, blizzard after blizzard, that cattle had been buried alive in the monstrous drifts, almost 80 percent of the millions of animals dying on the prairie. In any case livestock was becoming less profitable. In their greed the cattle ranchers had overgrazed, damaging the shortgrass prairie so that the amount of land needed to feed and raise a single steer had gone up and up, reaching almost fifty acres by the time of the snows of '86. The industry collapsed; and in their place came hundreds of thousands of migrants, some from the cities, some off the ship, with just an ox team, mules, and perhaps a gold dollar to start them up growing corn and beans. If the land approaching the 100th meridian and beyond it seemed too dry to make anything of, they had been told by respected persons that "rain followed the

plow"; that just by being there in their numbers and stirring the dirt
around a bit, they could make the air cooler and more humid. Prices
for the untillable prairie in the west section of the Indian Territory
began to climb and never stopped.

The Cherokee had their doubts about white men making rain, even
with the help of the Dance and the spirits. Their pessimism seemed
borne out by the brutal drought of 1890, which was followed by
another three years in a row of rainless heat. So when the offer of
$8 million to sell their remaining acres on the Strip came along, why
would they not take it? The council met, and the offer was accepted
without much regret.

It was approaching noon. According to the land offices there were
42,000 parcels of 160 acres each available for those who made the Run.
Seth Humphrey and his brother were not about to start as farmers,
however bad the times were; they had come with their bicycles for the
high jinks of it. Nothing had prepared them for the multitudes; some
"Sooners" who had got there ahead of the Run, staked a claim and
dared everyone else to get hopping mad about it; thousands of men
sleeping on blankets laid out on the buffalo grass or in makeshift tents;
everywhere the whinnying and the smell of more horses than you had
ever seen outside of an army camp; the creak and clang of wheels and
axles, buggies and carts. And now they were there and the locomotive
too, with its crazy passengers hanging out of the cattle cars and hollering
as though they could move the iron horse along quick if only they had
iron whips! The line of horses and carts and men and women went
on for miles. Some hundreds of soldiers had been stationed close by
to make sure the Run didn't turn into some sort of lawless battle. "I
casually wondered," wrote Seth, "how they would manage to dodge
the onrush; perhaps they were wondering that too."

At five before noon, the standing troops were given an order and
pointed their rifles to the sky. The idea was that a cannon stationed at
the eastern end of the Strip would fire and then the troops strung out
across the line would take up the report. As stopwatches moved toward
noon, Humphrey managed somehow to sneak out about fifty yards in
front of the crowd, between the horses and the train, and watched around
8,000 in his sight line of around two miles begin to stir with their reins
and spurs. "While we stood, numb with looking, the rifles snapped and

the line broke with a huge crackling roar. That one thundering moment of horseflesh by the mile quivering in its first leap forward was a gift of the gods, and its like will never come again. The next instant we were in a crush of vehicles, whizzing past us like a calamity."

At the end of the Run in the new townships of Enid, Perry, and Woodward, men were stationed in makeshift land offices, pen and ink and ledgers set on their trestle tables; men in little hats, their coats off for it was hot that September day. Long lines snaked out into the dirt street behind them, the "Boomers" who had made the Run, surprisingly few of them cursing or drinking or firing off, but waiting their turn to stake a claim to their 160 acres of America. Never mind that the acres were of hard-packed buffalo grass, not the kind of dirt a hoe could do anything with; they couldn't see much of it anyway, for the army of horsemen had kicked up such a storm of dust that it clouded the view, powdery dirt flying fine into the sky and down the streets of Enid, scratching the eyes of hopeful men.

37. The church of irrigation

One hundred and sixty acres were too many and yet not enough, John Wesley Powell told anyone who would listen to him at the Irrigation Congress that convened in Los Angeles in October 1893, about a month after the Cherokee Strip Run. It was the 100th meridian that made the difference. Farms in the "humid zone" to the east, or those in drier regions that benefited from intensive irrigation, hardly needed the sacred 160. But those in the west ought to content themselves with pasture and would need upwards of 2,500 acres to have any chance of success. The main thing, Powell said over and again, was that farming should go where the water was. Much as he was in favor of utilizing river waters through dams and ditches, on-stream reservoirs could only do so much. Where water was in short supply, like the prairie beyond the 100th meridian, the sensible thing was to create extensive "pasturage ranches" with water for not more than twenty acres of crops. And Powell believed that water resources in the arid West were too precious to leave to the market, which in all likelihood meant the kind of companies that had been smashed up in the economic train wreck of 1893. (In fact companies founded during the irrigation boom to produce and sell water were themselves notoriously unstable and insolvent, often underestimating fixed capital costs and the time it would take their customers to become viable farmers.)

Powell's views were heresy many times over; defiance of the plutocratic gods of the Gilded Age whose fortunes were built on economies of scale, vertical integration, and cartelized price fixing. The notion that government—either local or federal—should now be the trustee of water, just when it had at last offloaded millions of acres of unsettled land, appalled the company men. But coming as they did from Major John Wesley Powell, the views had to be given some credence because he was an all-American hero, soldier, explorer, and scientist, the director of the United States Geological Survey, someone who made no distinction between physical and intellectual bravery; a breed that was becoming as rare as the Great Plains bison.

In 1890, 90 percent of the population of the United States lived in its eastern half. With cities bursting at the seams, it was impossible for any far-sighted American statesman—like the young Theodore Roosevelt, for example, who had had personal experience of the West—not to see the region's transformation into settler-friendly farmland as the answer to so many of the nation's problems. But they had to listen to Major Powell. The son of a circuit-riding preacher in Illinois, John Wesley was largely self-taught in the natural history that allowed him, after the battle of Shiloh had blown off an arm, to become first a schoolteacher and then a university professor. In 1869 he had done the unthinkable by taking an expedition of four twenty-foot boats, three oak, one pine, a thousand miles down the notoriously lethal Colorado from its deceptively placid tributary, the Green River in northern Wyoming, through the vertiginous magnificence and terror of the Grand Canyon, hitherto seen only from the rim. Powell survived an ordeal of multiple capsizings, violent storms, and near starvation to make it through the canyon, though three members of his company decided, after a particularly harrowing time on the water, not to press on. Their bodies, probably killed by Shivwits Indians, were found at the top of the cliffs.

So Powell had a ripping yarn to tell, and he told it superlatively. *The Exploration of the Colorado River and Its Canyons* became an instant classic, its peculiarity calculated in the shifting tones of its voices. Sometimes our one-armed hero wants to present himself as the dry impersonal geologist, ethnographer, and naturalist: "The low desert, with its desolate mountains, which has thus been described is plainly separated from the upper region of plateau by the Mogollon Escarpment"; but he slips easily into Bartramesque prose-poetry: "thousands of these little lakes with deep, cold emerald waters, are embosomed among the crags of the Rocky Mountains." When he comes to his traveling companions, Powell turns ironist as mordant as Twain, and with the same observant touch of a master novelist. He describes his own brother "Captain Powell" (to his major) as "silent, moody and sarcastic, though sometimes he enlivens the camp at night with a song. He is never surprised at anything, his coolness never deserts him and he would choke the belching throat of a volcano if he thought the spitfire meant anything but fun. We call him 'Old Shady.'" When he needs to, at the moments of high drama, Powell knows how to push and pull the syntax of the writing as if it were carried away by the violent current. He has just gone over two falls in the river, the first a mere ten feet, then bounced to a fifty-foot drop: "I pass around a great crag just in time to see the boat strike a rock and, rebounding from the shock, careen and fill its open compartment with water. Two of the men lose their oars; she swings around and is carried down at a rapid rate, broadside on, for a few yards when, striking amidships on another rock with great force, she is broken quite in two and the men are thrown in the river."

More than simply an American odyssey through the watery jaws of hell and out again, Powell's story of his successful navigation of the river suggested that there might be more than one way to master the unruly Colorado. After the survey and the accumulation of knowledge came practical speculation. Perhaps this river might even be dammed, forced back into reservoirs from which water could feed the irrigation canals that, in the cliché of the time, would make the desert bloom. Powell's own head was full of hydraulic possibilities, but his *Report on the Lands of the Arid Region of the United States*, originally written for the General Land Office of the government and published by order of Congress in 1878, warned that only 3 percent of the West could ever be turned into farmland without irrigation. Even if the Colorado and other western rivers like the Snake could be dammed and canalized, only a tiny fraction of the immensity of the West would ever be viable for agriculture, much less capable of supporting new cities in Arizona, Nevada, or Southern California.

Powell did not mean to dash the hopes of the irrigation visionaries, but since water was such a precious resource in the West, he thought the only way it could be rationally utilized was through relatively small-scale and cooperative farming, of the kind he had seen working in the Mormon communities in Utah—the very definition of an unpromising arable landscape. The Mormons had kept the marketplace out of their irrigation projects, had allotted resources carefully and equitably, and as a result had been a model of how to do well in harsh circumstances that Powell thought the rest of America could learn from. The trouble was that his eminently sensible communally based approach to thinking of water as a publicly shared resource went directly against the spirit of the time, which was commercial, technological, and ruggedly individualist.

Powell thought of water as the refreshment of corrupt and jaded democracy: local, accountable; a bond that would hold communities together. What was the alternative: the extension of the bitter war between capital and labor into the West? But to politicians like big Bill Stewart of Nevada, this made Powell, who was evidently "drunk with power and deaf to reason," an enemy of the American way. For the sake of the "pioneers who are developing the country," Stewart declaimed from under his white Stetson, Powell needed to be chastened and his Irrigation Survey shut down before he could do more damage. It was all very simple for men like Stewart. Apply the mastery of hydraulic engineering to even the most intractable country, *tame* the rivers, and there would be water galore for western farmers in the hot valleys of Southern California—Sacramento and San Joaquin. They could produce not just corn but alfalfa for cattle feed; peaches, plums, and oranges; and vegetables in quantities as yet unimaginable. On their success, cities would indeed rise in the plain. Their tables would be bounteously served, and their gardens would forever be green.

This, at any rate, was the view of a newspaperman, William Ellsworth Smythe, for whom modern America was all about solving impossible problems. Smythe was the son of a wealthy Massachusetts shoe manufacturer but had lost his share of the family fortune in a failed publishing venture. To get back on his feet Smythe went to work for a land company that sent him to New Mexico, where he witnessed some of the same communally organized approach to water that the Mormons had used, for the good reason that the Latinos in the West had treated their wells and ditches in the same way. As editor of the *Omaha Bee,* Smythe traveled to Nebraska during the first year of the great drought of 1890, where scenes of farmers shooting livestock that were dying of thirst, and abandoning farms for want of any kind of water, cut themselves into his memory. "The spectacle of the landless man and the manless land," he wrote, "is enough to make the angels weep."

The next year Smythe established a journal he called *Irrigation Age* and attended an initial congress of all interested parties—farmers, engineers, and people from government—at Salt Lake City, which he promoted as a model for the rest of the West. He had already become more of a crusader than a promoter. His policy, he stated in the *Age,* would be "to champion the cause of irrigation; to keep step with the swift progress of science in discovering the water resources of the west, to encourage the settlement of the beautiful valleys, the wide-stretching plains and fertile tablelands of the arid west as fast as reclaimed; to explain and illustrate to the farmer the uses and benefits of water . . . from Kansas to California and from Manitoba to Mexico." Smythe's vision of what irrigation might be able to do for 100 million acres of the West, and thus for the future of the United States, was epic, but like Powell, he thought in social rather than macroeconomic terms. The great thing about irrigation projects was the amount of close attention they needed, to prevent canals and ditches from silting up and getting choked with debris and vegetation. A big commercial outfit using day-wage labor would never make for the same kind of success that family farms or small-town co-operatives would assure through their personal stake in the new farming.

Two years later in Los Angeles, a much bigger event was deliberately timed to coincide with the last month of the Chicago Exposition, in which America's agrarian past, present, and future had been prominently on display. Smythe printed 100,000 copies of a special issue of *Irrigation Age* for distribution at Chicago, and since delegates came to Los Angeles from Russia, Australia, and Peru it was not entirely vain to call the occasion an International Irrigation Congress. At the entrance to the Opera House where the convention was held, two massive pumps stood like guardians of the American transformation Smythe believed was about to happen. The cascade of disastrous news from business only made Smythe more resolved to promote irrigation as a way forward; and more controversially, to transfer some of the responsibility for it from private business to government. He said in his opening address: "we are laying today the cornerstone of the Republic of Irrigation. It shall not be laid in avarice and cemented with greed. That would not be fitting; for a people living in sunlit valleys guarded by eternal mountains have ever been the defenders of liberty. We will lay the superstructure of this edifice by the plumb line of justice and equity. We will write upon its white cornerstone 'Sacred to the Equality of Man.' We inscribe upon its especially massive arch those two synonymous terms 'Irrigation and Independence.'"

This was grand opera indeed. How Powell, who had been asked to address the congress not once but twice, felt about Smythe's ecstatic rhetoric can be guessed, especially in view of what he was to do: pour cold water on all the rhetorical heat. Powell's first address was mostly a reminiscence of his epic journey down the Colorado; and the intensity of those memories was such that the performance wore him out so badly that there was some doubt whether he would deliver his second lecture on present prospects. And when Smythe had heard what Powell had to say, it's possible he wished he had stayed in his hotel room. For there could not have been a more glaring contrast between the bullish mood of the *Irrigation Age* delegates and the beetle-browed Powell's cautionary words. Even if all the waters of the West were harnessed as they should like, he said, they could do no more than irrigate a minute proportion of the country. It was criminal to bring settlers into the West on promises that would not and could not be kept, and that in all these matters his golden rule that farms should go to where the water was, not the other way around, should be heeded. "Gentlemen," he went on, "it may be unpleasant for me to give you these facts. I hesitated a good deal but finally concluded to do so. I tell you gentlemen you are piling up a heritage of conflict and litigation of water rights, for there is not sufficient water to supply the land."

This is not what the delegates had come to Los Angeles to hear. Powell's unanticipated pessimism seemed at odds not just with their own temper of invincible achievement but with his own past record of enthusiasm about damming the rivers and creating on-stream reservoirs from which fields and homes could be supplied. The patriarch's comments were interrupted by booing and vocal outbursts of angry dissent. Though taken aback by Powell's attack on premature boosterism, Smythe himself had too much respect for the old major to pick a public quarrel. But Smythe's impassioned faith in a transformational irrigation of the West remained fierce. This would be America's true answer to the panic, anger, and defeatism that had gripped the country in that year of economic collapse. Messianic in his nationalist optimism, Smythe told readers of his book *The Conquest of Arid America* that it was "for all optimistic Americans . . . for homeseekers who under the leadership of the paternal Nation are to grapple with the desert, translate its gray barrenness into green fields and gardens, banish its silence with the laughter of children. This is the breed of men who make the Republic possible, who keep the lamp of faith burning through the night of corrupt commercialism." It was about "what is being done by the partnership of God and mankind in finishing one important corner of the world." The outlook in the 1890s might seem bleak for such grand projects, but the armies of the marching unemployed, the fear and gloom surrounding America's cities, were all the more reason to forge ahead, for "when Uncle Sam puts his hand to a task we know it will be done. Not even the hysteria of hard times can frighten him away from the work. When he waves his hand at the desert and says 'Let there be water!' we know that the stream will obey his command. We know more than that—know when the water will come, how much land will be reclaimed, how many homes will be built. We can even calculate with precision how many towns will spring up and where they will be."

Powell died in 1902, and in that same year, Theodore Roosevelt signed into law the National Reclamation Act, which owed perhaps more to Smythe's irrepressible optimism than to the major's emphasis on local conservation. For as was often the case with Roosevelt, the act brought the federal government deep into one of the most critical sectors of the economy. Though TR had to back off from the excited sponsor of the bill, Senator Francis Newlands, who claimed that with a stroke the government had "nationalized water," a new agency would now use tax revenues to undertake a sweeping range of hydraulic projects all over the West. Most of them would involve the creation of dams and reservoirs from which a systemic supply of irrigation could be provided to farmers in areas—like Southern California's parched Imperial Valley—that would otherwise never have been able to grow cereals and fruit. The Bureau of Reclamation would then provide the water and be repaid from the profits of the improved and productive farms, though at heavily subsidized rates. In the sense that they had hoped that irrigation was too precious to leave to the usual forces of the market, and that some sort of public intervention would be needed just to get the immense work of damming and creating reservoirs and canals done successfully, both Powell and Smythe might have felt satisfied that they had turned the United States in a new direction. For a generation at least it was no shame to be working for the government, just as in Montgomery Meigs's day, working for the U.S. Army Corps of Engineers was thought an honorable vocation.

Talented engineers in the years before and after World War I flocked to the reclamation agency, full of exactly the kind of can-do energy that Smythe had hoped would rekindle the spirit both of scientific ingenuity and public service in the United States. More than 600 dams were constructed in the first thirty years of the agency's existence, culminating in the spectacular achievements of Grand Coulee and Hoover, the last of which brought work to thousands of unemployed and created an entire city around the construction site. Millions of acres were brought under cultivation. Peaches, grapefruit, wheat, and alfalfa all prospered. And in that sense Smythe's vision had been realized, and Americans were now eating fruit and vegetables from places that had no natural business to be growing them.

Eventually deep fossil-bed aquifers—unknown to Powell—would be tapped for water, and almost nothing seemed to be beyond the powers of the church of irrigation. But then you look at the "bath rings" around the rock walls of Lake Mead that measure the bald fact of the reservoir being at barely 60 percent of capacity, and Powell doesn't seem so much off the mark after all. Something has to be done, and both Powell and Smythe would have agreed that the best you can say for a desperate situation is that it makes citizens reconsider what the government, which acts in their name and with their money, should, or should not, do.

38. Ghost house

We were bowling along the backroads of east Colorado looking for locations like a hound tracks game, side to side, sniffing for scat: "There"; "No there"; "How about *there?*" These days it takes a long while to get out of the Denver "boomburbs" that string themselves along the freeway, the beautiful lavender mountains always on the western horizon, running parallel to the road. That's where the tourists go, heading for the ski resorts of the Rockies, or in summer country hikes to the snow-capped Continental Divide. Pike's Peak will put you a good 12,000 feet up, and you feel, because you are, on top of the world.

I'd been there myself more than once and even got on a horse; at least I think it was a horse, so swayback was the poor creature with the burden of unnumbered urban cowboys that her belly almost touched the dirt. We were professors in the saddle; on parole from a hard morning's workout with business executives Needing to Know about Locke and Hobbes, and why not? Up into the Maroon Bells we went with our trusty steeds, when the ancient but courteous nag had a sudden notion that she was, for one last afternoon, the frisking filly she must once have been and took off from the trail of PhDs, galloping like she never wanted to hear another word about John Maynard Keynes ever again. Off she went over hill and down dale, and mountain meadows, taking me with her. Reins? What reins? I hung on, mostly to her old generous black mane, and after a bit the "for dear life" part of the sentence fell away. Eventually—five minutes later—the horse felt she had put enough distance between her and the philosophy rodeo, and Peggy (Pegasus to you) and I were Alone at last in Colorado. We both whinnied happily. Take me where you will, o winged one, I thought and possibly said (it had been a long morning), and obligingly she cantered a bit more through a glade of quaking aspen and out onto a bluff that looked down steeply to the valley below. I applied the brakes as best as old John Wayne movies had taught. Amazingly, "Whoa" seemed to work. Then without warning another rider appeared, perhaps thirty feet away on a little rise in the field, darkly silhouetted against the late afternoon Rocky Mountain light. To my amazement he then did that thing, leaning back in the saddle, taking the cowboy hat off, and waving it in salutation in the air. "HIYA," came the shout. This was it. The West and I were One. "HOWDY," I yelled back, getting into the spirit of the thing. There was a pause. Then came the fatal blow as deadly as if he'd beaten me to the draw. "That point you were making about Hobbes's theory of sovereignty this morning . . . Really *insight-ful!*" It was over. I had been gunned down by the political theory of contract. I slumped in the saddle. Peggy and I headed dejectedly back to the trail where the riders were chatting animatedly about Ingmar Bergman. "Where have you been?" one of them kindly inquired. "Oh just off west a bit," I said, accepting my place in the scheme of things. "What is it tonight, Schubert or Bartók?" Peggy snorted and emitted the richly foul aroma only horses can. I knew how she felt.

But we were filming a different Colorado, beyond the endless line of lumberyards and fixtures-and-fittings depots that go along with suburban sprawl. As you drive east about a hundred miles, the mountains become no more than a sawtooth edge to the far horizon, and out before and around you, 360 degrees, is nothing but an ocean of dark dirt, the Great Plains extending hundreds of miles through Kansas and south into the Oklahoma Panhandle, once the Cherokee Outlet, the land that the Runners of 1893 had galloped to buy. Not much goes on any longer around here—there are more abandoned farms than working ones—but the fields are plowed and will bear some wheat later in the season. Occasionally the skeletal sails of a small pump mill, still there hunting water long after the farmers have gone, clatter and moan. The silence gets broken only by the rumble of a distant truck that you smell before you see it; the unforgettably bad whiff of crated poultry coming from the vehicle as it rattles by, leaving a trail of atmospheric nastiness in the sweet spring air behind it.

Despite their name the plains aren't dead flat. Every so often the ground swells and rises; nothing more than a gentle wave but enough for the road to divert itself with a curve or two. Around one of those bends near a village called Hereford we came, and there, on the brow of one of those risers, was what once had been a farmhouse. From a distance the basics looked more or less intact, freeze-dried by the prairie winter: the pitched roof to let the snow slide away; the dark overlapping shingles along the side; the scatter of smaller wooden huts out back: a hog pen; a henhouse. But when you stepped through the balls of scratchy tumbleweed that had come to rest against the broken fence, you could see the place was held together by nothing more than the debris of its own ruin; the splintered wreck of a life that was hanging on in the middle of nowhere, so its reproach would endure against the Colorado sky like someone who wouldn't or couldn't stop crying.

Well, there was a lot to cry about. The dust storms of the 1930s that swept over the plains came like a biblical plague and in their gray-brown mountains of pulverized aerial dirt, took with them every hope of a share in American plenty the farmers in west Oklahoma, Kansas, the Texas Panhandle, and east Colorado ever had. Digging out after one of the dusters had gone through must have been like the aftermath of a nuclear attack, the bodies of cattle, blinded and choked, lying around in dunes of blowing dust. The flying grime settled in layers everywhere inside your house, and no matter how much you scrubbed it down, back it would come, turning the linen gray, covering the oilcloth on your table just as soon as you'd wiped it down, lodging in your children's ears till they hurt. At night you had to get up and down a glass of foul-tasting water just to make sure you didn't choke asleep what with all the dirt coming in as you slept.

The farm people of the western plains had seen a lot of trouble but never anything like this. One minute the sky overhead was clear blue; that washed-out, thin blue of the prairies; the next a monstrous wall of darkness made its appearance on the horizon, sometimes 30,000 feet high, gathering force from the wind. Often enough there would be no sound as it moved relentlessly toward your town or your house; only sometimes, from deep within the citadel of flying filth, there would be a low rumble of thunder and a shrouded bolt of lightning. There was nowhere you could get to quick enough; no good bundling the children in the old Model T as you might all be entombed by the wall of dust. All you could do was get everyone inside as the darkness rolled over you, blotting out the light, the suffocating blanket of dust hissing through every crack and cavity, falling with the quietness of snow on your grandma's chair, your kitchen sink, your marriage bed.

Outside your animals were dying and your fields were buried. Your life on the High Plains was over.

How had it happened? How had the "bowl" that its excited booster Thomas Hart Benton had insisted was *not* the Great American Desert, but a true dish of impending fruitfulness, turned into the Dust Bowl? The first generation of sodbusters had had a hard time making a go of it; the promised irrigation never really came, and no one knew about deep subterranean aquifers, much less had the means to get at them. Many had gone to the wall along with the early irrigation companies. Back came the Texas cattlemen, who this time managed not to overstock the shortgrass range. But then "dry farming"—deep plowing, mulching the soil with its own dust, then keeping some of the fields fallow through summer to retain moisture—seemed to offer whole new possibilities of making do with less water. Freshly ripped and broken sod, it turned out, was receptive to wheat, a crop no one had thought possible in the "semi-arid" western plains. What was more, international market conditions were perfect, with Europe's domestic production disrupted by World War I and postwar havoc. Demand rose steeply. The farmers who had hung on through the hard times of the 1890s and the first decade of the twentieth century found themselves in the unfamiliar position of cashing in from soaring cereal prices.

But if they were to profit properly from the opportunity, they needed help from heavy machinery and big capital. The old plowing days were a thing of the past. Now what was needed were tractors and the big disk plows that moved through the packed shortgrass sod, shallow and fast, tearing out the roots and leaving a fine dirt, the loose topsoil making sowing that much easier. At the other end of the process, a new generation of combine harvesters, especially the McCormicks, could get through a field in a fraction of the time needed in the old farming days. Scenting a killing, in came the banks offering the credit needed to rent or—in partnership with neighboring farms—buy the machines. Mortgages could be arranged now that the wheat boom had appreciated the price of their land, or advances given against future crops. The good times were so close. It was all so possible. But why should the rubes get all the action? So thought main chancers watching the price of wheat go up faster than the price of land. They had the wherewithal to buy huge tracts of land, plow and harvest on an industrial scale, the scale the agribusiness deserved. Never mind Jefferson's old dream of the husbandman-democrat and his hundred acres. This was the twentieth century.

Enter outsiders. They didn't get any more outside than Hickman Price, who in 1931 ditched his $50,000 salary from Fox Film Corps. in Hollywood to buy up 25,000 acres of Texas Panhandle prime. Everything was on a scale no one else could quite match: the half million bushels of wheat; the twenty-five combines painted a shiny silver, with jumbo lettering spelling out the name of the great man; the hundred trucks that took the wheat to the elevators at Kress; the motorcycle-riding harvest patrol roaring around the fields to check all was as it should be and reporting back to Price HQ if it wasn't; the 250 mobile maintenance units, carrying their sleeping gear with them, who could be deployed day and night if any of the combines and tractors broke down.

Some were even bigger. James Jelinek of Greeley County, Kansas, had 28,000 acres producing 620,000 bushels a year, a mere fourteen combines, and twenty tractors. But he also had his very own grain elevator. And then there was the prodigious Simon Fishman, who had come from Lithuania in the 1880s, aged twelve, had reckoned there were too many little Fishmans running around the Lower East Side, and moved on all the way to the High Plains, where he traveled around as Jews did, selling a bit of this and a bit of that. But there was a cult of the Land among Jews from the Pale around that time that produced a whole number who wanted to try their luck on the High Plains rather than Palestine: Benny Goldberg in North Dakota, or Samuel Kahn of Holt County, Nebraska, who mixed up wheat with livestock and other crops enough for him to be hailed far and wide as Kahn the Onion King. Encouraged, Simon Fishman turned farmer in Nebraska and did well enough to be elected mayor of Sydney in that state before moving on to Tribune, Kansas, where he established a wheat empire that shipped a million bushels a year from both his own and his neighbors' farms.

And, as generally happens, the wheat barons paid the price of their own success: overproduction, steeply falling prices; many of them going broke, starting with Hickman Price, who after only a few years of megaproduction was forced into bankruptcy by a hardware store to which he was in hock for 600 lousy bucks that somehow still couldn't get paid. But it was the small owners who got into deep trouble, mortgaged to the banks for their equipment against a harvest and their land that was now worth a fraction of its value in the boom years of the 1920s. Many were foreclosed, reduced to laboring.

And then, starting in 1931, there was nothing much to labor at. Droughts of the kind that no one could remember in their lifetime hit the High Plains. Such moisture as had been held beneath the broken sod was a distant memory. Wheat blackened and died. The shortgrass prairie turned into an ashland of burned-out prospects stretching as far as the Oklahoma horizon. And then, starting in 1933, it got worse. Winds started to blow, and there was now no vegetation to break their force, or to stop them sucking up the brown-black powder that was all that was left of the topsoil. The black blizzards were born.

The worst of all arrived on 14 April 1935 when an estimated 300,000 tons of flying dirt darkened skies all the way from eastern Colorado to Washington D.C., where Franklin Roosevelt's soil conservation specialist Hugh Bennett was about to testify before a congressional committee on restoring the integrity of the shortgrass prairie before nothing was left of it. On Friday the 19th, Bennett made his appearance announcing, as the sky over Washington turned dirty copper, that they were about to witness what had killed a whole age of farming. As the storm dirtied the windows of the Capitol, Bennett, nature helping his case, announced, "This, gentlemen, is what I'm talking about. There goes Oklahoma."

A week later the Soil Conservation Act was passed, funding an army of 20,000 in the plains who would attempt a restoration of the land; plant trees as windbreaks (a futile exercise in the semi-arid zone) and offer relief to the multitudes of destitute and homeless. To the Cherokee Strip, that just forty or so years before had seen the rush of horsemen to stake their claim to breadbasket prosperity, now came the chroniclers of America's agrarian agony: film-makers like Pare Lorentz, whose *The Plow That Broke the Plains* is still a documentary of tragic beauty; photographers like Walker Evans and Dorothea Lange, whose images of heroically lined begrimed faces, mothers and children standing before broken-down frame houses half buried in dune-dirt, can never leave anyone who has seen them; writers like John Steinbeck and Archibald MacLeish, who grasped that what had gone was more than just a moment in America's farm history, but rather a simple ideal that had been born with the republic itself, and with Jefferson's dream of a democratic republic of citizens of the soil.

"I think we should do a piece to camera inside," said the director, smiling one of his most disarming smiles. Behind him was the ruined house on the plains sunset. "Take a look; see what you think." I approached gingerly and with mixed feelings. It seemed—as it often does—indecent to pore over the wreckage of someone's misery; like making oneself comfortable sitting on their gravestone. But I was pulled by the kind of forensic curiosity that no self-respecting historian can do without. Any sort of step up to what was left of the front room had long ago vanished. Using an equipment case to hoist oneself up the two feet or so needed to get inside meant reaching out amid broken glass, shredded wood, and three-inch nails. I did it anyway. And there I was inside the dead and broken body of the dream. The terrible part about it was not the chunks of scrofulous plaster hanging from the remnant of the ceiling, or the areas of rotting wood floor that were now open to the dirt below, and beneath which something was scuttling. The worst part was that, through all the debris, it was so very easy to see the life that had once called this place home. Ragged and filthy fragments of wallpaper had made a cheerful splash in the front room. Through the doorway, its frame intact, had been a kitchen, the chimney still standing, and opposite the brick cavity that must have been a cold pantry. At the back of the shell was the ancient rusted remains of a spring mattress, small pieces of stuffing hanging on the wire. I stood there listening to the pet dog bark outside, a mother calling her kids back home in the sunset, a window framing a farmer high on his tractor before the light went; listening to sounds that weren't there.

We were packing up our gear when a pickup rolled by and stopped. "You guys know so and so?" said the voice from the truck, belonging to an open-faced friendly woman in her forties. I confessed we did not, wondering if the next question was what we were doing there, so and so owning the land we were standing on. But on the High Plains, once they have a sense you're not there to do any harm, they don't interrogate. I asked about its history, and she was happy to tell me, yes, the house was long abandoned and she didn't rightly know why it hadn't been leveled before now; and, yes, it had somehow survived the dusters and got fixed up "a bit" only to get knocked around again by all the bad times, and now it was just a family place where kids came and rode their horses during summer vacations, the parents making sure the kids didn't crawl all through the house and do themselves mischief. "So the family stayed after the storms?" I asked. "Oh sure; somehow you know they made it through; around here we're not the kind that gives up easily. But I tell you what, it was a close thing, nothing much between my husband's family—they settled here in the last years of the old century— and starving half to death. Now my husband he says, Katie, we could live on the buffalo grass if we had to, but that's just big talk. Now look around; you wouldn't know all that trouble now, would you?"

Katie drove off, a petulant little puff of dust raised by the back wheels of her truck as if warning us against undue complacency. But I looked through the gold radiance of the sunset at the plowed fields, the same ones that had risen in the sky, like a malevolent genie. I had had a sample of what they could do earlier when I picked up a handful and before I knew it, the dust had blown straight into my eyes, blinding me in seconds and taking hours and many glasses of water thrown in my face to get clean again. Now the fields looked like every farmer's dream of happiness. Swallows were doing showy aerobatics in the sky; a jackrabbit was scampering toward a distant fence; the early evening air smelled of germinating abundance. And then I thought of Woody Guthrie, the Oklahoma boy singing a road song:

> I'm blowin down this old dusty road
> I'm blowin down this old dusty road
> I'm blowin down this old dusty road, Lord, Lord
> And I ain't gonna be treated this way

> I'm goin where the dust storms never blow
> I'm goin where them dust storms never blow
> I'm goin where them dust storms never blow, blow, blow
> And I ain't gonna be treated this way

39. Roll up that lawn

I was ten and in the front stalls of the Golders Green Hippodrome. The finale to act 1: the whole cast, Curly and Laurey; cowboys and farmers, no more feuding; all pals now, holding hands marching to the front of the stage bathed in brilliant light; the orchestra ecstatic in the pit; the hairs on the back of my neck standing up.

> O---o---k---la ho ma,
> Where the wind comes sweepin' down the plain
> And the wavin' wheat can sure smell sweet,
> When the wind comes right behind the rain . . .
> We know we belong to the land
> And the land we belong to is grand!
> And when we say:
> Yeeow! Ayipioeeay!
> We're only sayin'
> "You're doin' fine, Oklahoma!"
> Oklahoma, O.K.!

This was just the ticket for Broadway and for America in 1943, two years into the war, the Beautiful Morning and the Surrey with the Fringe on Top wiping out, for an evening at least, the memory of the burning hulks of Pearl Harbor, and the anxiety about the Italian invasion; a fairy tale of the High Plains, provided by Rodgers and Hammerstein, two New York Jews who, it's safe to speculate, knew little of the Onion King of Nebraska or Simon Fishman the wheat baron of Kansas triumphally riding the combine in his snap-brim fedora and fancy suspenders.

Many of the real Oklahomans were living among the Cherokee who at last had got a break from history by having moved to the relatively dust-storm-free eastern end of the state, multiplying enough so they would become, in the twenty-first century, the most populous Native American nation in the United States. Or if they were really lucky, they might have lived near Oklahoma City and the gusher that blew in 1930, burning for eleven days before it was capped. But the Oklahomans whom the world knew about from Steinbeck and Dorothea Lange were the "Okies," for whom the High Plains were no longer a place to scratch a living. One of them, Babe Henry, still the boss of a tire business in the Imperial Valley of Southern California, a silver-haired octogenarian with a twinkle in his eye, confirmed that the Joads in The Grapes of Wrath and what they endured was not just a figment of Steinbeck's sympathetic imagination. All hope of work had gone, farming or otherwise; so off the Henrys went, down Route 66, the paralyzed mother and Babe's oldest brother, who was disabled with polio, riding in the cabin while the two other boys rode all the way on the flatbed, making sure the ropes on what furniture they had stayed secure.

It was known that the farmers needed labor for fruit and vegetable picking and since the stricter immigration laws had staunched the flow of Mexicans (even those who were actually Chinese), the Okies and refugees from the dusters would have to do. They were paid a pittance, a dollar a day, and lived in conditions of great poverty; but at least they ate. Like thousands of others in the valley, Babe Henry's family survived in a "tent house"; the floor and walls of wood up to about four feet, the rest of the space roofed with canvas. Even as they needed their labor, the Californians were not especially thrilled to have the migrants among them. Babe remembers realizing that his bib overalls gave him away in the schoolyard and getting into fights with boys who called him "Okie" as if it meant someone who was barely familiar with the basics of the twentieth century. "That's fine if they wanted to call us names. I wasn't ashamed of where I came from," and the glint of the tough little boy in the schoolyard comes back. "One thing that's good about the Imperial Valley, we have sun and we have water. We can grow crops all year round here."

On the day we talked to him, there was no argument about the sun. It was 120°F (around 50°C), but along the road from the Henry Tire Store, the irrigation sprays were moving smoothly through the fields. A bumper crop of corn was being harvested so the rest of America could get it in spring, not summer. It was the last of the fantasy lands of American plenty; the end of the road trip. After Bartram's Cherokee Elysian Fields, after the waving wheat dreams of the Land Runners, came the permanent year-round nursery garden of the valley, courtesy of the All-American Canal, one of those made possible by the Reclamation Act of 1902. But a look at the landscape on either side of the canal—to the left a dunescape that could be Mauritania; to the right, something more like the Kentish Weald or Vermont—is to see the crazy impossibility of it all, the kind of thing that got Powell worked up. The canal is fed from the Colorado just before it enters the Pacific; and though they are growing some of the most irrigation-intensive produce in the world so that America can have its winter strawberries and green beans, the farmers have more heavily subsidized water than they can use. Upstream in Nevada, years of drought, thin snowmelt, and massive urban use have taken Lake Mead, the great reservoir created by the Hoover Dam, down to less than 60 percent of optimal capacity. Evaporation in the ever more broiling summers has taken still more of the water into the blue Nevada sky. The jet-skiers on the lake don't care, though if they take a look they will see the abandoned hulks of old pleasure boats trapped in hard-caked mud where once there was water. A garden of frondlike reeds and grass has grown over the dry bed covering the embarrassment of the boats with their delicate greenery.

Something, someone, has to give: either the farmers sell or relinquish some of their quota from the Colorado Lower Basin allotments, or the pipes of Phoenix, Los Angeles, and San Diego will run dry. "Our water is not for sale" was the steely answer I got from one old Imperial Valley farmer, even though, like most of the farmers, his land is leased to agribusiness. The zero-sum game between competing interests was exactly what the prophetic Powell wanted to avoid. He had spent all his life on, or thinking about, the waters of the Colorado and the other western rivers, and saw in the prudent economies they prescribed a way for the American West to bring a sense of community to a nation he believed badly needed it. The refreshment in every sense of local self-determination.

Which is why I would have loved him to have been sitting with me in Las Vegas. I would even have taken the one-armed major to the Strip and into the twenty-four-hour night of the fantasy hotels, and before he could flee from the robot-chatter of the slots, I would have pointed him to the bodies of water that are everywhere in Vegas—the Pirates of the Caribbean Lagoon, the mock-Venetian Canals, the swimming pools the size of Brooklyn—and explained that they represent the triumph of recycling, every drop constituted from multiply used "gray water," and without me going into details he would have known exactly what I meant.

And then, if he had been reasonable, I would have taken the major to the real Las Vegas, miles away from the Strip, the one place he would have felt at home: the spring of "the meadows" (*vegas*) itself, around which the little town was originally founded; the water that the Indians and the Spanish used to feed their corn and beans and drank from. That it is still there is something of a wonder, but around it now is the Las Vegas Springs Preserve, one of the most beautiful and inspiring places in the United States: a few acres of desert garden; succulents, cacti, and other blooming plants that can be grown with the minimum of water; species entirely natural to south Nevada. Through the garden wind handsome trails; and gently rising over them is the educational Desert Living Center, where thousands of Las Vegas schoolkids learn something about the past and future of their water; the possibility of life in a heated-up world. On Fridays in summer there is Mozart and white wine, but the cultural trimmings should fool no one. At the heart of this whole enterprise is the South Nevada Water Authority, and the person who runs it is Pat Mulroy, who, one is absolutely certain, would have given John Wesley Powell and William Ellsworth Smythe a run for their money.

Being half Irish American and half German, Mulroy has steel and charm in exactly balancing proportions. She came to Las Vegas from Germany in 1974 before the mob got out of town, arriving too late one night to argue about rooms. The one she got boasted a circular bed and a mirror on the ceiling. Mulroy, who is middle-aged and still strikingly beautiful, had no idea there were such things. In the morning she rose from the round bed, took a look at the barely developed desert outside her window, and thought she was on Mars.

Her Mars definitely had water, and since 1989 she has been running the Water Authority, which, under the guise of its dull civic name, represents the most hopeful course for the American future, dominated by neither the raw power of the market nor the overbearing and remote authority of federal government. In the deals Mulroy has made for her metropolitan boom area, with others in the Lower Colorado she has managed to make local and common interests converge. She thinks the states of the Lower Colorado Basin (plus Mexico, which under the terms of the 1922 agreement gets a share) have no choice. Certainly Las Vegas, which now has a resident population of 400,000 and an annual tourist invasion of *forty million*, has no alternative but to be ecologically sound if it's going to survive. It starts with the principle "if it hits the sewers it gets recycled," but Mulroy has also pioneered the construction of pipes that take any storm runoff (for there are storms) and extra waste straight back to Lake Mead. In the 1990s it was possible to broker "banking" agreements with states like Arizona in which their unused groundwater allotment would be bought by needy Nevada. Put into action, Arizona would in a given year take less of its groundwater, and the corresponding amount of its share would then be available for Nevada at Lake Mead.

But that was before the first modern drought of near catastrophic magnitude in 2002. After that neither California nor Arizona had anything spare to trade in a water banking deal. Drastic measures were needed in Las Vegas if it was to survive. Mulroy then decided that since domestic and business interior use was almost all from gray water, a little educational incentive was needed to cope with exterior waste. Her targets were two of the most sacred spaces in American life: the golf course and the lawn. After 2002, Las Vegas residents would be paid a dollar a square foot to take up their lawn and replace it with desert "xerigraphic" landscaping; the species and rock gardens on show in her preserve, all of which require only the most minimal drip watering. Xerigraphic landscaping is now big business throughout the Southwest. But each week it's possible to see trucks from Pat Mulroy's Authority carrying away sod in cylinders that look like Swiss rolls. Golf clubs all over the metropolitan area have ripped out anything beyond the immediate playing areas on and close to the fairway and likewise replaced them with either artificial turf or desert landscape. If it sounds like symbolic rather than substantive action, it isn't. Mulroy reminds me that 70 percent of all Southern Californian water use goes on into-the-ground exterior use, overwhelmingly gardens, parks, and golf courses, where it irrecoverably disappears.

Outside the city, leading down from Lake Mead, Mulroy and her hydraulic engineers have built a series of stepped "minidams," eleven of them, to regulate the flow from the precious reservoir in keeping with whatever conditions in a given season or year require. The masonry and concrete used for the dams are debris from the demolition of the old hotels of the great days of the mob. So, those wicked dens of iniquity, the Dunes and the Desert Inn and the Sands, where Sinatra crooned and Martin drank and all of the Rat Pack behaved rattishly, have been given a second life as the enablers of the green life. The thought of all this saintliness may make their shades reach for a stiff JD on the rocks; but as long as the rocks are frozen gray water, who cares?

I ask Pat if she thinks America will come to Vegas to learn, of all things, about how to survive global warming rather than how to win big at the poker tables. She looks at me for a moment from those clear Nordic-Irish eyes, relaxes, smiles and says, "It's the next generation, our children." I can tell she has her own, pictured that very moment in her mind's eye. "They want to survive. And we want them to survive, don't we?"

40. Windmills

Flying back home to New York from San Diego and passing over the southern Sierra Nevada, I remembered a journey in the opposite direction spent beside Grigory the Russian. It was in the early days of glasnost, and this was Grigory's first time in the United States. He was on his way to a mathematics congress, I think at Stanford. After some polite mutual introductions, he trying out his fractured English, we relapsed into the usual things: reading, poking around in the meal tray for something that approximated food. Every so often Grigory, who had the window seat, would look out the window at whatever bit of America was passing below. And since it was one of those miraculously clear days, coast to coast, there was a whole lot of America to look at. Beyond the Great Lakes over Nebraska and its irrigation circles, the habit became more than looking. The Russian turned his body three-quarters in the seat and pressed his sandy-haired face against the window like a child trying to get at a Christmas display in a department store. As the scenery became not just picturesque but spectacular and the torn-off snowcaps of the Rockies climbed into the window frame, Grigory became fidgety, almost agitated. On and on he stared at Colorado and Utah with such strange intensity that I began to feel guilty for not looking quite so hard myself, all the more since I was then writing a book about landscape. It was a morning flight, due to land at San Francisco around one in the afternoon, but as the jet flew west, the interior of the plane had grown dark, most of the passengers opting for sleep or closing their shades to watch movies. It

turned out that it was this airborne darkness at noon that was upset-
ting Grigory and which he found incomprehensible when there was
the beauty of the earth to witness beyond those pockmarked little
windows. When we crossed Lake Tahoe and the pinewoods that Mark
Twain had accidentally set alight, it was finally too much and he broke
from his silent agitation. "AH," he said very loudly, turning to me and
waving a contemptuous hand at the dozers, "these people, why do
they SLEEP? Why don't they SEE? Why do they not UNDERSTAND?
This belongs to THEM! They must TREASURE like gold; you tell
them, you must tell them now, wake them up and tell them, go on,
yes, see, TREASURE." "I wish I could," I said, "but they're really into
Back to the Future."

Grigory, who perhaps had been too deeply immersed in Dostoyevsky
at some point, was tormented that the docile planeload of passengers,
their necks craned to the video monitors, were trapped in some sort of
moral and aesthetic neverland; that they might take America, its history,
and its geography for granted, preferring to opt instead for a snooze.
The president at that time was also famously partial to his daily naps,
and inevitably Grigory (who probably was more on the insomniac side)
passed animatedly from the topographic to the political. He wanted a
democracy that was wide awake, and if it was necessary to shake people
out of their childish slumber, so be it. But what I couldn't tell him was
that, though sometimes seeming to be lost in torpor, a visitation from
big trouble will always bring about an American awakening. Nothing
like hitting an air pocket to make the passengers vividly aware of the
scenery below.

Is that the case in 2008? Right now, airplane America has lost alti-
tude; the startle reflex has kicked in; and if the passengers are not
screaming in terror, they are certainly not dozing either. What they
are doing is *looking* hard at America—the whole bundle of history,
economy, geography, power—as though their life depends on it, which
it does. And they are considering which of the two men competing
for their vote feels more like the president who can somehow embody
that whole American bundle and by so doing call the country back to
a sense of common purpose, as all the great occupants of the White
House have done before. On the way to recovering that precious,
easily squandered sense of national community, a lot of hard knocks
will be given and taken, which is exactly what the Founding Fathers

prescribed: the storm of argument about whither and whence America without which elections are just so many exchanges of advertising techniques. Something else is going on this time: the republic shouting to be remade. Can it be done? Can the lumbering beast of American power, so big and clumsy, so taken by surprise when its good intentions go awry and the world takes offense, manage the latest act of self-transformation? I've lived in the United States half my life, and take it from me, it can, though nothing is a sure bet anymore. But—and this, I hope, may not come as a complete shock to any reader who has made it this far—the glory of American life is its complexity, not a word usually associated with the United States, but true nonetheless. From the richness of that complexity come, always, rejuvenating alternatives. The Hamiltonians have done you wrong for the last eight years? Well, you know where to go for redress. Anglo-America thinks it's going down to a Hispanic *reconquista*? Try remembering that America has always been shared between Latino and Anglo cultures.

But what this wealth of alternatives means is that, however dire the outlook, it's impossible to think of the United States at a dead end. Americans roused can turn on a dime, abandon habits of a lifetime (check out the waiting list for Smart cars), convert indignation into action, and before you know it there's a whole new United States in the neighborhood.

That too, the Founding Fathers hoped for: that nothing would be beyond American reinvention except their Constitution, and that too of course could be amended. But if the country is to come charging out of the gates of its several calamities with a fire-driven sense of national renewal, it will be because its people draw so ceaselessly on the lives and wisdoms of their ancestors. The history habit in America has nothing to do with reverence and everything to do with the timelessness they attach to their stories; moments that do have dates and dead people attached to them but which somehow leak into the present. They feel about Lincoln the way we feel about Shakespeare: the sound is old, the fury right now. It was striking that almost everyone I spoke to along this trip through American time and country sooner or later invoked Jefferson or DuBois, Teddy Roosevelt or FDR, Reagan or Hamilton, as though there were no distance at all between them and the YouTubers, which, in the long haul, there isn't. It's as though, at the most urgent moments of American decision, historical time folds in on itself and all of its shaping protagonists are there, like some ghostly chorus, to witness and instruct.

Sometimes past and future trip over each other in the most unlikely ways. Lately, the scourge of Democrats, the billionaire oilman T. Boone Pickens, born in Holdenville, Oklahoma, not all that far from the Cherokee, a devotee of George W. Bush and the bankroller of the Swift Boatmen who spent the campaign of 2004 besmirching John Kerry's honorable war record in Vietnam, has gone, to the horror of most of his natural allies, bright green. It is not necessarily a burgeoning sense of the planet in peril that accounts for this conversion. Pickens expects to make a bundle from the water reserves his Mesa Water company is stockpiling for the future; precisely the kind of aggressive capitalism that would have had Powell in a rage about "avarice." But Pickens has declared in public that the age of oil is over and America had better wise up fast to that fact. Drilling in the Arctic National Wildlife Refuge or anywhere else, he says, won't make a real dent in the country's energy deficit. So instead, the oilman is developing in Texas what will be the continent's biggest wind farm, thousands upon thousands of windmills to generate power for the entire South and Southwest. Plumb crazy, say the critics; the old boy has lost it.

And then considering this improbable turn, I think of someone else I encountered in 1964 on my first trip to America. After the Atlantic City convention, my coeditor and I headed back south again to Washington and then on by bus to Williamsburg, Virginia, or as most of America calls it, Colonial Williamsburg. It was all I expected and less: redcoats doing musket practice, men in periwigs, a lot of pewter tankarding. Too much eighteenth century, I thought, and yet not enough. So my friend and I escaped to where we were staying the night, which turned out to be—for reasons I can't remember (a bad sign)—the local mental hospital. Not among the patients but with the hospitable director and his wife, who, on learning I was reading history at Cambridge, peppered me with questions about Oliver Cromwell I was woefully unprepared to answer, though I did hazard a view that the rumor of Cromwell being a secret Jew was probably unfounded. But the price of hospitality was staying up late and talking a lot of Long Parliament.

The morning, though, was an American gift, pancakes and glittering sunshine streaming through the mercifully unleaded windows. Before departing we took a stroll in the grounds. Some of the patients were happily employed looking after the garden; sweeping a bit, weeding a lot. One of them was bending over something that wasn't in Billy Bartram's botany at all: a child's plastic windmill; a whirligig. He was a little man with a perfectly round, bald head, rimless glasses, and an impish grin which he turned to me as he patted the whirligig on the head, gave its little red sails a turn, rose, and came right over. "Would you like to see my windmills?" He said this with such elfin brightness that I said of course, and around the garden he led me, where amid the geraniums and buddleia were more and more; scores of his little whirligigs. With each one rediscovered, his beams grew sunnier until at length he came up very close, smelling strongly of institutional soap, and in a confessional whisper said, "Say, I bet you want to know why I'm planting my windmills." No reply was needed. He went on, "See, no one knows this, but I can tell you 'cause you look like a clever kid, I can tell you, that one night when the wind is"—and he put a chubby index finger to a pink thumb—"just SO, it'll catch, catch all my windmills and you see if this whole place doesn't take off and land somewhere where we will all be HAPPY." And he did a little giggle and skipped off down the sandy path to another shrubbery. And I thought, there goes Jefferson's kind of American, someone who reckons that "the care of human happiness not the destruction of life is the first and only object of good government." And what, really, is so crazy about that?

EPILOGUE:
THE IMPOTENT ANGEL?

Shortly after the outbreak of war in 1939, the Marxist philosopher Walter Benjamin, then living in Paris, offered a reading of a Paul Klee drawing he had bought eighteen years earlier. Klee had called his figure, suspended in a roughly sketched fiery void, *Novus Angelus*, the "New Angel." And while ever since, this particular angel has borne the weight of a lot of social theory, the truth is that he is a funny little thing, crowned with grade-school cut-out curls and three-toed birds' feet. Feeling swept into the vortex of European destruction, Benjamin burdened the peach-colored seraph with his own philosophical agonies, turning him into "The Angel of History." Noticing that the figure's head and torso seem to face opposite directions, Benjamin declared that the angel looks backward (toward us) into the immediate past, where a spectacle of mounting calamity registers in a bug-eyed stare. Why? Because the angel sees too much; the whole bloody mess, start to finish, while we poor mortals, the shortsighted historians of the contemporary, can only apprehend one damned thing after another; a chain of discrete episodes. In the meantime, the angel watches power-lessly as the wreckage rises into the sky.

"The angel," his custodial interlocutor writes, "would like to stay, to awaken the dead and make whole again what has been smashed. But a storm is blowing in from heaven, and has caught his outstretched wings with such force that he can no longer close them." Unable to edge forward into the smoking wind, prevented from ministering to the disaster, the angel is blown helplessly backward into the future by the violence of the unrelenting gale. And it is true; seen this way, the figure looks hopelessly wind-tossed, like dandelion fluff. "The storm," Benjamin added, unconvincingly, "is what we call progress."

Is this what has happened in the United States in the bleak fall of

2008? Is Barack Obama doomed to be the impotent angel, struggling against the tempest, but blown backward into the future? Whatever now unfolds, an American Tacitus, surveying one of the most momentous years in the annals of the Republic, is certain to note the sharp differences of political season. As light returned to America after the winter solstice, so did voters, flocking to caucuses and primaries in numbers that caught "seasoned observers" off guard. American democracy, trapped in a deep freeze of alienation, mistrust, and indifference, began to thaw. The waters of spring moved, and with them, the current of the future, pulling citizens along into the fast-flowing stream of democratic renewal. High summer was the zenith of optimism. Both parties had settled on candidates not cut from the conventional cloth of political fabric. Either of them was easily imaginable as a presidential improvement over the woeful performance of the incumbent and his hapless maladministration. The rhetoric of benign alteration, of an American resurrection, rose into the dog star sky along with the fireworks over Denver and the balloons of St. Paul. But the shadows lengthened, the dusk came sooner; the leaves colored and dropped along with the stock market. By the time the November consummation much of the country craved actually arrived, History, in the shape of financial catastrophe, had set a parenthesis around the verdict. The outcome matters, but not as much as you thought, was the message from the Angel of History, as a cascade of insolvencies, each more vertiginous than the last, drove the campaign from the pages. In the craters marked by the fall of the titans, lay the rubble of American economic security: retirement income, bank deposits, car warranties, university endowments, state and municipal revenues, consumption, employment. Confidence in a great American renewal, which had bounded through the spring and summer, disappeared into a sinkhole of debt that gaped wider and wider as the year moved to its merciful quietus.

Does all this mean that what the country sensed as a moment of historical truth was actually an illusion? The election was supposed to be unlike the travesties that have masqueraded as popular choice, in which hard questions were diverted to the politics of gossip, the tactics of takedown, a competition of inane flag waving. This one would be different—an election that, unlike many in the recent past, would face the troubles of America without without choking its dynamism or discounting its perennial capacity for reinvention. And debates there certainly were, on everything that ails the United States: health care, outsourcing, Afghanistan, NAFTA, FEMA, NATO. But in the end there has been TARP. So has the determination of the American people wanting to catch the wind, rather than be blown along by it, been set at naught by the debacle on Wall Street and Detroit? Does it mean the issues that have been the subject of this book have all turned out to be more marginal to American destiny than anyone, including this writer, anticipated? Will 2008 ultimately go down in American history as the year of traumatic reckoning, a systemic collapse of the national metabolism in which the election result that pulled crowds into the streets on the night of 4–5 November will be seen in retrospect as nothing more than a brief spike of joy, the feverish euphoria of the dying beast?

No. For, in the four o'clock twilight in which I'm writing, on a day when the memory of jubilant Chicago crowds has been replaced by the return of familiar stories of Illinois, it's possible to overcorrect. The historian of the American future will point out that there were *two* momentous alterations happening simultaneously in the country as the old regime left the White House: one boding well, the other ill. The two phenomena—the rehabilitation of governance from the odium and insignificance into which Reagan's demonizing doctrine thrust it, and the disintegration of economic security—are inescapably interlocked. The capacity of democratic government, with a greater degree of public trust behind it than at any time since the 1940s, to contain, arrest, and reverse economic atrophy and the demoralization of the markets is about to be tested. And while it would be naive to discount the magnitude of the disaster, I suspect that for that historian, the moment on the Capitol steps on January 20, 2009, will turn out to have been at least as formative for American destiny as the certainty of yet another panic on Wall Street. The administration of the forty-fourth president and the hydra-headed monster of recession will be tightly meshed together, as if in a gladiatorial net. On whether the former can marshal all the goodwill and resolution his campaign and election generated to contain and defeat the brute, the American future will depend.

So this is a moment in which the two forces that made America formidable—capitalist energy and democratic liberalism—get weighed in the balance. That America can depend on the robustness of its free society, and the demonstration of political inclusiveness supplied by the election, to arm itself against the social unhappiness to come, is just as well. For in most other respects, it's impossible for any attentive historian not to notice that the proliferation of multiple crises looks suspiciously like a classic prerevolutionary moment. We have seen this before, in eighteenth-century France and twentieth-century Russia. In those two overstretched empires, insupportable wars stretched state credit so badly that it became dependent on foreign bondholders who in a shrinking economy wondered whether there might come a time when revenues might not be enough to service the debt, and became ominously restive. Those were the times when anger and hunger moved together in the bad temper of the people, when rackety frauds besmirched the good standing of rulers, so that the only solution seemed to be to have done away with everything and everyone that had brought the country to that sorry pass. But America had an altogether different kind of revolution, and its consequences are still its saving grace. In France and Russia, absolutist governments were without the lightning rod of credibly representative political institutions able to conduct the electrical force of popular fury into constructive energy. The genius of America's revolution, and the constitution that came from it, has always been to draw the sting of rage with the promise of just and benevolent alteration.

Until this year, and this election, Lincoln's "government for the people, of the people, and by the people" was a promise incompletely realized. In hard times, the unequal distribution of fortune (which in the last eight years has grown wider than at any time in the past half century), was certain to make the disfranchised, those most distant from the center of power, feel their impotence most bitterly. With no stake in American democracy and not much to lose, councils of despair could, and did, turn incendiary. Cities burned. So this last benign and peaceful revolution that has brought an African American of mixed race and culture to the presidency is of incalculable importance to the resilience of the American democratic experiment. Obama's breathtaking ascent from congressional insignificance (much like his cynosure Lincoln) to the White House does not in itself, of course, guarantee that he might not squander the reserves of national goodwill he carries with him across its threshold, nor that he or his administration might not yet be defeated by the severity of the multiple disasters facing them.

But if there is certain hardship in the offing and if calamities mostly not of the people's doing have eaten away at the faith in an American future, it is just as well that it falls to a new president to begin the daunting work of reassurance and repair with a clean sheet and, for the time being, with a full measure of the people's trust. Nor could any designer of a national future come up with someone better suited to blunt the smarting sense of grievance that comes from the irrefutable truth that a few have inflicted disaster on the many. In the person he is, in the America he has come from, the forty-fourth president seems to embody the possibility of common purpose over the obstinacy of sectionalism.

The American story has always been a dialogue between Jefferson's unbounded faith in heroic individualism and the obligations of mutual community voiced by Lincoln and Franklin Roosevelt. But the supremacy of self-interest, of which corporatism was its mani-festation writ monstrous, is for the moment at any rate well and truly over, the casualty of its wildest ride. The Enron and Madoff frauds, both breathtaking in scale; the criminally tardy response of government to the devastation wrought by Hurricane Katrina, the abandonment of the people of New Orleans to their fate for days on end; the shaming sense that the American people were sold a bill of goods in Iraq, and that the country's "reconstruction" was merely an opportunity for outsourced freebooting have buried the era of "best of luck, pal." Instead, an alternative America has been recovered, one that was actually there all along. In this America, "government" is no longer the enemy of freedom but its guardian, no longer the bogeyman of enterprise but its honest conscience and forthright guide, no longer shrouded in furtive entanglements but vigorously transparent. A government that need not blush at its claim to be "for, of, and by the people."

And in this recovered United States, the matters that have threaded through American history in the pages of this book—the judgment of right and sufficient cause for war, the assertion of natural resources as public trust rather than sold property, the culturally inclusive nation, the bravely stubborn Jeffersonian belief that a moral society is also a tolerant one—will all be brought into the light of day, irrespective of how pressing the needs of economic rescue and repair become. Indeed it is at just such moments that the opportunity for righting a course can best be tackled. Whatever the American future might look like, it had better not resemble the version that has brought it down in the first place. Nor will it be. There are times in the Republic's history when it might suffice for the country to be constituted from a free but random agglomeration of individuals all looking out for themselves, the result being, somehow, by some social alchemy, a shared common fortune. But this is not one of them. In shared misfortune, the need to look out for one another seems at the very least unsentimental, a necessary condition of society holding together. American independence will not be jeopardized by American interdependence.

That's what the man in the red plaid shirt said he wanted from his government, his president, when he spoke to the Democratic Convention just before Obama's acceptance of the nomination. His face was pink, his passion was high, and he came from Marion, Indiana—a state that symbolized just how far things had come in America, by going for Obama in the election. He had been, he said, a lifelong Republican and had assumed, in his modest way, that the American way would allow him, as Jefferson had promised, his usual allowance of happiness: steady job, mortgage affordable on one income, decent health care. And then the consumer electronics plant closed and he was laid off with just ninety days' severance pay. He had been outsourced. It was the end of his America. Dismay turned to despair and then to anger. And in the city that had lynched blacks in 1930, the churchgoing Republican decided that his kind of American, his kind of president, was the African American candidate for the White House. And so he stood up before the Cecil B. DeMille fake-antique columns that were the backdrop for the revival of republican civism and said so. His name was Barney Smith, and what he wanted, he said, was a president who would "put Barney Smith before Smith Barney."

You have to hope that Barney gets what he wants; not least because a collapse of General Motors would leave Marion, Indiana—a city where unemployment of the eighteen- to twenty-five-year-olds and of older people is already running at near 25 percent—on the floor.

But perhaps Barney Smith will have celebrated at least an election in which politics conceived of as business as usual—negative campaigning, the pornography of patriotic paranoia—abjectly failed to deliver and the country was instantly the better for it. Washington on the night of 4 November, around eight in the evening, at any rate, was eerily quiet. Traffic was sparse; foot traffic in the rainy streets meager. There was little sign of people on their way to election-night parties; not a single placard or poster as the polls shut. The capital had shut its eyes, crossed its fingers tight enough to constrict its blood flow, and held its breath. Early data began to chatter in from exit polling and in the states across the Potomac that mattered, John McCain's numbers rose and rose; the silence in the city was so thick, you could cut it. Nothing by way of saying "it's the rural counties, south and west" made much difference; the intestinal knots got knottier.

Until somewhere around midnight, when at a fell swoop, Pennsylvania, Florida, and Ohio all went blue, and a great cork popped from the pent-up city. Block parties improvised, Washingtonians usually not given to dancing in the streets did just that. Hip-hop and salsa ruled on the damp sidewalks of Adams Morgan. All around the cities of the United States, an effusion of relief and almost incredulous glee poured through crowds, as if the country that had not quite dared to believe that it could be possible for someone so unlike the usual specifications for occupancy of 1600 Pennsylvania Avenue would nonetheless be taking up residence there, come 20 January. At long last, the ignominious betrayal of the American promise, inherent in slavery, had been effaced; the moral vileness of segregation wiped clean. Perfect strangers in Washington coffee shops and diners high-fived and hugged. Around the world, disbelief was swamped by joyous relief. The America the world wanted but assumed it had forever lost had returned. The Statue of Liberty was no longer a bad joke. Conceding, John McCain looked happy for the first time since he accepted the Republican nomination and went out of his way to garland the victor with heartfelt appreciation, as if he had been secretly wanting to do that for some time. Even the incumbent, whose presidency was being repudiated, understood that America had suddenly become better for what had happened and had the decency, in so many words, to say so.

At the Lincoln Memorial the following morning, every so often, people would arrive with bunches of flowers and set them at the foot

of the statue. Some were paying homage to the memory of Martin
Luther King and the day when his rhetoric rang out down the Mall
like a great cathedral bell, calling to God for a time he said when the
promise of America would finally be redeemed. Some were certainly
acknowledging the stand Lincoln himself had taken and the mortal
price he and the country had paid for it. Altogether, there was a
mysterious but unmistakably budding sense of reconnection; a country
remade through a simple, majestic act of popular will.

Acknowledging the magnitude of the disaster that has overcome
the economy, and the frightening scale of what needs to be accomplished to restore even a semblance of normality to its prospects, I
remain convinced that the American future, shaped by the epic of its
past, will turn fair once again. On a London street six weeks after the
election, a tall young man approached me, smiling. He was film-star
handsome, in a purely American way: square jawed and open faced.
He reminded me he had been my student many years before, at
Harvard, right at the outset of the Reagan presidency. He had done all
the things, chalked up all the points he had needed to be an American
success. The doors of corporate law had been thrown open. He could
be a deal maker. But, he said, he was going into government; into one
of the institutions that would determine where the country's money
went. He felt good about that, and so did I, since, anecdotally, I am
hearing this news from all over the place: an unapologetic return by
smart women and men, in their thirties, taking pay cuts to work for
the people's government. It is as if a call had been answered, even
though no one has yet thought to make it; a call to service that has
been made so often in the American past and will be again, if the
republic stays true to itself. So how bad can the American future be,
when it is in their strong, young, hardworking hands?

BIBLIOGRAPHY

Part One: American War

Bonner, Robert. *The Soldier's Pen: Firsthand Impressions of the Civil War.* Hill and Wang, New York, 2006.

Chernow, Ron. *Alexander Hamilton.* Penguin Press, New York, 2004.

Crackel, Theodore J. *West Point: A Bicentennial History.* University Press of Kansas, Lawrence, 2002, and Eurospan, London, 2003.

East, Sherrod E., "Montgomery Meigs and the Quartermaster Department." *Military Affairs* 25, 4 (Winter 1961–62).

Ellis, Joseph. *School for Soldiers: West Point and the Profession of Arms.* Oxford University Press, New York, 1974.

Faust, Drew Gilpin. *This Republic of Suffering: Death and the American Civil War.* Knopf, New York, 2008.

Giunta, Mary A., ed. *A Civil War Soldier of Christ and Country: The Selected Correspondence of John Rodgers Meigs, 1859–64.* University of Illinois Press, Urbana, 2006.

Hancock, Cornelia. *South after Gettysburg, Letters of Cornelia Hancock from the Army of the Potomac.* University of Pennsylvania Press, Philadelphia, 1937.

Harper, John Lamberton. *American Machiavelli: Alexander Hamilton and the Origins of U.S. Foreign Policy.* Cambridge University Press, Cambridge, 2004.

Hofstadter, Richard. *The Paranoid Style in American Politics and Other Essays.* Harvard University Press, Cambridge, Mass., 1965.

Kagan, Robert. *Dangerous Nation: America's Place in the World, from its Earliest Days to the Dawn of the 20th Century.* Knopf, New York, 2006.

Kaplan, Justin. *Mr. Clemens and Mark Twain*. Simon & Schuster, New York, 1991.

Karnow, Stanley. *In Our Image: America's Empire in the Philippines*. Random House, New York, 1989.

Kramer, Paul A. *The Blood of Government: Race, Empire, the United States, & the Philippines*. University of North Carolina Press, Chapel Hill, 2006.

Linn, Brian McAllister. *The U.S. Army and Counterinsurgency in the Philippine War, 1899–1902*. University of North Carolina Press, Chapel Hill, and London, 1989

McDonald, Robert M. S., ed. *Thomas Jefferson's Military Academy: Founding West Point*. University of Virginia Press, Charlottesville and London, 2004.

Miller, David C. *Second Only to Grant, Quartermaster General Montgomery C. Meigs*. White Main Books, Shippensburg, Pa., 2000.

Miller, Stuart Creighton. *Benevolent Assimilation: The American Conquest of the Philippines, 1899–1903*. Yale University Press, New Haven, Conn., and London, 1984.

Morris, Edmund. *Theodore Rex*. HarperCollins, London, 2002.

Onuf, Peter S. *The Mind of Thomas Jefferson*. University of Virginia Press, Charlottesville and London, 2007.

Phibbs, Brendan. *The Other Side of Time: A Combat Surgeon in World War II*. Little, Brown, Boston, 1987, and Hale, London, 1989.

Samet, Elizabeth D. "Great Men and Embryo Caesars: John Adams, Thomas Jefferson and the Figures-in-Arms." In Robert M. S. McDonald, *Thomas Jefferson's Military Academy: Founding West Point*. University of Virginia Press, Charlottesville and London, 2004.

Silbey, David J. *A War of Frontier and Empire: The Philippine-American War 1899–1902*. Hill and Wang, New York, 2007.

Stuart, Reginald C. *The Half-Way Pacifist: Thomas Jefferson's View of War*. University of Toronto Press, Toronto and London, 1978.

Thomas, Emory. *Robert E. Lee: A Biography*. W. W. Norton, New York and London, 1995.

Twain, Mark, "To the Person Sitting in Darkness." In *The Complete Essays of Mark Twain*, ed. Charles Neider. Da Capo Press, New York, 2000.

Weigley, Russell F. *Quartermaster General of the Union Army*. Columbia University Press, New York, 1959.

Yalom, Marilyn, photographs by Reid Yalom. *The American Resting Place*. Houghton Mifflin, Boston, 2008.

Part Two: American Fevor

Ahlstrom, Sydney E. *A Religious History of the American People.* Yale University Press, New Haven, Conn., and London, 2004.

Billingsley, Andrew. *Mighty Like a River: The Black Church and Social Reform.* Oxford University Press, Oxford and New York, 1999.

Blue, Frederick J. *No Taint of Compromise: Crusaders in Antislavery Politics.* Louisiana State University Press, Baton Rouge, 2005

Boles, John B. *The Great Revival: Beginnings of the Bible Belt.* University Press of Kentucky, Lexington, 1996.

Boyer, Paul S. *When Time Shall Be No More: Prophecy Belief in Modern American Culture.* Belknap Press of Harvard University Press, Cambridge, Mass., and London, 1992.

Bremer, Lenni, ed. *Jefferson and Madison on the Separation of Church and State: Writings on Religion and Secularism.* Barricade, Fort Lee, N.J., 2004.

Cross, Whitney R. *The Burned Over District: Social and Intellectual History of Enthusiasm.* Octagon Books, New York, 1981.

Dreisbach, Daniel L. *Thomas Jefferson and the Wall of Separation between Church and State.* New York University Press, New York and London, 2002.

Fairclough, Adam. *Better Day Coming: Blacks and Equality 1890–2000.* Penguin, New York and London, 2002.

Finney, Charles G. *Lectures on Revivals of Religion.* Simpkin and Marshall, London, 1840.

Frey, Sylvia, and Betty Wood. *Come Shouting to Zion: African American Protestantism in the American South and British Caribbean to 1830.* University of North Carolina Press, Chapel Hill and London, 1998.

Higginson, May Thacher, ed. *Letters and Journals of Thomas Wentworth Higginson, 1846–1906.* Da Capo Press, New York, 1969.

Higginson, Thomas Wentworth. *Army Life in a Black Regiment.* Dover Publications Mineola, N.Y., and David & Charles, Newton Abbot, Devon, 2002.

Hofstadter, Richard. *Anti-Intellectualism in American Life*. Jonathan Cape, London, 1964.

Lambert, Frank. *The Founding Fathers and the Place of Religion in America*. Princeton University Press, Princeton, N.J., and Oxford, 2003.

Levy, Leonard W. *The Establishment Clause: Religion and the First Amendment*. University of North Carolina Press, Chapel Hill and London, 2004.

Marsh, Charles. *God's Long Summer: Stories of Faith and Civil Rights*. Princeton University Press, Princeton, N.J., 1997.

Matthews, Donald G. *Religion in the Old South*. University of Chicago Press, Chicago and London, 1977.

McKivigan, John, and Mitchell Snay, eds. *Religion and the Antebellum Debate over Slavery*. University of Georgia Press, Athens and London, 1998.

Meacham, Jon. *American Gospel: God, the Founding Fathers, and the Making of a Nation*. Random House, New York, 2006.

Morgan, Edmund S. *Roger Williams: The Church and the State*. W. W. Norton, New York, 2007.

Noll, Mark A. *The Civil War as a Theological Crisis*. University of North Carolina Press, Chapel Hill, 2006.

Noonan, John T. *The Lustre of our Country: The American Experience of Religious Freedom*. University of California Press, Berkeley, and London, 1998.

Peterson, Merrill, and Robert C. Vaughan, eds. *The Virginia Statute for Religious Freedom: Its Evolution and Consequences in American History*. Cambridge University Press, Cambridge, 1988.

Raboteau, Albert J. *Canaan Land: A Religious History of African Americans*. Oxford University Press, Oxford, 2001.

Raboteau, Albert J. *Slave Religion: The Invisible Institution in the Antebellum South*. Oxford University Press, Oxford, 2004.

Snay, Mitchell. *Gospel of Disunion: Religion and Separatism in the Antebellum South*. Cambridge University Press, Cambridge, 1993.

Williams, Heather Andrea. *Self-Taught: African American Education in Slavery and Freedom*. University of North Carolina Press, Chapel Hill and London, 2005.

Wilson, Charles Reagan. *Baptized in Blood: The Religion of the Lost Cause 1865–1920*. University of Georgia Press, Athens, 1980.

Yellin, John Fagan, and John C. Van Horne, eds. *The Abolitionist Sisterhood: Women's Political Culture in Antebellum America.* Cornell University Press, Ithaca, N.Y., and London, 1994.

Part Three: What Is an American?

Abbott, Grace. *The Immigrant and the Community.* The Century Co., New York, 1917.

Allen, Gay Wilson, and Roger Asselineau. *St. John de Crèvecoeur: The Life of an American Farmer.* Viking, New York and London, 1987.

Anbinder, Tyler. *Nativism and Slavery: The Northern Know Nothings and the Politics of the 1850s.* Oxford University Press, New York, 2002.

Bailyn, Bernard. *The Peopling of British North America: An Introduction.* Knopf, New York, 1986.

Baldwin, Neil. *Henry Ford and the Jews: The Mass Production of Hate.* Public Affairs, New York, and Oxford Publicity Partnership, Oxford, 2001.

De Voto, Bernard Augustine. *The Year of Decision, 1846.* Eyre & Spottiswoode, London, 1957.

Francaviglia, Richard V., and Douglas W. Richmond, eds. *Dueling Eagles: Reinterpreting the U.S.–Mexican War, 1846–1848.* Texas Christian University Press, Fort Worth, 2000.

Glazer, Nathan, and Daniel Patrick Moynihan. *Beyond the Melting Pot: The Negroes, Puerto Ricans, Jews, Italians, and Irish of New York City.* M.I.T. Press, Cambridge, Mass., and London, 1970

Gómez, Laura E. *Manifest Destinies: The Making of the Mexican American Race.* New York University Press, New York, 2007.

Gonzalez, Gilbert G., and Paul A. Fernandez. *A Century of Chicano History: Empire, Nations, and Migration.* Routledge, New York and London, 2003.

Handlin, Oscar. *The Uprooted: The Epic Story of the Great Migration that Made the American People.* Little, Brown, Boston, 1973.

Hardin, Stephen L. *Texian Iliad: A Military History of the Texas Revolution, 1835–1836.* University of Texas Press, Austin, 1994.

Haynes, Samuel W., and Christopher Morris, eds. *Manifest Destiny and Empire: American Antebellum Expansion.* Texas A & M University Press, College Station, 1997.

Higham, John. *Strangers in the Land: Patterns of American Nativism, 1860–1925.* Atheneum, New York, 1963.

Hudson, Linda S. *Mistress of Manifest Destiny: A Biography of Jane McManus Storm Cazneau, 1807–1878.* Texas State Historical Association, Austin, 2001.

Kwong, Peter, and Dwsanka Miscevic. *Chinese America: The Untold Story of America's Oldest New Community.* New Press, New York, 2005.

Lee, Erika. *At America's Gates: Chinese Immigration During the Exclusion Era, 1882–1943.* University of North Carolina Press, Chapel Hill and London, 2003.

Martínez, Oscar J. *U.S.-Mexico Borderlands: Historical and Contemporary Perspectives.* University of Arizona Press, Tucson and London, 1994.

Massey, Douglas S.,ed. *New Faces in New Places: The Changing Geography of American Immigration.* Russell Sage Foundation, New York, 2008.

Montejano, David. *Anglos and Mexicans in the Making of Texas, 1836–1986.* University of Texas Press, Austin, 1987.

Ngai, Mae M. *Impossible Subjects: Illegal Aliens and the Making of Modern America.* Princeton University Press, Princeton, N.J., 2004.

Olmsted, Frederick Law. *A Journey through Texas; or, A Saddle-trip on the Southwestern Frontier.* Dix, Edwards, New York, 1857.

Pfaelzer, Jean. *Driven Out: The Forgotten War Against Chinese Americans.* Random House, New York, 2007.

Philbrick, Thomas. *St. John de Crèvecoeur.* Twayne Publishers, New York, 1970.

Portes, Alejandro, and Rubén G. Rumbaut. *Immigrant America: A Portrait.* University of California Press, Berkeley and London, 2006.

Saxton, Alexander. *The Indispensible Enemy: Labor and the Anti-Chinese Movement in California.* University of California Press, Berkeley, 1975.

Solnit, Rebecca. *Storming the Gates of Paradise: Landscapes for Politics.* University of California Press, Berkeley and London, 2007.

St. John de Crèvecoeur, J. Hector. *Letters from an American Farmer,* ed. Susan Manning. 1782, reprint Oxford University Press, Oxford, 1997.

Telles, Edward E., and Vilma Ortiz, eds. *Generations of Exclusion: Mexican Americans, Assimilation and Race.* Russell Sage Foundation, New York, 2008.

Truett, Samuel. *Fugitive Landscapes: The Forgotten History of the US–Mexico Borderlands*. Yale University Press, New Haven, Conn., and London, 2006.

Truett, Samuel, and Elliott Young, eds. *Continental Crossroads: Remapping U.S.–Mexico Borderlands History*. Duke University Press, Durham, N.C., 2004.

Waldinger, Roger, ed. *Strangers at the Gates: New Immigrants in Urban America*. University of California Press, Berkeley, 2001.

Watts, Stephen. *The People's Tycoon: Henry Ford and the American Century*. Knopf, New York, 2005.

Weber, David J. *The Mexican Frontier, 1821–1846: The American Southwest under Mexico*. University of New Mexico Press, Albuquerque, 1982.

Weber, David J., ed. *Foreigners in their Native Land: Historical Roots of the Mexican Americans*. University of New Mexico Press, Albuquerque, 2003.

Wheelan, Joseph. *Invading Mexico: America's Continental Dream and the Mexican War, 1846–1848*. Carroll & Graf, New York, 2007.

Zeh, Frederick. *An Immigrant Soldier in the Mexican War*. Trans. William J. Orr. Texas A & M University Press, College Station, 1997.

Zolberg, Aristide. *A Nation By Design: Immigration Policy in the Fashioning of America* Harvard University Press, Cambridge, Mass., and London, 2006.

Part Four: American Plenty

Brands, H. W. *Andrew Jackson: His Life and Times*. Doubleday, New York, 2005.

Dunar, Andrew J., and McBride, Dennis. *Building Hoover Dam: An Oral History of the Great Depression*. Twayne, New York, 1993.

Egan, Timothy. *The Worst Hard Time: The Untold Story of Those Who Survived the Great American Dust Bowl*. Houghton Mifflin, Boston, Mass., 2006.

Hine, Robert V., and John Mack Faragher. *The American West: A New Interpretive History*. Yale University Press, New Haven, Conn., 2000.

Lamarr, Howard, ed. *The New Encyclopedia of the American West*. Yale University Press, New Haven, Conn., and London, 1998.

McLoughlin, William. *After the Trail of Tears: The Cherokees' Struggle for Sovereignty, 1839–1880.* University of North Carolina Press, Chapel Hill, N.C., 1993.

McLoughlin, William. *Cherokee Renascence in the New Republic.* Princeton University Press, Princeton, N.J., 1986.

Perdue, Theda, ed. *Cherokee Editor: The Writings of Elias Boudinot.* Tennessee University Press, Knoxville, 1983.

Powell, John Wesley. *The Exploration of the Colorado River and Its Canyons.* 1875, reprint Dover, Mineola, N.Y., 1961.

Powell, John Wesley. *Report on the Lands of the Arid Region of the United States.* 1878, reprint Harvard University Press, Cambridge, Mass., 1983.

Reisner, Marc. *Cadillac Desert: The American West and Its Disappearing Water.* Pimlico, London, 2001.

Smythe, William E. *The Conquest of Arid America.* Harper, New York and London, 1900.

Stegner, Wallace Earle. *Beyond the Hundredth Meridian: John Wesley Powell and the Second Opening of the West.* Houghton Mifflin, Boston, 1954.

Worster, Donald. *A River Running West: The Life of John Wesley Powell.* Oxford University Press, Oxford, 2001.

Worster, Donald. *Dust Bowl: The Southern Plains in the 1930s.* Oxford University Press, New York and Oxford, 2004.

.

ACKNOWLEDGMENTS

This project has been a trip, through American space as well as time, and it has been informed by the willingness of many Americans; some in the thick of political life, some on the outside of it, to talk to me about their own perceptions of the historical moment in the life of their country. Without their engagement the book would have been an altogether poorer offering. An exhaustive list would fill the phone book of a small town, but I want in particular to thank the family of the late Staff Sergeant Kyu-Chay; Cadets Larry and Amber Choate of the United States Military Academy at West Point; Mark Anthony Green of Morehouse College; Katrina and Fred Gross; Vergie Hamer; Richard "Babe" Henry; Pastor Johnny Hunt; Dana Cochrane and Lou Stoker; Retired Generals Montgomery C. Meigs, Fernando Valenzuela, and Ricardo Sanchez; David Plylar; Ruth Malhotra; Jack and Jim McConnell; Charles McLaurin, Pat Mulroy of the South Nevada Water Authority; Representative Rick Noriega; Epifanio Salazar; Reverend Raphael Warnock of Ebenezer Baptist Church, Atlanta; Reverend Jim White.

Two good friends, Andrew Arends and Alice Sherwood, have been exceptionally generous with time given to close readings of the manuscript that have made the book so much better than it would have been without the gift of their critical sympathy for the work.

The inexorable nature of the election calendar meant that bringing both a writing and television project to fruition was always going to be a tall order. So I am even more grateful than usual to my literary agent and friend, Michael Sissons, for his unswerving belief that the work could get done and his sympathetic excitement on reading the manuscript as it went along. I am grateful too for enthusiasm shown by Caroline Michel of PFD for the book and for her kindness in reading sections of it as it progressed. My publisher, Will Sulkin of the Bodley Head Press, has been heroic in his willingness to adjust the usual timetable of production so that films and chapters could somehow get done in tandem, and I am deeply appreciative of his excitement about the project throughout the extended period of its conception and execution. I must also thank many others at the Bodley Head for their forbearance and friendly efficiency in accommodating themselves to a challenging schedule, in particular, Lizzie Dipple, Tessa Harvey, David Milner, Drummond Moir, and Laura Hassan. Gail Rebuck already knows how much I appreciate her acts of faith in this writer. My thanks also to Juliet Brightmore for her invaluable help with the illustrations. In the United States, I am once again grateful to my editor at Ecco Books, Dan Halpern, for his warmhearted sympathy for the project and his encouraging belief that it would have something fresh to say about the connections between past and present. Ginny Smith has been the kindest of abettors in seeing the book through to publication. My agent, Michael Carlisle, has been the usual tower of strength when the author seemed to totter. At BBC America, Melissa Green and Amy Mulcaire have done wonders to get the television series to a wide audience.

Alex Cummings and Ester Murdukhayeva of Columbia University were exceptionally helpful in providing some initial research on the religion and immigration portions of the book. Alan Brinkley, the provost of Columbia University, was kind enough to give his university professor the leave needed to complete the project, and in a more general sense I am grateful to many of my colleagues and friends in the History Department of Columbia for their collegial help and wisdom over the past few years, especially those in American history who have been hospitable to an intruder in their discipline, particularly Elizabeth Blackmar, Eric Foner, and Kenneth Jackson.

At the BBC, Glenwyn Benson, Roly Keating, George Entwistle, and Eamon Hardy have been enthusiastic supporters of the project since its inception, and Eamon has been an exceptionally constructive critic of early cuts of the films. My television agent, Rosemary Scoular, has been a tower of strength as well as a dear friend in getting me through the rougher patches of creating this work in two different media, and without her steady support the entire enterprise might not have come to fruition. I have been lucky to have worked with a gifted and supportive team at Oxford Films and Television, including two exceptionally talented directors in Sam Hobkinson and Ricardo Pollack. So much thanks are also due to Hilary Grove, Susannah Price, Matt Hill, Dirk Nel, Paul Nathan, Merce Williams, Glynis Robertson, and my irrepressible buddy and partner in crime behind the eyepiece, Neil Harvey.

In my office at Columbia University, Julina Rundberg has kept the ship afloat with efficient aplomb, even when asked to work beyond the call of duty—and has kept it from being swamped. Even more than usual, I am grateful for the forbearance of my family—Ginny, Gabriel, and Chloe—during the long and uneven seasons of the author's work on this book and television series. They know it could never have been begun, much less completed, without their tolerant affection. I am also grateful to Mike Pyle for his own contribution to the sum of energy and enthusiasm for the daunting project.

I owe more than even I can put into words to my dear pal and colleague Nick Kent, of Oxford Films, whose thought that I might want to tackle a big American television project happily coincided with my own less-well-shaped notion along the same lines, and who never flinched as my much more idiosyncratic idea of linking the past with the contemporary took shape. Nick has been the necessary partner and collaborator through this whole work: intellectually ebullient, creatively sympathetic, and a human cooling cloth on the often fevered brow of the writer-presenter, both when things went wrong and when things went right. Charlotte Sacher has also been the heart and soul of *The American Future*: a prodigious and brilliant researcher; a collaborator and good friend on location; a discriminating critic of both film and written prose. The finished product in both forms owes her an immeasurable debt, though she is not to be held responsible for any of its inevitable shortcomings. To both Nick and Charlotte—equally indispensable partners in this adventure into the past and future—this book is lovingly dedicated.

August 2008

PICTURE
ACKNOWLEDGMENTS

INDEX